T0136852

Progress in Nonlinear Differential Equations and Their Applications: Subseries in Control

Volume 92

More information about this series at http://www.springer.com/series/15137

Gengsheng Wang • Lijuan Wang • Yashan Xu
Yubiao Zhang

Time Optimal Control of
Evolution Equations

 Birkhäuser

Gengsheng Wang
Center for Applied Mathematics
Tianjin University
Tianjin, China

Lijuan Wang
School of Mathematics and Statistics
Wuhan University
Wuhan, China

Yashan Xu
School of Mathematical Sciences
Fudan University
Shanghai, China

Yubiao Zhang
Center for Applied Mathematics
Tianjin University
Tianjin, China

ISSN 1421-1750   ISSN 2374-0280   (electronic)
Progress in Nonlinear Differential Equations and Their Applications
Subseries in Control
ISBN 978-3-030-07021-2   ISBN 978-3-319-95363-2   (eBook)
https://doi.org/10.1007/978-3-319-95363-2

Mathematics Subject Classification (2010): 37L05, 49-02, 49J15, 49J20, 49J30, 93B05, 93C15, 93C20

This book is published under the imprint Birkhäuser, www.birkhauser-science.com by the registered company Springer Nature Switzerland AG part of Springer Nature.
The registered company address is: Gewerbestrasse 11, 6330 Cham, Switzerland

Preface

The studies on time optimal controls started with finite dimensional systems in the 1950s, and then extended to infinite dimensional systems in the 1960s. Time optimal control is a very important field in optimal control theory, from perspectives of both application and mathematics. Optimal control theory can be viewed as a branch of optimization theory or variational theory. In general, an optimal control problem is to ask for a control from an available set so that a certain optimality criterion, related to a given dynamic system, is achieved. When the above-mentioned criterion is elapsed time, such an optimal control problem is called a time optimal control problem. From the perspective of application, there are many practical examples which can be described as time optimal control problems. For instance, reach the top fastest by an elevator, warm the room by the stove quickly, and stop the harmonic oscillator in the shortest time. Systematic studies on time optimal control problems may help us to design the best strategy to solve these practical problems. Theoretically, time optimal control problems connect with several other fields of control theory, such as controllability, dynamic programming, and so on. The studies on time optimal controls may develop studies on these related fields. These can be viewed as the motivations to study time optimal controls.

The time optimal control problems studied in this monograph can be stated roughly as follows: Given a controlled system over the right half time interval, we ask for such a control (from a constraint set) that satisfies the following three conditions: (i) It is only active from a starting time point to an ending time, point; (ii) It drives the solution of the controlled system from an initial state set (at the starting time) to a target set (at the ending time); (iii) The length of the interval, where the control is active, is the shortest among all such candidates. Thus, each time optimal control problem has four undetermined variables: starting time, initial states, ending time and controls. They form a family of tetrads. When the starting time point is fixed, such a time optimal control problem is called a minimal time control problem. It is to initiate control (in a constraint set) from the beginning so that the corresponding solution (of a controlled system) reaches a given target set in the shortest time. When the ending time point is fixed, such a time optimal control problem is called a maximal time control problem. It is to delay the initiation of active control (in a constraint set) as late as possible, so that the corresponding solution (of a controlled system) reaches a given target set in a fixed ending time.

Most parts of this monograph focus on the above-mentioned two kinds of time optimal control problems for linear controlled evolution equations. In this monograph, we set up a general mathematical framework on time optimal control problems governed by some evolution equations. This framework has four ingredients: a controlled system, a control constraint set, a starting set, and an ending set. We introduce five themes on time optimal control problems under this framework: the existence of admissible controls and optimal controls, Pontryagin's maximum principle for optimal controls, the equivalence of several different kinds of optimal control problems, and bang-bang properties. The aim of this monograph is to summarize our, our seniors' and our collaborators', ideas, methods, and results in the aforementioned themes. Many of these ideas, methods, and results are the recent advances in the field of time optimal control. These may lead us to comprehensive understanding of the field on time optimal control. We tried our best to make the monograph self-contained and wish that it could interest specialists and graduate students in this field, as well as in other related fields. Indeed, the prerequisite is to know a little about the functional analysis and differential equations. There are a lot of literatures on time optimal control problems for evolution systems. We have not attempted to give a complete list of references. All the references in this monograph are closely related to the materials introduced here. It may possibly happen that some important works in this field have been overlooked.

More than thirteen years ago, H. O. Fattorini in his well-known book:

H. O. Fattorini, Infinite dimensional linear control systems, the time optimal and norm optimal problems, North-Holland Mathematics Studies, 201, Elsevier Science B. V., Amsterdam, 2005.

introduced minimal time control problems and minimal norm control problems. The controlled system there reads: $y' = Ay + u$, where A generates a C_0-semigroup in a Banach space. (Hence, the state space and the control space are the same, and the control operator is the identity.) The book is mainly concerned with three topics: First, the Pontryagin Maximum Principle with multipliers in different spaces which are called the regular space and the singular space; second, the bang-bang property of minimal time and the minimal norm controls; and third, connections between minimal time and minimal norm control problems. Also, the existence of optimal controls, based on the assumption that there is an admissible control, is presented. In the period after the book, the studies on time optimal control problems were developed greatly. We believe that it is the right time to summarize some of these developments. Differing from the book, the objects in the current monograph are the minimal time and the maximal time control problems; the controlled system in the current monograph reads: $y' = Ay + Bu$, where A generates a C_0-semigroup in a Hilbert space (which is the state space), B is a linear and bounded operator from the state space to another Hilbert space (which is the control space). Though many topics in the current monograph, such as the Pontryagin Maximum Principle, the bang-bang property, and the equivalence of different optimal control problems, are quite similar to those in the book, we study these issues from different perspectives and by using different ways. Moreover, studies on some topics are developed a lot, for instance, the equivalence of the minimal time control problems and the minimal norm control problems, and the bang-bang property for minimal time controls. The topic on the existence of admissible controls was not touched upon in the book, while in the current monograph we introduce several ways to study it.

Let us now be more precise on the contents of the different chapters of this monograph.

Chapter 1. This chapter introduces some very basic elements of preliminaries such as functional analysis, evolution equations, controllability, and observability estimates.

Chapter 2. In this chapter, we first set up a general mathematical framework of time optimal control problems for evolution equations and present four ingredients of this framework; we next introduce connections between minimal and maximal time control problems; we then show main subjects (on time optimal control problems) which will be studied in this monograph. Several examples are given to illustrate the framework, its ingredients, and main subjects.

What deserves to be mentioned is as follows: At the end of Section 2.1, we present a minimal blowup time control problem for some nonlinear ODEs with the blowup behavior. It is a special time optimal control problem (where the target is outside of the state space) and is not under our framework.

Chapter 3. The subject of this chapter is the existence of admissible controls and optimal controls for time optimal control problems. The key of this subject is the existence of admissible controls. We study it from three perspectives: the controllability, minimal norm problems, and reachable sets. Consequently, we provide three ways to obtain the existence of admissible controls. We also show how to derive the existence of optimal controls from that of admissible controls.

We end this chapter with the existence of optimal controls for a minimal blowup time control problem governed by nonlinear ODEs with the blowup behavior. This problem is very interesting from the viewpoints of application and mathematical theory. However, the studies of such a problem are more difficult than those of problems under our framework.

Chapter 4. This chapter is devoted to the Pontryagin Maximum Principle, which is indeed a kind of first-order necessary conditions on time optimal controls. In general, there are two methods to derive the Pontryagin Maximum Principle. They are the analytical method and the geometric method. We focus on the second one and show how to use it to derive the Pontryagin Maximum Principle for both the minimal time control problems and the maximal time control problems.

Three different kinds of maximum principles are introduced. They are the classical Pontryagin Maximum Principle, the local Pontryagin Maximum Principle, and the weak Pontryagin Maximum Principle. We find a way consisting of two steps to derive these maximum principles. In step 1, using the Hahn-Banach Theorem for different cases, we separate different objects in different spaces. In step 2, we build up different representation formulas for different cases. Besides, we give connections among the above-mentioned maximum principles.

Chapter 5. This chapter develops two equivalence theorems for several different kinds of optimal control problems. The first one is concerned with the equivalence of the minimal time control problem and the minimal norm control problem. The second one deals with the equivalence of the maximal time control problem, the minimal norm control problem, and the optimal target control problem. Some applications of these equivalence theorems are referred to in some examples and the miscellaneous note.

The above equivalence theorems allow us to obtain desired properties for one optimal control problem through studying another simpler one. Two facts deserve to be mentioned: First, between the minimal norm control problem and the minimal time control problem, the first one is simpler. Second, among the maximal time control problem, the minimal norm control problem, and the optimal target control problem, the first one is the most complicated one and the last one is the simplest one.

Chapter 6. In this chapter, we present the bang-bang property for time optimal control problems. In plain language, this property means that each optimal control reaches the boundary of a control constraint set at almost every time. It can be compared with such a property of a function that its extreme points stay only on the boundary of its domain. We begin with finite dimensional cases and then turn to infinite dimensional cases. We introduce two ways to approach the bang-bang property. The first way is the use of the Pontryagin Maximum Principle, together with a certain unique continuation. The second way is the use of a certain null controllability from measurable sets in time.

In this monograph, we give some miscellaneous notes at the end of each chapter. In these notes, we review the history and the related works of the involved study or point out some open problems in the related fields. Besides, after most theorems, we provide examples to help readers to understand the theorems better.

Tianjin, China   Gengsheng Wang
Wuhan, China   Lijuan Wang
Shanghai, China   Yashan Xu
Tianjin, China   Yubiao Zhang
July 2018

Acknowledgments

The authors wish to thank Professor Jean-Michel Coron for the kind invitation to write this book for the Control/PNLDE series. The following colleagues deserve special acknowledgments: Ping Lin (Northeast Normal University, China), Kim Dang Phung (Université d'Orléans, France), Shulin Qin (Wuhan University, China), Can Zhang (Wuhan University, China), and Enrique Zuazua (Universidad Autónoma de Madrid, Spain). Some materials in this monograph are based on our joint works with them. Finally, the authors would like to acknowledge support from the Natural Science Foundation of China under grants 11371285, 11471080, 11571264, 11631004, and 11771344.

Contents

Acronyms

$\mathbb{R}$	the set of all real numbers
$\mathbb{C}$	the set of all complex numbers
y	a state variable
Y	a state space
u	a control variable
U	a control space
$\mathbb{U}$	a control constraint set
$\mathrm{co}\mathbb{U}$	the convex hull of the set $\mathbb{U}$
$\mathscr{A}_{ad}$	a set of admissible tetrads
Q_S	a starting set
Q_E	an ending set
Y_E	a target
$Re\, z$	the real part of the complex number z
$\mathbb{R}^n$	the n-dimensional real Euclidean space
$\mathbb{R}^{n \times m}$	the set of all $n \times m$ real matrices
$\langle \cdot, \cdot \rangle$	an inner product in some Hilbert space
$\langle \cdot, \cdot \rangle_{X^*, X}$	the duality pairing between X^* and X
$\| \cdot \|$	a norm in some normed space
A^{-1}	the inverse of the reversible matrix A
2^Ω	the set of all subsets of Ω
$|E|$	the Lebesgue measure of the set E
$V^{\perp}$	the orthogonal complement space of the subspace V in a Hilbert space
$L(a, b; U)$	the set of Lebesgue measurable functions $\varphi : (a, b) \to U$
$L(a, b; \mathbb{U})$	the set of Lebesgue measurable functions $\varphi : (a, b) \to U$ with values on $\mathbb{U}$
$L^p(a, b; U)$	the set of Lebesgue measurable functions $\varphi : (a, b) \to U$ so that $\int_a^b \|\varphi(t)\|_U^p \, \mathrm{d}t < +\infty \ (p \in [1, +\infty))$
$L^\infty(a, b; U)$	the set of essential bounded Lebesgue measurable functions $\varphi : (a, b) \to U$

$\mathscr{L}(U, Y)$	the space of linear continuous operators from U to Y
$\mathscr{L}(Y)$	the space of linear continuous operators from Y to Y
$\mathscr{D}(A)$	the domain of the operator A
$\{e^{tA}\}_{t \geq 0}$	the C_0 semigroup generated by A
A^*	the adjoint of the operator A
$\mathscr{D}(A^*)'$	the dual space of $\mathscr{D}(A^*)$
$C([a, b]; Y)$	the space of continuous functions in the interval $[a, b]$ with the value in Y
$C(\overline{\Omega})$	the space of continuous functions in the closure of Ω
$C(\cdots)$	a positive constant dependent on those enclosed in the bracket
$\sigma(A)$	the spectrum of the operator A
$\mathscr{N}(A)$	the kernel of the operator A
$\mathscr{R}(A)$	the range of the operator A
$\eta^{\top}$	the transpose of the Euclidean vector η
X^*	the dual space of the normed space X
$\mathbb{N}$	the set of all natural numbers
$\mathbb{N}^+$	the set of all positive integers
$Int\, S$	the interior of the set S
$\overline{S}$	the closure of the set S
∂S	the boundary of the set S
$B_r(x_0)$	the closed ball with center at x_0 and of radius r
$\mathcal{O}_r(x_0)$	the open ball with center at x_0 and of radius r
χ_ω	the characteristic function of a nonempty set ω

Chapter 1
Mathematical Preliminaries

In this chapter, we recall some basic concepts and results which are necessary for the presentation of the theories in the later chapters.

1.1 Elements in Functional Analysis

In this section, some basic results of functional analysis are collected. Most proofs for the standard results will be omitted.

1.1.1 Spaces and Operators

We begin with the following definition.

Definition 1.1 Let X be a linear space over F ($F = \mathbb{R}$ or $\mathbb{C}$).

(i) A map $\varphi : X \mapsto \mathbb{R}$ is called a norm on X if it satisfies the following:

$$\begin{cases} \varphi(x) \geq 0 \text{ for all } x \in X; \quad \varphi(x) = 0 \Longleftrightarrow x = 0; \\ \varphi(\alpha x) = |\alpha|\varphi(x) \text{ for all } \alpha \in F \text{ and } x \in X; \\ \varphi(x + y) \leq \varphi(x) + \varphi(y) \text{ for all } x, y \in X. \end{cases} \quad (1.1)$$

(ii) A map $\psi : X \times X \mapsto F$ is called an inner product on X if it satisfies the following:

© Springer International Publishing AG, part of Springer Nature 2018
G. Wang et al., *Time Optimal Control of Evolution Equations*, Progress
in Nonlinear Differential Equations and Their Applications 92,
https://doi.org/10.1007/978-3-319-95363-2_1

$$\begin{cases} \psi(x, x) \geq 0 \text{ for all } x \in X; \quad \psi(x, x) = 0 \Longleftrightarrow x = 0; \\ \psi(x, y) = \overline{\psi(y, x)} \text{ for all } x, y \in X; \\ \psi(\alpha x + \beta y, z) = \alpha \psi(x, z) + \beta \psi(y, z) \text{ for all } \alpha, \beta \in F \\ \qquad\qquad\qquad\qquad\qquad\qquad\qquad \text{and } x, y, z \in X, \end{cases} \tag{1.2}$$

where $\overline{\psi(y, x)}$ is the complex conjugate of $\psi(y, x)$. Hereafter, we denote a norm on X (if it exists) by $\|\cdot\|$. Sometimes, $\|\cdot\|_X$ is used to indicate a norm defined on X. Similarly, we denote an inner product on X (if it exists) by $\langle\cdot, \cdot\rangle$ or $\langle\cdot, \cdot\rangle_X$ if the underlying space X needs to be emphasized. If X has a norm $\|\cdot\|$, $(X, \|\cdot\|)$ is called a normed linear space. In most time, we simply write X for $(X, \|\cdot\|)$.

Definition 1.2 A normed linear space $(X, \|\cdot\|)$ is called a Banach space if it is complete, i.e., for any Cauchy sequence $\{x_n\}_{n\geq 1} \subseteq X$, there exists an $x \in X$, so that $\lim_{n\to+\infty} \|x_n - x\| = 0$.

Definition 1.3 Let X be a linear space over F with an inner product $\langle\cdot, \cdot\rangle$ and let $\|\cdot\|$ be the induced norm (i.e., $\|x\| = \sqrt{\langle x, x\rangle}$ for any $x \in X$). When $(X, \|\cdot\|)$ is complete, X is called a Hilbert space.

Let us now recall some standard terminologies in Banach spaces.

Definition 1.4 Let X be a Banach space and let G be a subset of X.

(i) G is open if for any $x \in G$, $\mathcal{O}_r(x) \triangleq \{y \in X \mid \|y - x\| < r\} \subseteq G$ for some $r > 0$.

(ii) G is closed if $X \setminus G \triangleq \{x \in X \mid x \notin G\}$ is open.

(iii) The set $\text{Int}G \triangleq \{x \in G \mid \exists r > 0, \ \mathcal{O}_r(x) \subseteq G\}$ is called the interior of G; the smallest closed set containing G is called the closure of G, denoted by $\overline{G}$; and $\partial G \triangleq \overline{G} \setminus \text{Int}G$ is called the boundary of G.

(iv) G is compact if for any family of open sets $\{G_\alpha \mid \alpha \in \Lambda\}$ with $G \subseteq \bigcup_{\alpha \in \Lambda} G_\alpha$, there is a subset Λ_1 (of Λ) with finite number of elements so that $G \subseteq \bigcup_{\alpha \in \Lambda_1} G_\alpha$.

(v) G is relatively compact if the closure of G is compact.

(vi) G is separable if it admits a countable dense subset. When X is separable, it is called a separable Banach space.

Proposition 1.1 *Let X be a Banach space over F and G be a subset of X. Suppose that G is separable. Then any nonempty subset of G is also separable.*

Let X and Y be two Banach spaces and let $\mathcal{D}$ be a subspace of X (not necessarily closed). A map $A : \mathcal{D} \subseteq X \mapsto Y$ is called a linear operator if the following holds:

$$A(\alpha x + \beta y) = \alpha Ax + \beta Ay \quad \text{for all } x, y \in \mathcal{D} \text{ and } \alpha, \beta \in F.$$

The subspace $\mathcal{D}$ is called the domain of A. Thereafter, we denote it by $\mathcal{D}(A)$. If $\mathcal{D}(A) = X$ and A maps any bounded subset (of X) into a bounded subset (of Y), we say that A is bounded.

For any linear operator $A : \mathscr{D}(A) \subseteq X \mapsto Y$, we define

$$\begin{cases} \mathscr{N}(A) \triangleq \{x \in \mathscr{D}(A) : Ax = 0\}, \\ \mathscr{R}(A) \triangleq \{Ax : x \in \mathscr{D}(A)\}, \\ \mathrm{Graph}(A) \triangleq \{(x, Ax) : x \in \mathscr{D}(A)\}. \end{cases}$$

They are called the kernel, the range, and the graph of the operator A, respectively.
Let us now state some important results concerning linear bounded operators.

Theorem 1.1 (Closed Graph Theorem) *Let X and Y be Banach spaces. Let $A :$
$X \mapsto Y$ be a linear operator with $D(A) = X$. Suppose that $\mathrm{Graph}(A)$ is closed in
$X \times Y$. Then $A \in \mathscr{L}(X, Y)$.*

Theorem 1.2 (Principle of Uniform Boundedness) *Let X and Y be Banach
spaces and let $\mathscr{A} \subseteq \mathscr{L}(X, Y)$ be a nonempty family. Then,*

$$\sup_{A \in \mathscr{A}} \|Ax\| < +\infty \text{ for each } x \in X \Longrightarrow \sup_{A \in \mathscr{A}} \|A\| < +\infty.$$

Let us introduce linear functionals and dual spaces. Let $X^* = \mathscr{L}(X, F)$. It is
called the dual space of X. Each $f \in X^*$ is called a linear bounded (or continuous)
functional on X. Sometimes, we denote $f \in X^*$ by

$$f(x) = \langle f, x \rangle_{X^*, X} \text{ for all } x \in X.$$

The symbol $\langle \cdot, \cdot \rangle_{X^*, X}$ is referred to as the duality pairing between X^* and X. The
following results are very important in later applications.

Theorem 1.3 (Riesz Representation Theorem) *Let X be a Hilbert space. Then
$X^* = X$. More precisely, for any $f \in X^*$, there is $y \in X$ so that*

$$f(x) = \langle x, y \rangle \text{ for all } x \in X; \tag{1.3}$$

on the other hand, for any $y \in X$, by defining f as in (1.3), one has $f \in X^*$.

Theorem 1.4 (Riesz Representation Theorem in L^p)

*(i) Let $\Omega \subseteq \mathbb{R}^n$ be a domain. Then, for any $1 \leq p < +\infty$ and $1/p + 1/q = 1$,
$(L^p(\Omega))^* = L^q(\Omega)$.*

*(ii) Let $-\infty \leq a < b \leq +\infty$ and X be a real separable Hilbert space. Then, for
any $1 \leq p < +\infty$ and $1/p + 1/q = 1$, $(L^p(a, b; X))^* = L^q(a, b; X)$.*

Theorem 1.5 (Hahn-Banach Theorem) *Let X be a Banach space and X_0 be
a subspace of X inherited the norm from X. Let $f_0 \in X_0^*$. Then there exists an
extension $f \in X^*$ of f_0 so that $\|f\|_{X^*} = \|f_0\|_{X_0^*}$.*

Let X be a Banach space and X^* be its dual space. We denote the dual of X^* by
X^{**}.

Definition 1.5 A Banach space X is said to be reflexive if $X = X^{**}$.

Let $\Omega \subseteq \mathbb{R}^n$ be a domain. For each $p \in (1, +\infty)$ and $m \in \mathbb{N}^+$, we denote

$$W^{m,p}(\Omega) = \left\{ f : \Omega \mapsto \mathbb{R} \text{ is measurable } \mid \frac{\partial^{\alpha_1 + \cdots + \alpha_n} f}{\partial x_1^{\alpha_1} \cdots \partial x_n^{\alpha_n}} \in L^p(\Omega), \right.$$
$$\left. \alpha_1, \ldots, \alpha_n \in \mathbb{N} \text{ and } \sum_{i=1}^{n} \alpha_i \leq m \right\}.$$

Then, the spaces $L^p(\Omega)$ and $W^{m,p}(\Omega)$ are reflexive. Moreover, any Hilbert space is reflexive. But $C(\overline{\Omega})$, $L^1(\Omega)$, and $L^\infty(\Omega)$ are not reflexive.

Theorem 1.6 (James Theorem) *Let X be a real Banach space. Then X is reflexive if and only if for any $f \in X^*$, $\sup\limits_{\|x\| \leq 1} f(x) = \max\limits_{\|x\| \leq 1} f(x)$.*

Let X and Y be Banach spaces and $A \in \mathscr{L}(X, Y)$. We define a map $A^* : Y^* \mapsto X^*$ by the following manner:

$$\langle A^* y^*, x \rangle_{X^*, X} \triangleq \langle y^*, Ax \rangle_{Y^*, Y} \text{ for all } y^* \in Y^* \text{ and } x \in X.$$

Clearly, A^* is linear and bounded. We call A^* the adjoint operator of A. When X is a Hilbert space and $A \in \mathscr{L}(X)$ satisfies that $A^* = A$, A is called self-adjoint.

We end this subsection with the following definition on compact operators.

Definition 1.6 Let X and Y be Banach spaces and $A \in \mathscr{L}(X, Y)$. We say that A is compact if A maps any bounded set of X into a relatively compact set in Y, i.e., if G is bounded in X, then the closure of $A(G)$ is compact in Y.

1.1.2 Some Geometric Aspects of Banach Spaces

In this subsection, we present some results concerning certain geometric properties of Banach spaces and their subsets.

Let X be a Banach space. A subset G of X is said to be convex if for any $x, y \in G$ and $\lambda \in [0, 1]$, one has $\lambda x + (1 - \lambda) y \in G$. Denote $\text{co} S$ to be the smallest convex set containing S. We call $\text{co} S$ the convex hull of the set S. Important examples of convex sets are balls and subspaces. It should be pointed out that the intersection of convex sets is convex; but the union of two convex sets is not necessarily convex. Also, if G_1 and G_2 are convex, then for any $\lambda_1, \lambda_2 \in \mathbb{R}$, the set $\lambda_1 G_1 + \lambda_2 G_2 \triangleq \{\lambda_1 x_1 + \lambda_2 x_2 : x_1 \in G_1, x_2 \in G_2\}$ is convex.

Theorem 1.7 (Carathéodory Theorem) *Let S be a subset of $\mathbb{R}^n$. Let $x \in \text{co} S$. Then there exist $x_i \in S$ and $\alpha_i \geq 0$ ($0 \leq i \leq n$) so that*

$$x = \sum_{i=0}^{n} \alpha_i x_i \text{ and } \sum_{i=0}^{n} \alpha_i = 1.$$

Before presenting the next result, we introduce the concept of weak closedness. Let G be a subset of Banach space X. We say G is weakly closed if for any weakly convergent sequence $\{x_n\}_{n \geq 1}$ in G, its weakly convergent limit belongs to G.

Theorem 1.8 (Mazur Theorem) *Let X be a Banach space and G be a convex and closed set in X. Then G is weakly closed in X. Consequently, if $\{x_n\}_{n \geq 1} \subseteq X$ satisfies that $x_n \to x$ weakly in X, then there exist $\alpha_{n,k} \geq 0$ ($k = 1, \ldots, K_n$) with $\sum_{k=1}^{K_n} \alpha_{n,k} = 1$ so that*

$$\lim_{n \to +\infty} \left\| \sum_{k=1}^{K_n} \alpha_{n,k} x_{k+n} - x \right\| = 0.$$

We now give the next definition on the separability of two sets in a Banach space.

Definition 1.7 Let S_1 and S_2 be two nonempty subsets in a Banach space X. They are said to be separable in X if there is $\psi \in X^* \setminus \{0\}$ and $c \in \mathbb{R}$ so that

$$\inf_{x \in S_1} Re\langle \psi, x \rangle_{X^*, X} \geq c \geq \sup_{x \in S_2} Re\langle \psi, x \rangle_{X^*, X}. \tag{1.4}$$

When (1.4) holds, we say that the hyperplane $\{x \in X \mid Re\langle \psi, x \rangle_{X^*, X} = c\}$ (or the vector ψ) separates S_1 from S_2 in X; we call this hyperplane as a separating hyperplane; and we call ψ as a vector separating S_1 from S_2 in X, or simply, a separating vector. (A separating vector is a normal vector of a separating hyperplane.) Sometimes, we say that the vector ψ separates S_1 from S_2 in X.

Let us introduce some basic results on the separability. First we introduce the following result for finite dimensional spaces:

Theorem 1.9 ([29]) *Let S be a nonempty convex subset in $\mathbb{R}^n$. Assume $z_0 \in \partial S$, where ∂S is the boundary of S in $\mathbb{R}^n$. Then, there exists $f \in \mathbb{R}^n$ with $\|f\|_{\mathbb{R}^n} = 1$ so that*

$$\inf_{s \in S} f(s) \geq f(z_0).$$

We next introduce the following result for infinite dimensional spaces:

Theorem 1.10 ([19]) *Let X be a Banach space and S be a nonempty convex subset of X. Assume $Int\, S \neq \emptyset$ and $z_0 \notin Int\, S$. Then there exists an $f \in X^*$ with $\|f\|_{X^*} = 1$ so that*

$$\inf_{z \in S} Re f(z) \geq Re f(z_0).$$

Theorem 1.11 (Hahn-Banach Theorem (Geometry Form) [24]) *Suppose that S_1 and S_2 are disjoint, nonempty, convex sets in a normed linear space X. The following statements hold:*

(i) *If S_1 is closed and S_2 is compact, then there exists $f \in X^*$ (with $\|f\|_{X^*} = 1$), $c_1 \in \mathbb{R}$ and $c_2 \in \mathbb{R}$ so that*

$$Re f(s_1) > c_2 > c_1 > Re f(s_2) \quad \text{for all } s_1 \in S_1 \ \text{ and } \ s_2 \in S_2.$$

(ii) *If S_1 is open, then there exists $f \in X^*$ (with $\|f\|_{X^*} = 1$) and $c \in \mathbb{R}$ so that*

$$Re f(s_2) \geq c > Re f(s_1) \quad \text{for all } s_1 \in S_1 \ \text{ and } \ s_2 \in S_2.$$

Remark 1.1 Theorem 1.11 is still true when X is a locally convex topological vector space (see Theorem 3.4 in [24]).

1.1.3   Semigroups and Evolution Equations

Through this subsection, X and Y are Banach spaces. We will present some basic properties on C_0 semigroups. We begin with the next definition.

Definition 1.8 Let $A : \mathscr{D}(A) \mapsto Y$ be a linear operator with $\mathscr{D}(A)$ a linear subspace of X (not necessarily closed).

(i) We say that A is densely defined if $\mathscr{D}(A)$ is dense in X.
(ii) We say that A is closed if its graph

$$\text{Graph}(A) = \left\{ (x, y) \in X \times Y \mid x \in \mathscr{D}(A), y = Ax \right\} \tag{1.5}$$

is closed in $X \times Y$.

We know that any linear bounded operator $A \in \mathscr{L}(X)$ is densely defined and closed. However, there are many linear, densely defined, closed operators that map some bounded sets to unbounded sets. These operators are called unbounded linear operators. We cannot define any norm for an unbounded operator. A typical linear, densely defined, closed, and unbounded operator is as: $A \triangleq d/dx$ on $X = L^2(0, 1)$ with $\mathscr{D}(A) = \{y$ is absolutely continuous on $[0, 1] \mid \dot{y} \in L^2(0, 1), y(0) = 0\}$.

Let us first introduce the following definition.

Definition 1.9

(i) Let $\{T(t)\}_{t \geq 0} \subseteq \mathscr{L}(X)$. We call it a C_0 semigroup on X if

$$\begin{cases} T(0) = I, \\ T(t + s) = T(t)T(s) \ \text{ for all } t, s \geq 0, \\ \lim_{s \to 0+} \|T(s)x - x\| = 0 \ \text{ for each } x \in X. \end{cases}$$

(ii) Let $\{T(t)\}_{t\in\mathbb{R}} \subseteq \mathscr{L}(X)$. We call it a C_0 group on X if

$$\begin{cases} T(0) = I, \\ T(t+s) = T(t)T(s) \ \text{ for all } \ t, s \in \mathbb{R}, \\ \lim_{s\to 0} \|T(s)x - x\| = 0 \ \text{ for each } \ x \in X. \end{cases}$$

Proposition 1.2 *Let $\{T(t)\}_{t\geq 0}$ be a C_0 semigroup on X. Then there exist two constants $M \geq 1$ and $\omega \in \mathbb{R}$, so that*

$$\|T(t)\|_{\mathscr{L}(X)} \leq Me^{\omega t} \ \text{ for all } \ t \geq 0.$$

Definition 1.10 Let $\{T(t)\}_{t\geq 0}$ be a C_0 semigroup on X. Let

$$\begin{cases} \mathscr{D}(A) \triangleq \left\{ x \in X \mid \text{s--}\lim\limits_{t\to 0+} \frac{T(t)-I}{t}x \text{ exists} \right\}, \\ Ax \triangleq \text{s--}\lim\limits_{t\to 0+} \frac{T(t)-I}{t}x \ \text{ for all } \ x \in \mathscr{D}(A), \end{cases}$$

where s–lim stands for the strong limit in X. The operator $A : \mathscr{D}(A) \subseteq X \mapsto X$ is called the generator of the semigroup $\{T(t)\}_{t\geq 0}$. In general, the operator A may not be bounded.

We now present some results for evolution equations. Let $T > 0$, $f(\cdot) \in L^1(0, T; X)$, and $y_0 \in X$. Consider

$$\begin{cases} \dot{y}(t) = Ay(t) + f(t), \quad t \in (0, T), \\ y(0) = y_0. \end{cases} \tag{1.6}$$

Here, $A : \mathscr{D}(A) \subseteq X \mapsto X$ generates a C_0 semigroup $\{e^{At}\}_{t\geq 0}$ on X. We first introduce some notions of solutions.

Definition 1.11 For the Equation (1.6),

(i) $y \in C([0, T]; X)$ is called a strong solution of (1.6) if $y(0) = y_0$; y is strongly differentiable almost everywhere on $(0, T)$; $y(t) \in \mathscr{D}(A)$ for a.c. $t \subset (0, T)$; and the equation in (1.6) is satisfied for almost every $t \in (0, T)$.
(ii) $y \in C([0, T]; X)$ is called a weak solution of (1.6) if for any $x^* \in D(A^*)$, $\langle x^*, y(\cdot) \rangle$ is absolutely continuous on $[0, T]$ and

$$\langle x^*, y(t) \rangle = \langle x^*, y_0 \rangle + \int_0^t \left[\langle A^*x^*, y(s) \rangle + \langle x^*, f(s) \rangle \right] ds \ \text{ for all } \ t \in [0, T].$$

(iii) $y \in C([0, T]; X)$ is called a mild solution of (1.6) if

$$y(t) = e^{At}y_0 + \int_0^t e^{A(t-s)} f(s) ds \ \text{ for all } \ t \in [0, T].$$

We next consider the following semilinear equation:

$$\begin{cases} \dot{y}(t) = Ay(t) + f(t, y(t)), & t \in (0, T), \\ y(0) = y_0. \end{cases} \tag{1.7}$$

where $A : \mathcal{D}(A) \subseteq X \mapsto X$ generates a C_0 semigroup $\{e^{At}\}_{t \geq 0}$ on X and $f : [0, T] \times X \mapsto X$ satisfies the following conditions:

(i) For each $x \in X$, $f(\cdot, x)$ is strongly measurable;
(ii) There exists a function $\ell(\cdot) \in L^1(0, T)$ so that

$$\begin{cases} \|f(t, x_1) - f(t, x_2)\| \leq \ell(t) \|x_1 - x_2\| & \text{for all } x_1, x_2 \in X \text{ and a.e. } t \in (0, T), \\ \|f(t, 0)\| \leq \ell(t) & \text{for a.e. } t \in (0, T). \end{cases}$$

We call $y \in C([0, T]; X)$ a mild solution of (1.7) if

$$y(t) = e^{At} y_0 + \int_0^t e^{A(t-s)} f(s, y(s)) \quad \text{for all } t \in [0, T]. \tag{1.8}$$

The following result is concerned with the existence and uniqueness of the solution to (1.8).

Proposition 1.3 *Let the above assumptions concerning A and f hold. Then, for any $y_0 \in X$, (1.8) admits a unique solution y. Moreover, if we let $y(\cdot; y_0)$ be the solution corresponding to y_0, and let the C_0 semigroup $\{e^{At}\}_{t \geq 0}$ satisfy*

$$\|e^{At}\|_{\mathcal{L}(X)} \leq M e^{\omega t} \quad \text{for all } t \geq 0,$$

where $M \geq 1$ and $\omega \in \mathbb{R}$, then

$$\|y(t; y_0)\| \leq M e^{\omega t + M \int_0^t \ell(s) ds} (1 + \|y_0\|) \quad \text{for all } 0 \leq t \leq T \quad \text{and } y_0 \in X,$$

and

$$\|y(t; y_{0,1}) - y(t; y_{0,2})\| \leq M e^{\omega t + M \int_0^t \ell(s) ds} \|y_{0,1} - y_{0,2}\|$$

$$\text{for all } 0 \leq t \leq T \quad \text{and } y_{0,1}, y_{0,2} \in X.$$

When we discuss perturbed linear evolution equations, i.e., the Equation (1.7) with $f(t, y) = D(t)y$, we have the following variation of constants formula:

Lemma 1.1 *Let $\{e^{At}\}_{t \geq 0}$ be a C_0 semigroup on X and $D(\cdot) \in L^1(0, T; \mathcal{L}(X))$. Then there exists a unique strongly continuous function $G : \overline{\Delta} \mapsto \mathcal{L}(X)$ with $\Delta = \{(t, s) \in [0, T] \times [0, T] : 0 \leq s < t \leq T\}$, so that*

$$\begin{cases} G(t, t) = I & \text{for all } t \in [0, T], \\ G(t, r)G(r, s) = G(t, s) & \text{for all } 0 \leq s \leq r \leq t \leq T, \end{cases} \tag{1.9}$$

and so that for any $0 \le s \le t \le T$ and $x \in X$,

$$G(t, s)x = e^{A(t-s)}x + \int_s^t e^{A(t-r)}D(r)G(r, s)x\,dr$$

$$= e^{A(t-s)}x + \int_s^t G(t, r)D(r)e^{A(r-s)}x\,dr.$$

Because of the property (1.9), we refer to the operator valued function $G(\cdot, \cdot)$ as the evolution operator generated by $A + D(\cdot)$. Next, we consider the following nonhomogeneous linear equations:

$$\xi(t) = e^{At}\xi_0 + \int_0^t e^{A(t-s)}D(s)\xi(s)\,ds + \int_0^t e^{A(t-s)}g(s)\,ds, \quad t \in [0, T], \qquad (1.10)$$

and

$$\begin{aligned}
\eta(t) &= e^{A^*(T-t)}\eta_T + \int_t^T e^{A^*(s-t)}D(s)^*\eta(s)\,ds \\
&\quad - \int_t^T e^{A^*(s-t)}h(s)\,ds, \quad t \in [0, T],
\end{aligned} \qquad (1.11)$$

where $\xi_0 \in X$ and $\eta_T \in X^*$. It is not hard to see that both (1.10) and (1.11) admit unique solutions. The next result gives the variation of constants formula.

Proposition 1.4 *Let $G(\cdot, \cdot)$ be the evolution operator generated by $A + D(\cdot)$. Then the solutions ξ of (1.10) and η of (1.11) can be represented respectively by*

$$\xi(t) = G(t, 0)\xi_0 + \int_0^t G(t, s)g(s)\,ds \quad \text{for all } t \in [0, T],$$

and

$$\eta(t) = G(T, t)^*\eta_T - \int_t^T G(s, t)^*h(s)\,ds \quad \text{for all } t \in [0, T].$$

Moreover, for any $0 \le s \le t \le T$, it holds that

$$\langle \eta(t), \xi(t) \rangle - \langle \eta(s), \xi(s) \rangle = \int_s^t \left[\langle \eta(r), g(r) \rangle + \langle h(r), \xi(r) \rangle \right]dr.$$

Finally, we give the well posedness for the following two heat equations:

$$\begin{cases}
\partial_t y - \Delta y + a(x, t)y = g & \text{in } \Omega \times (0, T), \\
y = 0 & \text{on } \partial\Omega \times (0, T), \\
y(0) = y_0 & \text{in } \Omega,
\end{cases} \qquad (1.12)$$

and

$$
\begin{cases}
\partial_t y - \Delta y + f(y) = g & \text{in } \Omega \times (0, T), \\
y = 0 & \text{on } \partial\Omega \times (0, T), \\
y(0) = y_0 & \text{in } \Omega.
\end{cases}
\tag{1.13}
$$

Here, $\Omega \subseteq \mathbb{R}^n (n \geq 1)$ is a bounded domain with a C^2 boundary $\partial\Omega$, $T > 0$, $a \in L^\infty(\Omega \times (0, T))$, $g \in L^2(0, T; L^2(\Omega))$, $y_0 \in L^2(\Omega)$, and $f : \mathbb{R} \mapsto \mathbb{R}$ is a globally Lipschitz function satisfying that $f(r)r \geq 0$ for all $r \in \mathbb{R}$.

Theorem 1.12 *The following statements are true:*

(i) *The Equation (1.12) has a unique solution* $y \in L^2(0, T; H_0^1(\Omega)) \cap W^{1,2}(0, T; H^{-1}(\Omega))$ *and*

$$
\|y\|_{L^2(0,T;H_0^1(\Omega))} + \|y\|_{W^{1,2}(0,T;H^{-1}(\Omega))} \leq C(\|y_0\|_{L^2(\Omega)} + \|g\|_{L^2(0,T;L^2(\Omega))}).
\tag{1.14}
$$

Here C is a constant independent of y_0 and g, but depending on $\|a\|_{L^\infty(\Omega \times (0,T))}$.

(ii) *The Equation (1.13) has a unique solution* $y \in L^2(0, T; H_0^1(\Omega)) \cap W^{1,2}(0, T; H^{-1}(\Omega))$ *and*

$$
\|y\|_{L^2(0,T;H_0^1(\Omega))} + \|y\|_{W^{1,2}(0,T;H^{-1}(\Omega))} \leq C(\|y_0\|_{L^2(\Omega)} + \|g\|_{L^2(0,T;L^2(\Omega))}).
\tag{1.15}
$$

Here C is a constant independent of y_0 and g, but depending on the Lipschitz constant of f. Moreover, if $g = 0$, then

$$
\|y(T)\|_{L^2(\Omega)} \leq e^{-\lambda_1 T} \|y_0\|_{L^2(\Omega)},
\tag{1.16}
$$

where λ_1 is the first eigenvalue of $-\Delta$ with the homogeneous boundary condition.

Proof We prove the conclusions one by one.

(i) The existence and the uniqueness of the solution to (1.12) follow from Theorem 3 and Theorem 4 of Section 7.1 in Chapter 7, [10]. We omit the detailed proofs.

The remainder is to show (1.14). For this purpose, we multiply the first equation of (1.12) by y and integrate it over $\Omega \times (0, t)$ ($t \in (0, T)$). Then we obtain that

$$
\|y(t)\|_{L^2(\Omega)}^2 + 2\int_0^t \|\nabla y(s)\|_{L^2(\Omega)}^2 ds
$$
$$
\leq \|y_0\|_{L^2(\Omega)}^2 + \|g\|_{L^2(0,T;L^2(\Omega))}^2 + 2(\|a\|_{L^\infty(\Omega \times (0,T))} + 1)\int_0^t \|y(s)\|_{L^2(\Omega)}^2 ds.
$$

This, along with Gronwall's inequality, implies that

$$\|y\|^2_{C([0,T];L^2(\Omega))} + \|y\|^2_{L^2(0,T;H^1_0(\Omega))} \leq C(\|y_0\|^2_{L^2(\Omega)} + \|g\|^2_{L^2(0,T;L^2(\Omega))}).$$

The above inequality and (1.12) lead to (1.14) immediately.

(ii) The existence and the uniqueness of the solution to (1.13) follow from Theorem 2.6 of Chapter 7, [4]. The remainder is to show (1.15) and (1.16). We start with the proof of (1.15). Since $f(r)r \geq 0$ for each $r \in \mathbb{R}$, by multiplying the first equation of (1.13) by y and then by integrating it over $\Omega \times (0, t)$ $(t \in (0, T))$, we obtain that

$$\|y(t)\|^2_{L^2(\Omega)} + 2\int_0^t \|\nabla y(s)\|^2_{L^2(\Omega)} ds$$
$$\leq \|y_0\|^2_{L^2(\Omega)} + \|g\|^2_{L^2(0,T;L^2(\Omega))} + \int_0^t \|y(s)\|^2_{L^2(\Omega)} ds.$$

This, along with Gronwall's inequality, implies that

$$\|y\|^2_{C([0,T];L^2(\Omega))} + \|\nabla y\|^2_{L^2(0,T;L^2(\Omega))} \leq 2e^T(T+1)(\|y_0\|^2_{L^2(\Omega)} + \|g\|^2_{L^2(0,T;L^2(\Omega))}). \tag{1.17}$$

Meanwhile, since $f(r)r \geq 0$ for each $r \in \mathbb{R}$ and $f : \mathbb{R} \mapsto \mathbb{R}$ is globally Lipschitz, we have that $f(0) = 0$ and $|f(r)| \leq L(f)|r|$ for each $r \in \mathbb{R}$. Here, $L(f)$ denotes the Lipschitz constant of the function f. Hence,

$$\|f(y)\|_{L^2(0,T;L^2(\Omega))} \leq L(f)\|y\|_{L^2(0,T;L^2(\Omega))}. \tag{1.18}$$

By (1.17), (1.18), and (1.13), we get (1.15).

We next prove (1.16). Since $g = 0$ and $f(r)r \geq 0$ for each $r \in \mathbb{R}$, by multiplying the first equation of (1.13) by y and then by integrating it over Ω, we see that

$$\frac{d}{dt}\|y(t)\|^2_{L^2(\Omega)} + 2\|\nabla y(t)\|^2_{L^2(\Omega)} \leq 0 \quad \text{for a.e. } t \in (0, T). \tag{1.19}$$

Noting that $\|\nabla w\|^2_{L^2(\Omega)} \geq \lambda_1\|w\|^2_{L^2(\Omega)}$ for each $w \in H^1_0(\Omega)$, by (1.19), we get that

$$\frac{d}{dt}\|y(t)\|^2_{L^2(\Omega)} + 2\lambda_1\|y(t)\|^2_{L^2(\Omega)} \leq 0 \quad \text{for a.e. } t \in (0, T). \tag{1.20}$$

Multiplying (1.20) by $e^{2\lambda_1 t}$ and integrating it over $(0, T)$, we obtain (1.16).

Thus, we finish the proof of Theorem 1.12. □

1.1.4 Minimization of Functionals

In this subsection, we introduce a theorem concerning with the existence of solutions for the following optimization problem:

$$\inf_{u \in X} F(u), \quad \text{with } F \text{ a functional over } X. \tag{1.21}$$

Theorem 1.13 *Assume that the following conditions hold:*

(i)   X is a real reflexive Banach space;
(ii) $F : X \mapsto (-\infty, +\infty)$ is convex, lower semicontinuous;
(iii) $F(u) \to +\infty$ when $\|u\|_X \to +\infty$.

Then the problem (1.21) has a solution.

Proof Let $\widehat{d} \triangleq \inf_{u \in X} F(u)$. We claim that

$$\widehat{d} > -\infty. \tag{1.22}$$

Indeed, if (1.22) was not true, then there would be a sequence $\{v_\ell\}_{\ell \geq 1} \subseteq X$ so that

$$F(v_\ell) \to -\infty \quad \text{when } \ell \to +\infty. \tag{1.23}$$

From (1.23) and the condition (iii), we find that $\{v_\ell\}_{\ell \geq 1}$ is bounded. This, together with the condition (i), implies that there exists a subsequence of $\{v_\ell\}_{\ell \geq 1}$, still denoted by itself, so that

$$v_\ell \to \widehat{v} \quad \text{weakly in } X \text{ for some } \widehat{v} \in X.$$

The above, along with the condition (ii) and (1.23), yields that

$$F(\widehat{v}) \leq \liminf_{\ell \to +\infty} F(v_\ell) = -\infty,$$

which leads to a contradiction, since we assumed that F takes values over $(-\infty, +\infty)$. Hence, (1.22) is true.

According to (1.22), there exists a sequence $\{u_\ell\}_{\ell \geq 1} \subseteq X$ so that

$$\widehat{d} = \lim_{\ell \to +\infty} F(u_\ell). \tag{1.24}$$

By the condition (iii) and (1.24), we find that $\{u_\ell\}_{l \geq 1}$ is bounded. This, together with the condition (i), implies that there is a subsequence of $\{u_\ell\}_{l \geq 1}$, denoted in the same manner, so that

$$u_\ell \to \widehat{u} \quad \text{weakly in } X \text{ for some } \widehat{u} \in X. \tag{1.25}$$

It follows from (1.24), the condition (ii), and (1.25) that

$$\widehat{d} = \liminf_{\ell \to +\infty} F(u_\ell) \geq F(\widehat{u}). \tag{1.26}$$

Since $\widehat{u} \in X$ and $\widehat{d} = \inf_{u \in X} F(u)$, it follows by (1.26) that $\widehat{u}$ is a solution to the problem (1.21).

In summary, we finish the proof of Theorem 1.13. $\qquad\qquad\qquad\qquad\square$

1.2 Set-Valued Map

Let X and Y be two Banach spaces. Let S be a nonempty subset of X. Denote by 2^Y the set consisting of all subsets of Y. We call any map $\Gamma : S \mapsto 2^Y \setminus \{\emptyset\}$ a set-valued map (see [3]).

We now present the concept of upper semicontinuous set-valued map, which was introduced by Bouligand and Kuratowski (see [6] and [18]).

Definition 1.12 Let X and Y be Banach spaces. Let $S \subseteq X$. Let $\Gamma : S \mapsto 2^Y \setminus \{\emptyset\}$ be a set-valued map. We say that Γ is upper semicontinuous at $x_0 \in S$, if for any neighborhood G of $\Gamma(x_0)$, there is $\eta > 0$ so that $\Gamma(x) \subseteq G$ for all $x \in \mathscr{O}_\eta(x_0) \bigcap S$. When Γ is upper semicontinuous at each point in S, it is said to be upper semicontinuous on S.

Let

$$\Gamma^{-1}(M) \triangleq \{x \in S \mid \Gamma(x) \bigcap M \neq \emptyset\}$$

and

$$\Gamma^{+1}(M) \triangleq \{x \in S \mid \Gamma(x) \subseteq M\}.$$

The subset $\Gamma^{-1}(M)$ is called the inverse image of M by Γ, and $\Gamma^{+1}(M)$ is called the core of M by Γ. They naturally coincide when Γ is single-valued. We observe that

$$\Gamma^{+1}(Y \setminus M) = S \setminus \Gamma^{-1}(M) \text{ and } \Gamma^{-1}(Y \setminus M) = S \setminus \Gamma^{+1}(M).$$

We can use the concepts of inverse images and cores to characterize upper semicontinuous maps (see Proposition 1.4.4 in [3]).

Proposition 1.5 *A set-valued map $\Gamma : S \mapsto 2^Y \setminus \{\emptyset\}$ is upper semicontinuous at $x \in S$ if the core of any neighborhood of $\Gamma(x)$ by Γ is a neighborhood of x. Hence, Γ is upper semicontinuous if and only if the core of any open subset by Γ is open. In another word, Γ is upper semicontinuous if and only if the inverse image of any closed subset by Γ is closed.*

Let $\Gamma : S \mapsto 2^Y \setminus \{\emptyset\}$ be a set-valued map. Its graph is defined by

$$\text{Graph}(\Gamma) \triangleq \{(x, y) \in S \times Y \mid y \in \Gamma(x)\}.$$

Then we have the following result (see Propositions 1.4.8 and 1.4.9 in [3]):

Proposition 1.6 *Let S be a nonempty closed subset of X and $\Gamma : S \mapsto 2^Y \setminus \{\emptyset\}$ be a set-valued map.*

(i) *If Γ is upper semicontinuous and of closed values (i.e., for each $x \in S$, $\Gamma(x)$ is a closed subset of Y), then $\text{Graph}(\Gamma)$ is closed.*

(ii) *Let K be a compact subset of Y. If $\Gamma(x) \subseteq K$ for each $x \in S$ and satisfies that $\text{Graph}(\Gamma)$ is closed, then Γ is upper semicontinuous.*

Theorem 1.14 (Kakutani-Fan-Glicksberg Theorem) *Let K be a nonempty, compact, and convex subset of a Banach space X. Let $\Gamma : K \mapsto 2^K \setminus \{\emptyset\}$ be a mapping with convex values (i.e., for each $x \in K$, $\Gamma(x)$ is a nonempty and convex subset of K). Then there is at least one $x \in K$ so that $x \in \Gamma(x)$ when either of the following two conditions holds:*

(i) *Γ is upper semicontinuous and of closed values;*

(ii) *$\text{Graph}(\Gamma)$ is closed.*

Proof By Proposition 1.6, we observe that if (i) holds, then (ii) is true. The converse is also true. Thus, we can assume, without loss of generality, that (i) is true. Arbitrarily fix $\varepsilon > 0$ and $x \in K$. Set

$$U_\varepsilon(x) \triangleq \{y \in K \mid \Gamma(y) \subseteq \Gamma(x) + \mathcal{O}_\varepsilon(0)\}.$$

Since Γ is upper semicontinuous, by Proposition 1.5, we have that $U_\varepsilon(x)$ is open. Consider the open cover $\{U_\varepsilon(x) \cap \mathcal{O}_\varepsilon(x)\}_{x \in K}$ of K. Since K is paracompact, there is a neighborhood-finite barycentric refinement $\{V_\lambda\}_{\lambda \in \Lambda}$, i.e., if $\cap V_{\lambda_i} \neq \emptyset$, then there is $x_0 \in K$ so that $\bigcup V_{\lambda_i} \subseteq U_\varepsilon(x_0) \cap \mathcal{O}_\varepsilon(x_0)$ (see Chapter 7 in [13]).

We now claim that for the above ε, there is a continuous function $f^\varepsilon : K \mapsto K$ so that

$$\left[\mathcal{O}_\varepsilon(x) \times \mathcal{O}_\varepsilon(f^\varepsilon(x))\right] \bigcap \text{Graph}(\Gamma) \neq \emptyset \quad \text{for all } x \in K, \tag{1.27}$$

i.e., for each $x \in K$, there is $(x_\varepsilon, y_\varepsilon) \in K \times K$ so that

$$y_\varepsilon \in \Gamma(x_\varepsilon), \quad \|x_\varepsilon - x\|_X < \varepsilon \ \text{ and } \ \|y_\varepsilon - f^\varepsilon(x)\|_X < \varepsilon. \tag{1.28}$$

To this end, let $\{\beta_\lambda\}_{\lambda \in \Lambda}$ be a partition of unity subordinate to the cover $\{V_\lambda\}_{\lambda \in \Lambda}$. For each $\lambda \in \Lambda$, we choose a $z_\lambda \in \Gamma(V_\lambda) \subseteq K$ and define a function

$$f^\varepsilon(x) \triangleq \sum_{\lambda \in \Lambda} \beta_\lambda(x) z_\lambda \quad \text{for all } x \in K. \tag{1.29}$$

It is obvious that $f^\varepsilon : K \mapsto K$ is continuous, where the convexity of K is used. Then the function $f^\varepsilon(\cdot)$ defined in (1.29) satisfies (1.27) (or (1.28)). Indeed, we arbitrarily fix $x \in K$ and let $V_{\lambda_1}, \ldots, V_{\lambda_m}$ be all the sets in $\{V_\lambda\}_{\lambda \in \Lambda}$ containing x. Since $\{V_\lambda\}_{\lambda \in \Lambda}$ is a barycentric refinement, there is $x_\varepsilon \in K$ so that

$$x \in \bigcup_{i=1}^{m} V_{\lambda_i} \subseteq U_\varepsilon(x_\varepsilon) \bigcap \mathscr{O}_\varepsilon(x_\varepsilon).$$

In particular, $x \in \mathscr{O}_\varepsilon(x_\varepsilon)$ and

$$\bigcup_{i=1}^{m} V_{\lambda_i} \subseteq U_\varepsilon(x_\varepsilon) = \left\{ y \in K \mid \Gamma(y) \subseteq \Gamma(x_\varepsilon) + \mathscr{O}_\varepsilon(0) \right\}. \tag{1.30}$$

From (1.30) it follows that

$$z_{\lambda_i} \in \Gamma(V_{\lambda_i}) \subseteq \Gamma(x_\varepsilon) + \mathscr{O}_\varepsilon(0). \tag{1.31}$$

Since $\Gamma(x_\varepsilon) + \mathscr{O}_\varepsilon(0)$ is convex, by (1.29) and (1.31), we see that $f^\varepsilon(x) \in \Gamma(x_\varepsilon) + \mathscr{O}_\varepsilon(0)$. Thus, there is $y_\varepsilon \in \Gamma(x_\varepsilon) \subseteq K$ so that $\|y_\varepsilon - f^\varepsilon(x)\|_X < \varepsilon$. From the above arguments, we see that (1.27) (or (1.28)) holds.

Since $f^\varepsilon : K \mapsto K$ is continuous and K is a compact subset of X, according to the Schauder Fixed Point Theorem, there is $\widetilde{x}_\varepsilon \in K$ so that $f^\varepsilon(\widetilde{x}_\varepsilon) = \widetilde{x}_\varepsilon$. This, together with (1.28), implies that there is $(\widehat{x}_\varepsilon, \widehat{y}_\varepsilon) \in K \times K$ so that

$$\|\widehat{x}_\varepsilon - \widetilde{x}_\varepsilon\|_X < \varepsilon, \quad \|\widehat{y}_\varepsilon - \widetilde{x}_\varepsilon\|_X < \varepsilon \text{ and } \widehat{y}_\varepsilon \in \Gamma(\widehat{x}_\varepsilon). \tag{1.32}$$

Since K is compact, without loss of generality, we can assume that

$$\widetilde{x}_\varepsilon \to \widetilde{x} \in K, \quad \widehat{x}_\varepsilon \to \widetilde{x} \text{ and } \widehat{y}_\varepsilon \to \widetilde{x} \text{ when } \varepsilon \to 0. \tag{1.33}$$

Since Graph(Γ) is closed, by (1.33) and the third conclusion in (1.32), we obtain that $\widetilde{x} \in \Gamma(\widetilde{x})$.

Hence, we finish the proof of Theorem 1.14. $\qquad\square$

Remark 1.2 Theorem 1.14 is a weakened version of Corollary 17.55 in [1].

Theorem 1.15 (Lyapunov Theorem) *Let* $y \in L^1(0, T; \mathbb{R}^n)$. *If* $\lambda : [0, T] \mapsto [0, 1]$ *is Lebesgue measurable, then there is a Lebesgue measurable set* $E \subseteq [0, T]$ *so that*

$$\int_0^T \lambda(t) y(t) \, \mathrm{d}t = \int_E y(t) \, \mathrm{d}t.$$

Theorem 1.15 is a weakened version of Lyapunov's theorem for vector measures (see [21] and [14]), which states that the set of values of an n-dimensional, nonatomic, and bounded vector-measure is a convex compact set. The proof of this theorem is adapted from [2] (see Pages 580 and 581 in [2]).

Proof We denote by $\mathscr{F}$ the σ-field, which consists of all Lebesgue measurable sets in $[0, T]$. Let

$$\mu(\mathscr{F}) \triangleq \left\{ \int_E y(t)\, \mathrm{d}t \mid E \in \mathscr{F} \right\} \tag{1.34}$$

and

$$\mathscr{T} \triangleq \left\{ \varphi \in L^\infty(0, T; \mathbb{R}) \mid 0 \le \varphi(t) \le 1 \ \text{a.e.} \ t \in (0, T) \right\}.$$

Furthermore, we define a continuous linear functional $L : L^\infty(0, T; \mathbb{R}) \mapsto \mathbb{R}^n$ by

$$L(\varphi) \triangleq \int_0^T \varphi(t) y(t)\mathrm{d}t \ \text{ for all } \ \varphi \in L^\infty(0, T; \mathbb{R}), \tag{1.35}$$

and denote

$$L(\mathscr{T}) \triangleq \left\{ \int_0^T \varphi(t) y(t)\mathrm{d}t \mid \varphi \in \mathscr{T} \right\}.$$

It is clear that $L(\mathscr{T})$ is a nonempty, convex, and compact subset of $\mathbb{R}^n$.

We shall conclude by proving that $\mu(\mathscr{F}) = L(\mathscr{T})$. Since $\mu(\mathscr{F})$ is obviously contained in $L(\mathscr{T})$, it remains to prove that $L(\mathscr{T}) \subseteq \mu(\mathscr{F})$. To this end, let $z \in L(\mathscr{T})$ and define

$$\mathscr{T}_z \triangleq \left\{ \varphi \in \mathscr{T} \mid L(\varphi) = z \right\}.$$

One can directly check that $\mathscr{T}_z$ is a nonempty, convex, and weakly star compact subset of $L^\infty(0, T; \mathbb{R})$. Then, by the Krein-Milman Theorem, $\mathscr{T}_z$ has an extremal point. If we could prove that

$$\text{any extremal point of } \ \mathscr{T}_z \ \text{ is a characteristic function,} \tag{1.36}$$

then there exists $E \in \mathscr{F}$ so that $z = L(\chi_E)$. This, together with (1.35) and (1.34), implies that $z \in \mu(\mathscr{F})$ and leads to the conclusion.

We next show (1.36). By contradiction, suppose that it was not true. Then there would exist an extremal point of $\mathscr{T}_z$, denoted by φ_0, which is not a characteristic function. Then, there is a constant $c_0 \in (0, 1/2)$ and a set $E_0 \in \mathscr{F}$ with $|E_0| > 0$ so that

$$c_0 \le \varphi_0(t) \le 1 - c_0 \ \text{ for all } \ t \in E_0. \tag{1.37}$$

Consider the subspace $V \triangleq \{\chi_{E_0}\varphi \mid \varphi \in L^\infty(0, T; \mathbb{R})\}$. Since the dimension of V is bigger than n, the kernel of the operator $L|_V : V \mapsto \mathbb{R}^n$ defined by

$$L|_V(\chi_{E_0}\varphi) \triangleq \int_{E_0} \varphi(t)y(t)\mathrm{d}t \quad \text{for all} \quad \chi_{E_0}\varphi \in V,$$

is nonzero. Therefore, there exists a function $\chi_{E_0}\varphi_1 \neq 0$ with $\varphi_1 \in L^\infty(0, T; \mathbb{R})$ so that

$$\int_{E_0} \varphi_1(t)y(t)\mathrm{d}t = 0. \tag{1.38}$$

Denote $\varphi_2(t) \triangleq c_0 \chi_{E_0}\varphi_1(t)/\|\chi_{E_0}\varphi_1\|_{L^\infty(0,T;\mathbb{R})}$ for a.e. $t \in (0, T)$. Since $\varphi_0 \in \mathscr{T}_z$, it follows from (1.38) and (1.37) that

$$\varphi_0 \pm \varphi_2 \in \mathscr{T} \quad \text{and} \quad L(\varphi_0 \pm \varphi_2) = z.$$

Hence, $\varphi_0 + \varphi_2$ and $\varphi_0 - \varphi_2$ belong to $\mathscr{T}_z$. These lead to a contradiction since φ_0 is an extremal point of $\mathscr{T}_z$.

Thus, we finish the proof of Theorem 1.15. □

Let us introduce the measurability of a set-valued map.

Definition 1.13 Let $D \subseteq \mathbb{R}^n$ be a nonempty, Lebesgue measurable set. Let X be a separable Banach space. A set-valued map $\Gamma : D \mapsto 2^X \setminus \{\emptyset\}$ is Lebesgue measurable if for any nonempty closed set $F \subseteq X$, the set $\Gamma^{-1}(F)$ is Lebesgue measurable in D.

Theorem 1.16 (Kuratowski and Ryll-Nardzewski Theorem) *Let $D \subseteq \mathbb{R}^n$ be a nonempty, Lebesgue measurable set and X be a separable Banach space. If $\Gamma : D \mapsto 2^X \setminus \{\emptyset\}$ is Lebesgue measurable and takes closed set values (i.e., $\Gamma(r)$ is closed for each $r \in D$), then it admits a measurable selection. That is, there is a Lebesgue measurable map $f : D \mapsto X$ so that $f(r) \in \Gamma(r)$ for a.e. $r \in D$.*

Proof Without loss of generality, we assume that

$$\Gamma(r) \bigcap \mathscr{O}_1(0) \neq \emptyset \quad \text{for any } r \in D. \tag{1.39}$$

Indeed, by setting

$$S_k \triangleq \left\{r \in D \mid \Gamma(r) \bigcap \mathscr{O}_k(0) \neq \emptyset\right\}, \quad D_1 \triangleq S_1$$

$$\text{and} \quad D_{k+1} = S_{k+1} \setminus S_k \quad \text{for all} \quad k \in \mathbb{N}^+,$$

we observe that Γ admits a measurable selection if and only if $\Gamma|_{D_k}$ admits a measurable selection for each $k \in \mathbb{N}^+$.

We will use the recursive method to define a sequence of Lebesgue measurable functions $\{f_n(\cdot)\}_{n\geq 1}$ with $f_n(\cdot) : D \mapsto X$ so that

$$d(f_n(r), \Gamma(r)) < 2^{-n} \text{ for all } r \in D, \tag{1.40}$$

and so that

$$d(f_n(r), f_{n-1}(r)) < 2^{-n+1} \text{ for all } r \in D. \tag{1.41}$$

(Here and throughout the proof of this theorem, $d(x_1, x_2)$ denotes the distance between x_1 and x_2 in X.)

When this is done, it follows from (1.41) that $\{f_n(r)\}_{n\geq 1}$ is a Cauchy sequence in the Banach space X. Thus, there is $f(r) \in X$ so that $f_n(r) \to f(r)$. From this, as well as the facts that $\Gamma(r)$ is closed and each f_n is Lebesgue measurable, we can pass to the limit for $n \to +\infty$ in (1.40) to see that $f(\cdot)$ is a measurable selection of Γ. This ends the proof of Theorem 1.16.

The remainder is to construct the desired $\{f_n\}_{n\geq 1}$. Let $f_0(\cdot), f_1(\cdot) : D \mapsto X$ be given by $f_0(\cdot) = f_1(\cdot) \equiv 0$. It is clear that $f_0(\cdot)$ and $f_1(\cdot)$ are Lebesgue measurable. Further, it follows from (1.39) that (1.40) and (1.41) hold for $n = 1$. Given $n \geq 2$, suppose that for each $m < n$, f_m with (1.40) and (1.41) has been defined.

We now define the desired $f_n(\cdot)$. Since X is a separable Banach space, there is a dense subset $\{x_k\}_{k\geq 1} \subseteq X$. For each $k \geq 1$, we set

$$E_k \triangleq \left\{ r \in D : d(x_k, f_{n-1}(r)) < 2^{-n+1} \text{ and } d(x_k, \Gamma(r)) < 2^{-n} \right\}. \tag{1.42}$$

Since $f_{n-1}(\cdot)$ and $\Gamma(\cdot)$ are Lebesgue measurable, we obtain that

$$E_k = f_{n-1}^{-1}(\mathscr{O}_{2^{-n+1}}(x_k)) \bigcap \Gamma^{-1}(\mathscr{O}_{2^{-n}}(x_k))$$

is Lebesgue measurable. We further claim that

$$\bigcup_{k\geq 1} E_k = D. \tag{1.43}$$

Indeed, given $r \in D$, $d(f_{n-1}(r), \Gamma(r)) < 2^{-n+1}$ and $\Gamma(r)$ is closed. Thus there is a $y \in \Gamma(r)$ so that

$$d(f_{n-1}(r), y) < 2^{-n+1}. \tag{1.44}$$

Since $\{x_k\}_{k\geq 1}$ is dense, it follows from (1.44) that there exists an $\ell \in \mathbb{N}^+$ so that

$$d(x_\ell, y) < \min\{2^{-n+1} - d(f_{n-1}(r), y), 2^{-n}\}.$$

This implies that $d(x_\ell, f_{n-1}(r)) < 2^{-n+1}$ and $d(x_\ell, \Gamma(r)) \leq d(x_\ell, y) < 2^{-n}$. Thus, we have that $r \in E_\ell \subseteq \bigcup_{k\geq 1} E_k$, which leads to (1.43).

Next, by (1.43), we define

$$f_n(r) \triangleq x_k \quad \text{when} \quad r \in E_k \setminus \bigcup_{i=1}^{k-1} E_i. \tag{1.45}$$

It is clear that $f_n(\cdot)$ is measurable. Moreover, (1.40) and (1.41) follow from (1.45) and (1.42).

Hence, we finish the proof of Theorem 1.16. $\qquad\square$

Remark 1.3 The proof of Theorem 1.16 is adapted from Theorem 5.2.1 in [26] and Theorem 2.23 in Chapter 3 of [19].

Let $D \subseteq \mathbb{R}^n$ be a nonempty and Lebesgue measurable set, and X and Y be two separable Banach spaces. Let the map $\varphi : D \times X \mapsto Y$ satisfy the following conditions:

(i) For any $x \in X$, the map $\varphi(\cdot, x)$ is measurable;

(ii) $\varphi(\cdot, \cdot)$ is uniformly continuous in u locally at any $(r_0, x_0) \in D \times X$,

i.e., for any $(r_0, x_0) \in D \times X$, there exists a constant $\delta > 0$ and a

modulus of continuity $\omega(\cdot) \triangleq \omega(\cdot, r_0, x_0, \delta)$ so that

$$\|\varphi(r, x) - \varphi(r, \widehat{x})\|_Y \leq \omega(\|x - \widehat{x}\|_X)$$

for all $(r, x), (r, \widehat{x}) \in \left(\mathscr{O}_\delta(r_0) \bigcap D\right) \times \mathscr{O}_\delta(x_0)$.

$$\tag{1.46}$$

Theorem 1.17 (Measurable Selection Theorem) *Let $D \subseteq \mathbb{R}^n$ be a nonempty, Lebesgue measurable set with $|D| < +\infty$. Let X and Y be two separable Banach spaces. Let $\Gamma : D \mapsto 2^X \setminus \{\emptyset\}$ be a measurable set-valued map taking closed set values (i.e., for each $r \in D$, $\Gamma(r)$ is closed). Let $\varphi : D \times X \mapsto Y$ satisfy the above condition (1.46). Then for every Lebesgue measurable function $h : D \mapsto Y$ satisfying*

$$h(r) \in \varphi(r, \Gamma(r)) \quad \text{for almost all } r \in D, \tag{1.47}$$

there exists a Lebesgue measurable map $f : D \mapsto X$ so that $f(r) \in \Gamma(r)$ for a.e. $r \in D$ and so that

$$h(r) = \varphi(r, f(r)) \quad \text{for almost all } r \in D. \tag{1.48}$$

Proof Let

$$\widehat{\varphi}(r, x) \triangleq \varphi(r, x) - h(r) \quad \text{for any } (r, x) \in D \times X.$$

It is clear that $\widehat{\varphi}$ satisfies the condition (1.46). Define a set-valued map $\Lambda : D \mapsto 2^X$ by

$$\Lambda(r) \triangleq \{x \in X \mid \widehat{\varphi}(r, x) = 0\} \quad \text{for all} \ \ r \in D. \tag{1.49}$$

For each $r \in D$, it follows from the continuity of $\widehat{\varphi}(r, \cdot)$, (1.47), and (1.49) that

$$\Lambda(r) \ \ \text{and} \ \ \Lambda(r) \bigcap \Gamma(r) \ \ \text{are nonempty and closed.} \tag{1.50}$$

Here, we used the fact that $\Gamma(r)$ is closed.

By (1.50), we can define a set-valued map $\Lambda \bigcap \Gamma : D \mapsto 2^X \setminus \{\emptyset\}$ in the following manner:

$$(\Lambda \bigcap \Gamma)(r) \triangleq \Lambda(r) \bigcap \Gamma(r) \quad \text{for all} \ \ r \in D. \tag{1.51}$$

We now claim that

$$\Lambda : D \mapsto 2^X \setminus \{\emptyset\} \ \ \text{is Lebesgue measurable.} \tag{1.52}$$

When (1.52) is proved, (1.48) follows from the Lebesgue measurability of $\Gamma(\cdot)$, (1.51), (1.50), Theorem 1.16, and (1.49).

To show (1.52), we arbitrarily fix a constant $\rho > 0$. Since X is a separable Banach space, there is a sequence $\{x_k\}_{k \geq 1}$, which is dense in X. For each k, according to the Lusin Theorem, there exists a compact set $D_{\rho,k} \subseteq D$ with $|D \setminus D_{\rho,k}| < \rho/2^k$, so that $\widehat{\varphi}(\cdot, x_k)$ is continuous on $D_{\rho,k}$. Set

$$D_\rho \triangleq \bigcap_{k=1}^{+\infty} D_{\rho,k}.$$

We see that

$$D_\rho \ \ \text{is closed,} \ \ |D \setminus D_\rho| < \rho, \tag{1.53}$$

and

$$\widehat{\varphi}(\cdot, x_k) \ \ \text{is continuous on} \ \ D_\rho \ \ \text{for all} \ \ k \in \mathbb{N}^+. \tag{1.54}$$

The rest of the proof will be carried out by the following two steps:

Step 1. We prove that $\widehat{\varphi}(\cdot, \cdot)$ is continuous on $D_\rho \times X$.

To this end, we arbitrarily fix $(r_0, x_0) \in D_\rho \times X$. Since $\widehat{\varphi}$ satisfies the condition (1.46), there is a constant $\delta > 0$ and a modulus of continuity $\omega(\cdot)$ so that

$$\|\widehat{\varphi}(r, x) - \widehat{\varphi}(r, \widehat{x})\|_Y \leq \omega(\|x - \widehat{x}\|_X)$$
$$\text{for all} \ \ (r, x), (r, \widehat{x}) \in \big(\mathscr{O}_\delta(r_0) \bigcap D\big) \times \mathscr{O}_\delta(x_0). \tag{1.55}$$

Moreover, for arbitrarily fixed $\varepsilon > 0$, there is $\widehat{\delta} \triangleq \widehat{\delta}(\varepsilon) \in (0, \delta/2)$ so that

$$\omega(2\widehat{\delta}) \leq \varepsilon/3. \tag{1.56}$$

Meanwhile, since $\{x_k\}_{k \geq 1}$ is dense in X, there is $k_\varepsilon \in \mathbb{N}^+$ so that

$$\|x_{k_\varepsilon} - x_0\|_X \leq \widehat{\delta}. \tag{1.57}$$

Moreover, according to (1.54), there is $\widetilde{\delta} \triangleq \widetilde{\delta}(\varepsilon) \in (0, \widehat{\delta})$ so that

$$\|\widehat{\varphi}(r, x_{k_\varepsilon}) - \widehat{\varphi}(r_0, x_{k_\varepsilon})\|_Y \leq \varepsilon/3 \text{ for all } r \in \mathscr{O}_{\widetilde{\delta}}(r_0) \bigcap D_\rho. \tag{1.58}$$

Now it follows from (1.57), (1.55), (1.56), and (1.58) that for any $(r, x) \in (\mathscr{O}_{\widetilde{\delta}}(r_0) \bigcap D_\rho) \times \mathscr{O}_{\widetilde{\delta}}(x_0)$,

$$\|x - x_{k_\varepsilon}\|_X \leq \|x - x_0\|_X + \|x_0 - x_{k_\varepsilon}\|_X \leq \widetilde{\delta} + \widehat{\delta} \leq 2\widehat{\delta} \leq \delta,$$

and

$$\|\widehat{\varphi}(r, x) - \widehat{\varphi}(r_0, x_0)\|_Y \leq \|\widehat{\varphi}(r, x) - \widehat{\varphi}(r, x_{k_\varepsilon})\|_Y$$
$$+ \|\widehat{\varphi}(r, x_{k_\varepsilon}) - \widehat{\varphi}(r_0, x_{k_\varepsilon})\|_Y + \|\widehat{\varphi}(r_0, x_{k_\varepsilon}) - \widehat{\varphi}(r_0, x_0)\|_Y$$
$$\leq \varepsilon/3 + \varepsilon/3 + \varepsilon/3 = \varepsilon.$$

Since (r_0, x_0) is taken arbitrarily from $D_\rho \times X$, we complete the proof of Step 1.

Step 2. We finish the proof of (1.52).

By Step 1, (1.49) and the first conclusion in (1.53), we find that

$$\text{Graph}(\Lambda|_{D_\rho}) \triangleq \{(r, x) \mid r \in D_\rho \text{ and } x \in \Lambda(r)\} \text{ is closed.} \tag{1.59}$$

Set

$$\widehat{D} \triangleq \bigcup_{m=1}^{+\infty} D_{1/m}. \tag{1.60}$$

It follows from (1.53) that

$$\widehat{D} \text{ is a Borel set and } |D \setminus \widehat{D}| = 0. \tag{1.61}$$

Meanwhile, by (1.59) and (1.60), we see that

$$\text{Graph}(\Lambda|_{\widehat{D}}) \triangleq \{(r, x) \mid r \in \widehat{D} \text{ and } x \in \Lambda(r)\}$$
$$= \bigcup_{m=1}^{+\infty} \text{Graph}(\Lambda|_{D_{1/m}}) \text{ is a Borel set.} \tag{1.62}$$

Let $P_D : D \times X \mapsto D$ be the projection operator defined by

$$P_D(r, x) \triangleq r \quad \text{for all} \quad (r, x) \in D \times X.$$

Then, for any nonempty and closed set $S \subseteq X$, we have that

$$\Lambda^{-1}(S) = P_D\big(\text{Graph}(\Lambda) \bigcap (D \times S)\big)$$
$$= P_D\big(\text{Graph}(\Lambda) \bigcap (\widehat{D} \times S)\big) \bigcup P_D\big(\text{Graph}(\Lambda) \bigcap ((D \setminus \widehat{D}) \times S)\big),$$

which indicates that

$$\Lambda^{-1}(S) = P_D\big(\text{Graph}(\Lambda|_{\widehat{D}}) \bigcap (\widehat{D} \times S)\big) \bigcup P_D\big(\text{Graph}(\Lambda) \bigcap ((D \setminus \widehat{D}) \times S)\big). \tag{1.63}$$

By the first conclusion of (1.61) and (1.62), we see that the first term in the right-hand side of (1.63) is Souslinian. Further, it is Lebesgue measurable. For the second term in the right-hand side of (1.63), we observe that

$$P_D\big(\text{Graph}(\Lambda) \bigcap ((D \setminus \widehat{D}) \times S)\big) \subseteq P_D\big((D \setminus \widehat{D}) \times S\big) = D \setminus \widehat{D}.$$

This, together with the second conclusion of (1.61), implies that

$$P_D\big(\text{Graph}(\Lambda) \bigcap ((D \setminus \widehat{D}) \times S)\big) \quad \text{is a Lebesgue measurable set.}$$

Hence, by (1.63), $\Lambda^{-1}(S)$ is Lebesgue measurable. This implies (1.52).

Thus, we finish the proof of Theorem 1.17.   □

Remark 1.4 The proof of Theorem 1.17 is based on Theorem 2.2 in [15] and Theorem 2.24 in Chapter 3 of [19].

Corollary 1.1 (Filippov Lemma) *Let Y and U be separable Banach spaces. Let $D \subseteq \mathbb{R}^n$ be a Lebesgue measurable set with $|D| < +\infty$. Let $\Gamma : D \mapsto 2^U \setminus \{\emptyset\}$ be a Lebesgue measurable set-valued map taking closed set values. Let $f : D \times Y \times U \mapsto Y$ satisfy the conditions:*

(i) f is measurable in (r, y, u);

(ii) f is uniformly continuous in u locally at (r_0, y_0, u_0) for a.e. $r_0 \in D$,

each $y_0 \in Y$ and each $u_0 \in U$;

(iii) For each $u \in U$ and for a.e. $r \in D$, $f(r, \cdot, u) : Y \mapsto Y$ is continuous.
$$\tag{1.64}$$

Then for every Lebesgue measurable function $h : D \mapsto Y$ and $z \in C(D; Y)$ satisfying

$$h(r) \in f(r, z(r), \Gamma(r)) \quad \text{for a.e. } r \in D,$$

there exists a measurable selection $u(r) \in \Gamma(r)$, a.e. $r \in D$, so that

$$h(r) = f(r, z(r), u(r)) \quad \text{for a.e. } r \in D.$$

Proof We define $\varphi : D \times U \mapsto Y$ by

$$\varphi(r, u) \triangleq f(r, z(r), u) \quad \text{for all } (r, u) \in D \times U.$$

By Theorem 1.17, we only need to check that φ satisfies the conditions in (1.46). To this end, we use the measurability of $z(\cdot)$ to find a sequence of simple functions $z_m(\cdot) : D \mapsto Y$ converging pointwisely to $z(\cdot)$. Then, according to (i) and (iii) in (1.64), for each $u \in U$, the function $r \to f(r, z_m(r), u)$ is measurable and

$$\lim_{m \to +\infty} f(r, z_m(r), u) = f(r, z(r), u) = \varphi(r, u) \quad \text{for a.e. } r \in D.$$

These imply that φ satisfies the first condition in (1.46).

Next, according to the second condition in (1.64), for a.e. $r_0 \in D$ and for each $u_0 \in U$, there is a constant $\delta > 0$ and a modulus of continuity $\omega(\cdot) \triangleq \omega(\cdot; r_0, z(r_0), u_0, \delta)$ so that

$$\|f(r, y, u) - f(r, y, \widehat{u})\|_Y \le \omega(\|u - \widehat{u}\|_U), \tag{1.65}$$

for all $r \in \mathscr{O}_\delta(r_0) \bigcap D$, $y \in \mathscr{O}_\delta(z(r_0))$, and $u, \widehat{u} \in \mathscr{O}_\delta(u_0)$. Since $z \in C(D; Y)$, there is $\widehat{\delta} \in (0, \delta)$ so that

$$\|z(r) - z(r_0)\|_Y < \delta \quad \text{for all } r \in \mathscr{O}_{\widehat{\delta}}(r_0) \bigcap D.$$

This, together with (1.65), implies that

$$\|f(r, z(r), u) - f(r, z(r), \widehat{u})\|_Y \le \omega(\|u - \widehat{u}\|_U)$$

for all $(r, u), (r, \widehat{u}) \in (\mathscr{O}_{\widehat{\delta}}(r_0) \bigcap D) \times \mathscr{O}_{\widehat{\delta}}(u_0)$, i.e., φ satisfies the second condition in (1.46).

Hence, we finish the proof of Corollary 1.1. □

1.3 Controllability and Observability Estimate

In this section, some results on controllability and observability estimates for linear differential equations are presented. We start with some basic concepts and results on controllability and observability.

Let X and U be two reflexive Banach spaces. Let $A : \mathscr{D}(A) \subseteq X \mapsto X$ be a linear operator, which generates a C_0 semigroup $\{e^{At}\}_{t \geq 0}$ on X. Consider the following controlled system:

$$y(t) = e^{At}x + \int_0^t e^{A(t-s)} f(s, y(s), u(s))\mathrm{d}s, \quad t \geq 0, \tag{1.66}$$

where f satisfies proper conditions so that for any $x \in X$ and $u \in L^p(0, +\infty; U)$ with $p \in [1, +\infty]$, (1.66) admits a unique solution $y(\cdot) \triangleq y(\cdot; 0, x, u) \in C([0, +\infty); X)$. This is the case if, for instance, $f(t, x, u)$ is Lipschitz continuous and grows at most linearly in x, and uniformly in (t, u).

Let $T > 0$, $x \in X$, and $p \in [1, +\infty]$. We define the reachable set of the system (1.66) with the initial state x at T and with the control set $L^p(0, T; U)$ by

$$Y_R(T, x; p) \triangleq \{y(T; 0, x, u) \mid u \in L^p(0, T; U)\}.$$

We now introduce definitions of various controllabilities.

Definition 1.14 The system (1.66) is said to be

(i) exactly controllable over $(0, T)$ with the control set $L^p(0, T; U)$ if

$$Y_R(T, x; p) = X \quad \text{for each } x \in X;$$

(ii) approximately controllable over $(0, T)$ with the control set $L^p(0, T; U)$ if

$$\overline{Y_R(T, x; p)} = X \quad \text{for each } x \in X;$$

(iii) null controllable over $(0, T)$ with the control set $L^p(0, T; U)$ if

$$0 \in Y_R(T, x; p) \quad \text{for each } x \in X.$$

The following theorem plays an important role in the study of the controllability.

Theorem 1.18 (Range Comparison Theorem) *Let W, V, and Z be Banach spaces where W is reflexive. Let $F \in \mathscr{L}(V, Z)$ and $G \in \mathscr{L}(W, Z)$. Then*

$$\mathscr{R}(F) \subseteq \mathscr{R}(G) \tag{1.67}$$

if and only if there is $\delta > 0$ so that

$$\|G^*z^*\|_{W^*} \geq \delta \|F^*z^*\|_{V^*} \text{ for all } z^* \in Z^*. \tag{1.68}$$

Furthermore, if there is a constant $C > 0$ so that (1.68) holds with $\delta = 1/C$, then for any $v \in V$, there is $w \in W$ with

$$\|w\|_W \leq C\|v\|_V \tag{1.69}$$

so that

$$F(v) = G(w). \tag{1.70}$$

Proof We first show that $(1.67) \Rightarrow (1.68)$. By contradiction, we suppose that it was not true. Then, for any $n \in \mathbb{N}^+$, there would exist $z_n^* \in Z^*$ so that

$$\|G^* z_n^*\|_{W^*} < \frac{1}{n} \|F^* z_n^*\|_{V^*}. \tag{1.71}$$

Denoting $\widehat{z}_n^* \triangleq \sqrt{n} z_n^* / \|F^* z_n^*\|_{V^*}$, by (1.71), we have that

$$\|G^* \widehat{z}_n^*\|_{W^*} \leq \frac{1}{\sqrt{n}} \to 0 \text{ and } \|F^* \widehat{z}_n^*\|_{V^*} = \sqrt{n} \to +\infty. \tag{1.72}$$

According to (1.67), for any $v \in V$, there exists a $w \in W$ so that $Fv = Gw$. Then by the first conclusion of (1.72), we obtain that

$$\langle F^* \widehat{z}_n^*, v \rangle_{V^*, V} = \langle \widehat{z}_n^*, Fv \rangle_{Z^*, Z} = \langle \widehat{z}_n^*, Gw \rangle_{Z^*, Z} = \langle G^* \widehat{z}_n^*, w \rangle_{W^*, W} \to 0.$$

This, together with Principle of Uniform Boundedness (see Theorem 1.2), implies that the sequence $\{F^* \widehat{z}_n^*\}_{n \geq 1}$ must be bounded, which leads to a contradiction with the second conclusion of (1.72).

We next show that $(1.68) \Rightarrow (1.67)$. For this purpose, we arbitrarily fix $v \in V$ and define $f_v : \mathcal{R}(G^*) \mapsto \mathbb{R}$ by

$$f_v(G^* z^*) \triangleq \langle F^* z^*, v \rangle_{V^*, V}, \quad z^* \in Z^*. \tag{1.73}$$

It is clear that $f_v : \mathcal{R}(G^*) \mapsto \mathbb{R}$ is well defined. Indeed, for any $z_1^*, z_2^* \in Z^*$ with $G^* z_1^* = G^* z_2^*$, by (1.68), we observe that

$$\|F^* (z_1^* - z_2^*)\|_{V^*} \leq \frac{1}{\delta} \|G^* (z_1^* - z_2^*)\|_{W^*} = 0,$$

which indicates that $F^* (z_1^*) = F^* (z_2^*)$. Moreover, it follows from (1.73) and (1.68) that $f_v : \mathcal{R}(G^*) \mapsto \mathbb{R}$ is linear and

$$|f_v(G^* z^*)| = |\langle F^* z^*, v \rangle_{V^*, V}| \leq \frac{1}{\delta} \|v\|_V \|G^* z^*\|_{W^*}.$$

Then, by the Hahn Banach Theorem (see Theorem 1.5) and the reflexivity of W, we can find $w \in W$ so that

$$\|w\|_W \leq \|v\|_V / \delta \tag{1.74}$$

and so that

$$f_v(G^*z^*) = \langle G^*z^*, w \rangle_{W^*, W} \quad \text{for all} \quad z^* \in Z^*. \tag{1.75}$$

From (1.73) and (1.75) it follows that

$$\langle z^*, Fv \rangle_{Z^*, Z} = \langle F^*z^*, v \rangle_{V^*, V}$$
$$= \langle G^*z^*, w \rangle_{W^*, W} = \langle z^*, Gw \rangle_{Z^*, Z} \quad \text{for each} \quad z^* \in Z^*.$$

This implies that

$$Fv = Gw. \tag{1.76}$$

Hence, $\mathscr{R}(F) \subseteq \mathscr{R}(G)$.

Finally, (1.69) and (1.70) follow from (1.74) and (1.76). Thus, we finish the proof of Theorem 1.18. □

Let $T > 0$, $D(\cdot) \in L^1(0, T; \mathscr{L}(X))$, and $B(\cdot) \in L^\infty(0, T; \mathscr{L}(U, X))$. Consider the following controlled linear equation:

$$\begin{cases} \dot{y}(t) = Ay(t) + D(t)y(t) + B(t)u(t), & t \in (0, T), \\ y(0) \in X. \end{cases} \tag{1.77}$$

Write $\{\Phi(t, s) \mid t \geq s \geq 0\}$ for the evolution operator generated by $A + D(\cdot)$ over X. We define two operators $F_T \in \mathscr{L}(X)$ and $G_T \in \mathscr{L}(L^p(0, T; U), X)$ ($p \geq 1$) by:

$$F_T(y_0) \triangleq \Phi(T, 0)y_0 \quad \text{for all} \quad y_0 \in X, \tag{1.78}$$

and

$$G_T(u(\cdot)) \triangleq \int_0^T \Phi(T, s)B(s)u(s)\,ds \quad \text{for all} \quad u \in L^p(0, T; U), \tag{1.79}$$

respectively. Then by Definition 1.14, one can easily check the following fact:

- The system (1.77) is exactly (or null, or approximately) controllable over $(0, T)$ with the control set $L^p(0, T; U)$ if and only if $X \subseteq \mathscr{R}(G_T)$ (or $\mathscr{R}(F_T) \subseteq \mathscr{R}(G_T)$, or $\overline{\mathscr{R}(G_T)} = X$).

Next, based on Theorem 1.18 and the above fact, we give another equivalent condition on the exact controllability (or the null controllability) of the system (1.77). To this end, we introduce the adjoint equation of (1.77):

$$\begin{cases} \dot{\varphi}(t) = -A^*\varphi(t) - D(t)^*\varphi(t), & t \in (0, T), \\ \varphi(T) \in X^*. \end{cases} \tag{1.80}$$

We write $\varphi(\cdot; T, \varphi_T)$ for the solution of (1.80) over $[0, T]$ with the terminal condition that $\varphi(T) = \varphi_T \in X^*$. It is clear that

$$G_T^*(\varphi_T) = B(\cdot)^* \varphi(\cdot; T, \varphi_T) \text{ for all } \varphi_T \in X^*. \tag{1.81}$$

Let $p > 1$, $W \triangleq L^p(0, T; U)$, $V = Z \triangleq X$, $G \triangleq G_T$, and $F \triangleq I$ (or $F \triangleq F_T$). Then, by Theorem 1.18 and (1.81), we have the following result:

Corollary 1.2 *Let $p, q \in (1, +\infty)$ and $1/p + 1/q = 1$.*

(i) The system (1.77) is exactly controllable over $(0, T)$ with the control set $L^p(0, T; U)$ if and only if

$$\|\varphi_T\|_{X^*} \leq C \|B(\cdot)^* \varphi(\cdot; T, \varphi_T)\|_{L^q(0,T;U^*)} \text{ for all } \varphi_T \in X^*, \tag{1.82}$$

where $C > 0$ is a constant. Furthermore, if (1.82) holds, then for any $z \in X$ and $y_0 \in X$, there is a control $u \in L^p(0, T; U)$ so that

$$y(T; 0, y_0, u) = z \quad and \quad \|u\|_{L^p(0,T;U)} \leq C\|z - \Phi(T, 0)y_0\|_X.$$

(ii) The system (1.77) is null controllable over $(0, T)$ with the control set $L^p(0, T; U)$ if

$$\|\varphi(0; T, \varphi_T)\|_{X^*} \leq C \|B(\cdot)^* \varphi(\cdot; T, \varphi_T)\|_{L^q(0,T;U^*)} \text{ for all } \varphi_T \in X^*, \tag{1.83}$$

where $C > 0$ is a constant. Furthermore, if (1.83) holds, then for any $y_0 \in X$, there is a control $u \in L^p(0, T; U)$ so that

$$y(T; 0, y_0, u) = 0 \quad and \quad \|u\|_{L^p(0,T;U)} \leq C\|y_0\|_X.$$

We call (1.82) and (1.83) as observability estimates. About the approximate controllability, we have the next theorem.

Theorem 1.19 *Let $p \in [1, +\infty]$. Then the following conclusions are equivalent:*

(i) The system (1.77) is approximately controllable over $(0, T)$ with the control set $L^p(0, T; U)$;
(ii) $\overline{\mathscr{R}(G_T)} = X$;
(iii) The operator G_T^* is injective, i.e., $\mathscr{N}(G_T^*) = \{0\}$;
(iv) $\varphi_T = 0$ if and only if $B(\cdot)^* \varphi(\cdot; T, \varphi_T) = 0$.

Proof It is clear that (i), (ii), and (iii) are equivalent. We now show that (iii)$\Rightarrow$(iv)$\Rightarrow$(ii).

(iii)$\Rightarrow$(iv). On one hand, if $\varphi_T = 0$, then it is obvious that $B(\cdot)^* \varphi(\cdot; T, \varphi_T) = 0$. On the other hand, let $\varphi_T \in X^*$ satisfy $B(\cdot)^* \varphi(\cdot; T, \varphi_T) = 0$. This, together with (1.81), implies that $G_T^*(\varphi_T) = 0$. Then by (iii), we have that $\varphi_T = 0$.

(iv)$\Rightarrow$(ii). By contradiction, we suppose that $\overline{\mathscr{R}(G_T)} \neq X$. Then, there would exist $z \in X$ so that $z \notin \overline{\mathscr{R}(G_T)}$. Since $\overline{\mathscr{R}(G_T)}$ is a nonempty, convex, and closed subset of X, by Theorem 1.11, we can find $y^* \in X^*$ with $\|y^*\|_{X^*} = 1$ so that

$$\langle y^*, z \rangle_{X^*,X} \leq \left\langle y^*, \int_0^T \Phi(T,s)B(s)u(s)\mathrm{d}s \right\rangle_{X^*,X} \quad \text{for all } u \in L^p(0,T;U).$$

Replacing u in the above inequality by ku $(k > 0)$, we find that

$$\int_0^T \langle B(s)^*\Phi(T,s)^*y^*, u(s)\rangle \mathrm{d}s = 0 \quad \text{for all } u \in L^p(0,T;U).$$

Consequently, $B(\cdot)^*\Phi(T,\cdot)^*y^* = 0$. Then, by (iv), we have that $y^* = 0$, which leads to a contradiction.

Hence, we finish the proof of Theorem 1.19. $\qquad\qquad\qquad\qquad\qquad\square$

We next introduce connections between L^∞-null controllability and L^1-observability in Hilbert spaces. Let Y and U be two real separable Hilbert spaces. The controlled system is as:

$$\dot{y}(t) = Ay(t) + D(t)y(t) + B(t)u(t), \quad t \in (0,+\infty), \tag{1.84}$$

where $D(\cdot) \in L^1_{loc}(0,+\infty;\mathscr{L}(Y))$, $A : \mathscr{D}(A) \subseteq Y \mapsto Y$ generates a C_0 semigroup $\{e^{At}\}_{t\geq 0}$ on Y and $B(\cdot) \in L^\infty(0,+\infty;\mathscr{L}(U,Y))$. Given $\tau \in [0,+\infty)$, $y_0 \in Y$ and $u \in L^\infty(\tau,+\infty;U)$, write $y(\cdot;\tau,y_0,u)$ for the solution of (1.84) over $[\tau,+\infty)$, with the initial condition that $y(\tau) = y_0$. Moreover, write $\{\Phi(t,s) : t \geq s \geq 0\}$ for the evolution operator generated by $A + D(\cdot)$ over Y.

Then we have the following equivalence between null controllability and observability estimate.

Theorem 1.20 *The following statements are equivalent:*

(i) *For any $T_2 > T_1 \geq 0$, there exists a constant $C_1(T_1,T_2) \in (0,+\infty)$ so that*

$$\|\Phi(T_2,T_1)^*z\|_Y \leq C_1(T_1,T_2)\|B(\cdot)^*\Phi(T_2,\cdot)^*z\|_{L^1(T_1,T_2;U)} \quad \text{for all } z \in Y. \tag{1.85}$$

(ii) *For any $T_2 > T_1 \geq 0$, there exists a constant $C_2(T_1,T_2) \in (0,+\infty)$ so that for each $y_0 \in Y$, there is a control $u \in L^\infty(T_1,T_2;U)$ satisfying that*

$$y(T_2;T_1,y_0,u) = 0 \quad \text{and} \quad \|u\|_{L^\infty(T_1,T_2;U)} \leq C_2(T_1,T_2)\|y_0\|_Y. \tag{1.86}$$

(iii) *For any $T_2 > T_1 \geq 0$ and $y_0 \in Y$, there exists a control $u \in L^\infty(T_1,T_2;U)$ satisfying that $y(T_2;T_1,y_0,u) = 0$.*

Furthermore, when one of the above three conclusions is valid, the constants $C_1(T_1,T_2)$ in (1.85) and $C_2(T_1,T_2)$ in (1.86) can be taken as the same number.

Proof We divide the proof into the following four steps.

Step 1. We show that (i)⟹(ii).

Suppose that (i) holds. Let $T_2 > T_1 \geq 0$ and let $C_1(T_1, T_2)$ be given by (1.85). Arbitrarily fix $y_0 \in Y$ and set

$$X_{T_1, T_2} \triangleq \{ B(\cdot)^* \Phi(T_2, \cdot)^* z|_{(T_1, T_2)} \mid z \in Y \} \subseteq L^1(T_1, T_2; U).$$

Define a map $\mathscr{F}_{T_1, T_2, y_0} : X_{T_1, T_2} \mapsto \mathbb{R}$ in the following manner:

$$\mathscr{F}_{T_1, T_2, y_0}\big(B(\cdot)^* \Phi(T_2, \cdot)^* z|_{(T_1, T_2)} \big) \triangleq \langle y_0, \Phi(T_2, T_1)^* z \rangle_Y$$

$$\text{for each } z \in Y. \qquad (1.87)$$

We first claim that $\mathscr{F}_{T_1, T_2, y_0}$ is well defined. In fact, if

$$z_1, z_2 \in Y \text{ so that } B(\cdot)^* \Phi(T_2, \cdot)^* z_1 = B(\cdot)^* \Phi(T_2, \cdot)^* z_2 \text{ over } (T_1, T_2),$$

then by (1.85), it follows that $\Phi(T_2, T_1)^* z_1 = \Phi(T_2, T_1)^* z_2$ in Y. Hence, $\mathscr{F}_{T_1, T_2, y_0}$ is well defined. Besides, one can easily check that $\mathscr{F}_{T_1, T_2, y_0}$ is linear. By making use of (1.85) again, we can find that

$$\big| \mathscr{F}_{T_1, T_2, y_0}\big(B(\cdot)^* \Phi(T_2, \cdot)^* z|_{(T_1, T_2)} \big) \big|$$

$$\leq C_1(T_1, T_2) \|y_0\|_Y \| B(\cdot)^* \Phi(T_2, \cdot)^* z \|_{L^1(T_1, T_2; U)} \text{ for all } z \in Y.$$

From this, we see that

$$\| \mathscr{F}_{T_1, T_2, y_0} \|_{\mathscr{L}(X_{T_1, T_2}, \mathbb{R})} \leq C_1(T_1, T_2) \|y_0\|_Y. \qquad (1.88)$$

Since X_{T_1, T_2} is a subspace of $L^1(T_1, T_2; U)$, we can apply the Hahn-Banach Theorem (see Theorem 1.5) to find a functional $\widetilde{\mathscr{F}}_{T_1, T_2, y_0} \in (L^1(T_1, T_2; U))^*$ so that

$$\| \mathscr{F}_{T_1, T_2, y_0} \|_{\mathscr{L}(X_{T_1, T_2}, \mathbb{R})} = \| \widetilde{\mathscr{F}}_{T_1, T_2, y_0} \|_{(L^1(T_1, T_2; U))^*}$$

and

$$\mathscr{F}_{T_1, T_2, y_0}(g) = \widetilde{\mathscr{F}}_{T_1, T_2, y_0}(g) \text{ for all } g \in X_{T_1, T_2}.$$

From these, we can apply the Riesz Representation Theorem (see Theorem 1.4) to find a function $u \in L^\infty(T_1, T_2; U)$ so that

$$\| \mathscr{F}_{T_1, T_2, y_0} \|_{\mathscr{L}(X_{T_1, T_2}, \mathbb{R})} = \|u\|_{L^\infty(T_1, T_2; U)} \qquad (1.89)$$

and so that

$$\mathscr{F}_{T_1, T_2, y_0}(g) = \int_{T_1}^{T_2} \langle g(t), -u(t) \rangle_U \, dt \text{ for all } g \in X_{T_1, T_2}. \qquad (1.90)$$

From (1.87) and (1.90), we see that for each $z \in Y$,

$$\langle \Phi(T_2, T_1) y_0, z \rangle_Y = \int_{T_1}^{T_2} \langle -u(t), B(t)^* \Phi(T_2, t)^* z \rangle_U \, dt$$

$$= -\left\langle \int_{T_1}^{T_2} \Phi(T_2, t) B(t) u(t) dt, z \right\rangle_Y,$$

which indicates that

$$\Phi(T_2, T_1) y_0 + \int_{T_1}^{T_2} \Phi(T_2, t) B(t) u(t) dt = 0.$$

This, along with (1.89) and (1.88), yields (1.86) with $C_2(T_1, T_2) = C_1(T_1, T_2)$.

Step 2. We prove that (ii)$\Rightarrow$(i).

Suppose that (ii) holds. Let $T_2 > T_1 \geq 0$ and let $C_2(T_1, T_2)$ be given by (ii). Arbitrarily fix $y_0 \in Y$. By (ii), there is $u \in L^\infty(T_1, T_2; U)$ so that

$$y(T_2; T_1, y_0, u) = 0 \quad \text{and} \quad \|u\|_{L^\infty(T_1, T_2; U)} \leq C_2(T_1, T_2) \|y_0\|_Y. \tag{1.91}$$

By the equality in (1.91), we find that

$$\langle y_0, \Phi(T_2, T_1)^* z \rangle_Y = -\int_{T_1}^{T_2} \langle u(t), B(t)^* \Phi(T_2, t)^* z \rangle_U \, dt \quad \text{for all } z \in Y.$$

This, along with the inequality in (1.91), yields that

$$|\langle y_0, \Phi(T_2, T_1)^* z \rangle_Y| \leq C_2(T_1, T_2) \|y_0\|_Y \|B(\cdot)^* \Phi(T_2, \cdot)^* z\|_{L^1(T_1, T_2; U)} \quad \text{for all } z \in Y.$$

Since y_0 is arbitrarily taken from Y, the above implies that for all $z \in Y$,

$$\|\Phi(T_2, T_1)^* z\|_Y \leq C_2(T_1, T_2) \|B(\cdot)^* \Phi(T_2, \cdot)^* z\|_{L^1(T_1, T_2; U)},$$

which leads to (1.85) with $C_1(T_1, T_2) = C_2(T_1, T_2)$.

Step 3. We show that (ii)$\Leftrightarrow$(iii).

It is clear that (ii)$\Rightarrow$(iii). We now show the reverse. Suppose that (iii) holds. Let $T_2 > T_1 \geq 0$. Define a linear operator $\mathscr{G}_{T_1, T_2} : L^\infty(T_1, T_2; U) \mapsto Y$ by setting

$$\mathscr{G}_{T_1, T_2}(u) \triangleq \int_{T_1}^{T_2} \Phi(T_2, t) B(t) u(t) dt \quad \text{for each } u \in L^\infty(T_1, T_2; U). \tag{1.92}$$

Then it is clear that $\mathscr{G}_{T_1, T_2}$ is bounded. By (iii), we know that for each $y_0 \in Y$, there is $u \in L^\infty(T_1, T_2; U)$ so that $y(T_2; T_1, y_0, u) = 0$, i.e.,

$$0 = \Phi(T_2, T_1) y_0 + \int_{T_1}^{T_2} \Phi(T_2, t) B(t) u(t) dt. \tag{1.93}$$

From (1.92) and (1.93), we see that

$$\text{Range } \Phi(T_2, T_1) \subseteq \text{Range } \mathscr{G}_{T_1, T_2}. \tag{1.94}$$

Write Q_{T_1, T_2} for the quotient space of $L^\infty(T_1, T_2; U)$ with respect to $\text{Ker } \mathscr{G}_{T_1, T_2}$, i.e.,

$$Q_{T_1, T_2} \triangleq L^\infty(T_1, T_2; U)/\text{Ker } \mathscr{G}_{T_1, T_2}.$$

Let $\pi_{T_1, T_2} : L^\infty(T_1, T_2; U) \mapsto Q_{T_1, T_2}$ be the quotient map. Then π_{T_1, T_2} is surjective and it holds that

$$\begin{aligned}
&\|\pi_{T_1, T_2}(u)\|_{Q_{T_1, T_2}} \\
&= \inf \left\{ \|w\|_{L^\infty(T_1, T_2; U)} : w \in u + \text{Ker } \mathscr{G}_{T_1, T_2} \right\} \quad \text{for each } u \in L^\infty(T_1, T_2; U).
\end{aligned} \tag{1.95}$$

Define a map $\widehat{\mathscr{G}}_{T_1, T_2} : Q_{T_1, T_2} \mapsto Y$ in the following manner:

$$\widehat{\mathscr{G}}_{T_1, T_2}(\pi_{T_1, T_2}(u)) \triangleq \mathscr{G}_{T_1, T_2}(u) \quad \text{for each } \pi_{T_1, T_2}(u) \in Q_{T_1, T_2}. \tag{1.96}$$

One can easily check that $\widehat{\mathscr{G}}_{T_1, T_2}$ is linear and bounded. By (1.96) and (1.94), we see that $\widehat{\mathscr{G}}_{T_1, T_2}$ is injective and that

$$\text{Range } \Phi(T_2, T_1) \subseteq \text{Range } \widehat{\mathscr{G}}_{T_1, T_2}.$$

From these, we find that for each $y_0 \in Y$, there is a unique $\pi_{T_1, T_2}(u_{y_0}) \in Q_{T_1, T_2}$ so that

$$\Phi(T_2, T_1)y_0 = \widehat{\mathscr{G}}_{T_1, T_2}\left(\pi_{T_1, T_2}(u_{y_0})\right). \tag{1.97}$$

We next define another map $\mathscr{T}_{T_1, T_2} : Y \mapsto Q_{T_1, T_2}$ by

$$\mathscr{T}_{T_1, T_2}(y_0) \triangleq \pi_{T_1, T_2}(u_{y_0}) \quad \text{for each } y_0 \in Y. \tag{1.98}$$

One can easily check that $\mathscr{T}_{T_1, T_2}$ is well defined and linear. We will use the Closed Graph Theorem (see Theorem 1.1) to show that $\mathscr{T}_{T_1, T_2}$ is bounded. For this purpose, we let $\{y_k\}_{k \geq 1} \subseteq Y$ satisfy that

$$y_k \to \widehat{y} \text{ in } Y \quad \text{and} \quad \mathscr{T}_{T_1, T_2}(y_k) \to \widehat{h} \text{ in } Q_{T_1, T_2} \text{ as } k \to +\infty. \tag{1.99}$$

Because $\widehat{\mathscr{G}}_{T_1, T_2}$ and $\Phi(T_1, T_2)$ are linear and bounded, it follows from (1.99), (1.98), and (1.97) that

$$\begin{aligned}
\widehat{\mathscr{G}}_{T_1, T_2}(\widehat{h}) &= \lim_{k \to +\infty} \widehat{\mathscr{G}}_{T_1, T_2}\left(\mathscr{T}_{T_1, T_2}(y_k)\right) = \lim_{k \to +\infty} \widehat{\mathscr{G}}_{T_1, T_2}\left(\pi_{T_1, T_2}(u_{y_k})\right) \\
&= \lim_{k \to +\infty} \Phi(T_2, T_1)y_k = \Phi(T_2, T_1)\widehat{y}.
\end{aligned} \tag{1.100}$$

Meanwhile, by (1.97) and (1.98), we find that

$$\Phi(T_2, T_1)\widehat{y} = \widehat{\mathscr{G}}_{T_1,T_2}\big(\pi_{T_1,T_2}(u_{\widehat{y}})\big) = \widehat{\mathscr{G}}_{T_1,T_2}\big(\mathscr{T}_{T_1,T_2}(\widehat{y})\big).$$

This, together with (1.100), yields that $\widehat{\mathscr{G}}_{T_1,T_2}(\widehat{h}) = \widehat{\mathscr{G}}_{T_1,T_2}\big(\mathscr{T}_{T_1,T_2}(\widehat{y})\big)$, which, combined with the injectivity of $\widehat{\mathscr{G}}_{T_1,T_2}$, indicates that $\widehat{h} = \mathscr{T}_{T_1,T_2}(\widehat{y})$. So the graph of $\mathscr{T}_{T_1,T_2}$ is closed. Now we can apply the Closed Graph Theorem (see Theorem 1.1) to see that $\mathscr{T}_{T_1,T_2}$ is bounded. Hence, there is a constant $C(T_1, T_2) \in (0, +\infty)$ so that $\|\mathscr{T}_{T_1,T_2}(y_0)\|_{Q_{T_1,T_2}} \leq C(T_1, T_2)\|y_0\|_Y$ for all $y_0 \in Y$. This, along with (1.98), implies that

$$\|\pi_{T_1,T_2}(u_{y_0})\|_{Q_{T_1,T_2}} \leq C(T_1, T_2)\|y_0\|_Y \quad \text{for each} \ \ y_0 \in Y. \tag{1.101}$$

Meanwhile, by (1.95), we see that for each $y_0 \in Y$, there is $\widetilde{u}_{y_0}$ so that

$$\widetilde{u}_{y_0} \in u_{y_0} + \operatorname{Ker}\mathscr{G}_{T_1,T_2} \quad \text{and}$$

$$\|\widetilde{u}_{y_0}\|_{L^\infty(T_1,T_2;U)} \leq 2\|\pi_{T_1,T_2}(u_{y_0})\|_{Q_{T_1,T_2}}. \tag{1.102}$$

From (1.97), (1.96), (1.102), and (1.101), we find that for each $y_0 \in Y$, there is a control $\widetilde{u}_{y_0} \in L^\infty(T_1, T_2; U)$ so that

$$\Phi(T_2, T_1)y_0 = \mathscr{G}_{T_1,T_2}(\widetilde{u}_{y_0}) \quad \text{and}$$

$$\|\widetilde{u}_{y_0}\|_{L^\infty(T_1,T_2;U)} \leq 2C(T_1, T_2)\|y_0\|_Y. \tag{1.103}$$

Then by (1.92) and (1.103), we see that for each $y_0 \in Y$, there is a control $\widetilde{u}_{y_0} \in L^\infty(T_1, T_2; U)$ so that

$$y(T_2; T_1, y_0, -\widetilde{u}_{y_0}) = 0 \quad \text{and} \quad \|\widetilde{u}_{y_0}\|_{L^\infty(T_1,T_2;U)} \leq 2C(T_1, T_2)\|y_0\|_Y.$$

These lead to (1.86) with $C_2(T_1, T_2) = 2C(T_1, T_2)$.

Step 4. About the constants $C_1(T_1, T_2)$ and $C_2(T_1, T_2)$.

From the proofs in Step 1 to Step 3, we find that the constants $C_1(T_1, T_2)$ in (1.85) and $C_2(T_1, T_2)$ in (1.86) can be taken as the same number, provided that one of the conclusions (i)–(iii) holds.

In summary, we end the proof of Theorem 1.20. $\square$

We next restrict our discussions on linear heat equations. Let $\Omega \subseteq \mathbb{R}^n$ ($n \geq 1$) be a bounded domain with a C^2 boundary $\partial\Omega$. Let $\omega \subseteq \Omega$ be an open and nonempty subset with its characteristic function χ_ω. Let $E \subseteq (0, T)$ be a subset of positive measure. Let χ_E be its characteristic function. Consider the following heat equation:

$$\begin{cases} \partial_t y - \Delta y + ay = \chi_\omega \chi_E u & \text{in } \Omega \times (0, T), \\ y = 0 & \text{on } \partial\Omega \times (0, T), \\ y(0) = y_0 & \text{in } \Omega. \end{cases} \tag{1.104}$$

Here, $y_0 \in L^2(\Omega)$ and $a \in L^\infty(\Omega \times (0, T))$, with the norm: $\|a\|_\infty \triangleq \|a\|_{L^\infty(\Omega \times (0,T))}$. The adjoint equation of (1.104) reads

$$\begin{cases} \partial_t \varphi + \Delta \varphi - a\varphi = 0 & \text{in } \Omega \times (0, T), \\ \varphi = 0 & \text{on } \partial\Omega \times (0, T), \\ \varphi(T) \in L^2(\Omega). \end{cases} \tag{1.105}$$

Using the similar arguments as those used in the proof of Theorem 1.20, we can obtain the following theorem:

Theorem 1.21 *The following statements are equivalent:*

(i) For any $T > 0$, there exists a constant $C_1(T) \in (0, +\infty)$ so that any solution φ to (1.105) satisfies that

$$\|\varphi(0)\|_{L^2(\Omega)} \le C_1(T)\|\chi_\omega \chi_E \varphi\|_{L^1(0,T;L^2(\Omega))}. \tag{1.106}$$

(ii) For any $T > 0$, there exists a constant $C_2(T) \in (0, +\infty)$ so that for each $y_0 \in L^2(\Omega)$, there is a control $u \in L^\infty(0, T; L^2(\Omega))$ satisfying that

$$y(T; 0, y_0, u) = 0 \quad \text{and} \quad \|u\|_{L^\infty(0,T;L^2(\Omega))} \le C_2(T)\|y_0\|_{L^2(\Omega)}. \tag{1.107}$$

Furthermore, when one of the above two conclusions is valid, the constants $C_1(T)$ in (1.106) and $C_2(T)$ in (1.107) can be taken as the same number.

The estimate in (i) of Theorem 1.21 can be derived from the interpolation inequality in the next lemma.

Lemma 1.2 ([23]) *There are two positive constants $C \triangleq C(\Omega, \omega)$ and $\beta \triangleq \beta(\Omega, \omega) \in (0, 1)$ so that for any $T > 0$ and $\varphi_T \in L^2(\Omega)$,*

$$\|\varphi(0)\|_{L^2(\Omega)} \le N\|\varphi_T\|_{L^2(\Omega)}^\beta \|\varphi(0)\|_{L^2(\omega)}^{1-\beta}, \tag{1.108}$$

where φ is the solution to Equation (1.105), with $\varphi(T) = \varphi_T$, and where

$$N \triangleq \exp\left(C\left(1 + T^{-1} + T\|a\|_\infty + \|a\|_\infty^{2/3}\right)\right).$$

Based on Lemma 1.2, we obtain the following estimate from measurable set in time:

Theorem 1.22 ([23]) *Any solution φ to Equation (1.105) satisfies the following observability estimate:*

$$\|\varphi(0)\|_{L^2(\Omega)} \le C(\Omega, \omega, E, T, \|a\|_\infty) \int_{\omega \times E} |\varphi(x, t)| dx dt, \tag{1.109}$$

where

$$C\left(\Omega, \omega, E, T, \|a\|_\infty\right) \triangleq \exp\left(C(\Omega, \omega, E)\right)$$
$$\exp\left(C\left(\Omega, \omega\right)\left(1 + T\|a\|_\infty + \|a\|_\infty^{2/3}\right)\right).$$

Remark 1.5 The inequality (1.108) is a unique continuation estimate at one time. It is a strong estimate from two perspectives as follows: (i) It gives the unique continuation for Equation (1.105). Indeed, if $\varphi(0) = 0$ in a nonempty open set of Ω, then from (1.108), we have that $\varphi(0) = 0$ in Ω. This, along with the backward uniqueness of parabolic equations (see [5]), implies that $\varphi = 0$ in $\Omega \times (0, T)$. (ii) It implies the observability estimate (1.109).

Finally, we consider the following controlled ODE:

$$\begin{cases} \dot{y} = Ay(t) + Bu(t), & t \in (0, T), \\ y(0) \in \mathbb{R}^n, \end{cases} \tag{1.110}$$

where $T > 0$, $A \in \mathbb{R}^{n \times n}$ and $B \in \mathbb{R}^{n \times m}$. Then we have the following result.

Theorem 1.23 ([8]) *For each $y_0 \in \mathbb{R}^n$ and $y_1 \in \mathbb{R}^n$, there exists a control $u \in L^\infty(0, T; \mathbb{R}^m)$ so that $y(T; 0, y_0, u) = y_1$ if and only if*

$$rank(B, AB, \dots, A^{n-1}B) = n \quad (Kalman\ controllability\ rank\ condition).$$

Here, $y(\cdot; 0, y_0, u)$ is the solution to (1.110) with the initial condition $y(0) = y_0$.

Miscellaneous Notes

The material of Section 1.1 is almost standard. Most of them are taken from [19]. We recommend the following references on functional analysis: [7, 24] and [28]. Some of the materials in Section 1.2 are adapted from [1–3, 6, 18, 19, 26], and [15].

The notion of controllability was introduced by R. E. Kalman for finite dimensional systems in [16] (see also [17]). Then it was extended to infinite dimensional systems. We would like to mention the following references on controllability for infinite dimensional systems: [9, 11, 12, 20, 25, 27], and [22].

References

1. C.D. Aliprantis, K.C. Border, *Infinite Dimensional Analysis, A Hitchhiker's Guide*, 3rd edn. (Springer, Berlin, 2006)
2. J. Aubin, *Mathematical Methods of Game and Economic Theory*, Reprint of the 1982 revised edition, with a new preface by the author (Dover Publications, Mineola, NY, 2007)

3. J. Aubin, H. Frankowska, *Set-valued Analysis*, Reprint of the 1990 edition. Modern Birkhäuser Classics (Birkhäuser Boston, Inc, Boston, MA, 2009)
4. V. Barbu, *Nonlinear Semigroup and Differential Equations in Banach Spaces* (Noordhoff International Publishing, Leyden, 1976)
5. C. Bardos, L. Tartar, Sur l'unicité rétrograde des équations parabliques et quelques questions voisines (French). Arch. Ration. Mech. Anal. **50**, 10–25 (1973)
6. G. Bouligand, Sur la semi-continuité d'inclusions et quelques sujets conn exes. Enseignement Mathématique **31**, 14–22 (1932)
7. J.B. Conway, *A Course in Functional Analysis*. Graduate Texts in Mathematics, vol. 96 (Springer, New York, 1985)
8. J.M. Coron, *Control and Nonlinearity*. Mathematical Surveys and Monographs, vol. 136 (American Mathematical Society, Providence, RI, 2007)
9. T. Duyckaerts, X. Zhang, E. Zuazua, On the optimality of the observability inequalities for parabolic and hyperbolic systems with potentials. Ann. Inst. H. Poincaré Anal. Non Linéaire **25**, 1–41 (2008)
10. L.C. Evans, *Partial Differential Equations*. Graduate Studies in Mathematics, vol. 19 (American Mathematical Society, Providence, RI, 1998)
11. H.O. Fattorini, Some remarks on complete controllability. SIAM J. Control **4**, 686–694 (1966)
12. A.V. Fursikov, O. Yu. Imanuvilov, *Controllability of Evolution Equations*. Lecture Notes Series, vol. 34 (Seoul National University, Seoul, 1996)
13. A. Granas, J. Dugundji, *Fixed Point Theory*. Springer Monographs in Mathematics (Springer, New York, 2003)
14. P.R. Halmos, The range of a vector measure. Bull. Am. Math. Soc. **54**, 416–421 (1948)
15. M.Q. Jacobs, Remarks on some recent extensions of Filippov's implicit functions lemma. SIAM J. Control **5**, 622–627 (1967)
16. R.E. Kalman, Contributions to the theory of optimal control. Bol. Soc. Mat. Mexicana **5**, 102–119 (1960)
17. R.E. Kalman, Mathematical description of linear dynamical systems. J. SIAM Control A **1**, 152–192 (1963)
18. K. Kuratowski, Les fonctions semi-continues dans l'espace des ensembles fermés. Fundam. Math. **18**, 148–159 (1932)
19. X. Li, J. Yong, *Optimal Control Theory for Infinite Dimensional Systems* (Birkhäuser, Boston, 1995)
20. J.-L. Lions, Exact controllability, stabilization and perturbations for distributed systems. SIAM Rev. **30**, 1–68 (1988)
21. A. Lyapunov, Sur les fonctions-vecteurs complètement additives (Russian). Bull. Acad. Sci. URSS Sér. Math. **4**, 465–478 (1940)
22. K.D. Phung, G. Wang, An observability estimate for parabolic equations from a measurable set in time and its application. J. Eur. Math. Soc. **15**, 681–703 (2013)
23. K.D. Phung, L. Wang, C. Zhang, Bang-bang property for time optimal control of semilinear heat equation. Ann. Inst. H. Poincaré Anal. Non Linéaire **31**, 477–499 (2014)
24. W. Rudin, *Functional Analysis*. International Series in Pure and Applied Mathematics, 2nd edn. (McGraw-Hill Inc., New York, 1991)
25. D.L. Russell, Controllability and stabilizability theory for linear partial differential equations: recent progress and open questions. SIAM Rev. **20**, 639–739 (1978)
26. S.M. Srivastava, *A Course on Borel Sets*. Graduate Texts in Mathematics, vol. 180 (Springer, New York, 1998)
27. M. Tucsnak, G. Weiss, *Observation and Control for Operator Semigroups* (Birkhäuser Verlag, Basel, 2009)
28. K. Yosida, *Functional Analysis*. Grundlehren der Mathematischen Wissenschaften, 6th edn., vol. 123 (Springer, Berlin, 1980)
29. X. Zhang, X. Li, Z. Chen, *Differential Equation Theory for Optimal Control Systems (Chinese)* (Higher Education Publishing House, Shanghai, 1989)

Chapter 2
Time Optimal Control Problems

2.1 Overview on Time Optimal Control Problems

2.1.1 Introduction

Optimization is a selection of the best "candidate" from a set of available alternatives, in virtue of some criterion. As a branch of optimization, optimal control problem is to ask for a control from an available set so that a certain optimality criterion, related to a given dynamic system, is achieved. When the criterion is the elapsed time, we call such problem a time optimal control problem. This book is concerned with time optimal control theory.

Let us give a practical example of time optimal controls. Consider a car traveling through a hilly road. How should the driver press the accelerator pedal in order to minimize the total traveling time? In this example, the control law refers to the way via which the driver presses the accelerator pedal and shifts the gears. The dynamic system consists of both the car and the road. The optimality criterion is the minimization of the total traveling time. In general, control problems include some ancillary constraints. For instance, in the above example, the amount of available fuel and speed of the car should be limited, and the accelerator pedal cannot be pushed through the floor of the car, and so on.

We next introduce a mathematical framework of time optimal control problems. Our framework contains the following four parts (or four ingredients):

The first part is a controlled system (or a state equation) which reads:

$$\dot{y}(t) = Ay(t) + f(t, y(t), u(t)), \quad t \in (0, +\infty), \tag{2.1}$$

where and throughout this monograph, the following **Basic assumptions** are effective:

© Springer International Publishing AG, part of Springer Nature 2018
G. Wang et al., *Time Optimal Control of Evolution Equations*, Progress
in Nonlinear Differential Equations and Their Applications 92,
https://doi.org/10.1007/978-3-319-95363-2_2

($\mathcal{H}_1$) State space Y is a real separable Hilbert space and control space U is another real separable Hilbert space.

($\mathcal{H}_2$) The operator $A : \mathcal{D}(A) \subseteq Y \mapsto Y$ generates a C_0 semigroup $\{e^{At}\}_{t \geq 0}$ on Y and $f : (0, +\infty) \times Y \times U \mapsto Y$ is a given map.

About the Equation (2.1), we would like to mention what follows:

(i) Recall that $L(a, b; U)$, with $0 \leq a < b < +\infty$, is the space of all Lebesgue measurable functions defined on $(a, b) \subseteq (0, +\infty)$ and taking values in U. If for each $y_0 \in Y$, each pair (a, b) (with $0 \leq a < b < +\infty$) and each $u \in L(a, b; U)$, the Equation (2.1), with the initial condition that $y(a) = y_0$ (or with the initial data (a, y_0)), has a unique mild solution $y(\cdot; a, y_0, u, b)$ over $[a, b]$, then we call this solution the state trajectory over $[a, b]$, corresponding to the control u and the initial data (a, y_0).

(ii) Recall that $y(\cdot) \triangleq y(\cdot; a, y_0, u, b)$ is a mild solution to (2.1), if $y \in C([a, b]; Y)$ and

$$y(t) = e^{A(t-a)}y_0 + \int_a^t e^{A(t-s)} f(s, y(s), u(s)) \, ds, \quad t \in [a, b]. \tag{2.2}$$

(iii) Given $y_0 \in Y$ and $a \geq 0$, we simply write $y(\cdot; a, y_0, u)$ for $y(\cdot; a, y_0, u, b)$, provided that $y(\cdot; a, y_0, u, b)$ exists for some $b > a$ and $u \in L(a, b; U)$.

(iv) When we write $y(T; a, y_0, u)$ (with $T > a \geq 0$, $y_0 \in Y$ and $u \in L(a, T; U)$), we agree that the Equation (2.1), with the initial condition that $y(a) = y_0$, has a unique mild solution $y(\cdot; a, y_0, u)$ on $[a, T]$. Hence, $y(T; a, y_0, u)$ denotes the value of this solution at time T.

The second part is a control constraint set (or control constraint). Generally, in time optimal control problems, controls are always constrained. Most kinds of control constraints can be put into the framework that $u(t) \in \mathbb{U}$ a.e. $t \in (0, +\infty)$, where $\mathbb{U}$ is a nonempty subset of U.

The third part is a starting set Q_S, which is a nonempty subset of $[0, +\infty) \times Y$. In many literatures, it is taken as $\{t_0\} \times \{y_0\}$, with $t_0 \in [0, +\infty)$ and $y_0 \in Y$.

The last part is an ending set Q_E, which is also a nonempty subset of $(0, +\infty) \times Y$. In many literatures, it is taken as $(0, +\infty) \times S$, with S a subset of Y.

With the aid of the above four parts, we define the following set:

$$\mathcal{A}_{ad} \triangleq \Big\{(t_S, y_0, t_E, u) \in Q_S \times (t_S, +\infty) \times L(t_S, t_E; \mathbb{U}) \; \Big|$$

$$(t_E, y(t_E; t_S, y_0, u)) \in Q_E \Big\}, \tag{2.3}$$

where $L(t_S, t_E; \mathbb{U})$ is the set of all functions v in $L(t_S, t_E; U)$ so that $v(t) \in \mathbb{U}$ for a.e. $t \in (t_S, t_E)$. Then the time optimal control problem studied in most parts of this monograph can be stated as follows:

Problem (TP):   Find a tetrad $(t_S^*, y_0^*, t_E^*, u^*)$ in $\mathcal{A}_{ad}$ so that

$$t^* \triangleq t_E^* - t_S^* = \inf_{(t_S, y_0, t_E, u) \in \mathscr{A}_{ad}} (t_E - t_S). \tag{2.4}$$

In this problem,

(a) Each (t_S, y_0, t_E, u) in $\mathscr{A}_{ad}$ is called an admissible tetrad, while $\mathscr{A}_{ad}$ is called the set of admissible tetrads (or simply the admissible set). When $(t_S, y_0, t_E, u) \in \mathscr{A}_{ad}$, u is called an admissible control.

(b) When the infimum in (2.4) is reached at $(t_S^*, y_0^*, t_E^*, u^*)$, the tetrad $(t_S^*, y_0^*, t_E^*, u^*)$ is called an optimal tetrad; the number $t_E^* - t_S^*$ is called the optimal time; u^* is called an optimal control, *which is a function, defined on* (t_S^*, t_E^*) *and taking values on* $\mathbb{U}$; the solution $y(\cdot; t_S^*, y_0^*, u^*)$ (to the Equation (2.1) over $[t_S^*, t_E^*]$) is called an optimal trajectory (or an optimal state).

Remark 2.1 (i) According to definitions of $\mathscr{A}_{ad}$ and the optimal tetrad, respectively, each admissible tetrad (t_S, y_0, t_E, u) satisfies that $t_E > t_S$, and that each optimal tetrad $(t_S^*, y_0^*, t_E^*, u^*)$ satisfies that $t_E^* > t_S^*$. (ii) When $\mathscr{A}_{ad} = \emptyset$, we agree that $t^* = +\infty$.

Two special cases will be introduced as follows:

(i) The first case is that t_S is fixed. We call the corresponding **Problem (TP)** a minimal time control problem. We use $(TP)_{min}^{Q_S, Q_E}$ to denote **Problem (TP)**. It covers the following important minimal time control problem:

$(P)_{min}$ $t^* \triangleq \inf\{t \in (0, +\infty) \mid y(t; 0, y_0, u) \in Y_E$ and $u \in L(0, t; \mathbb{U})\}$, $\tag{2.5}$

where $y_0 \in Y$ and $Y_E \subseteq Y$ are fixed. (The set Y_E is called the target of this problem.) The above problem can be treated as the following time optimal control problem: (It will be explained later.)

$(\widetilde{P})_{min}$ $\widetilde{t}^* \triangleq \inf\{t \in (0, +\infty) \mid y(t; 0, y_0, u) \in Y_E$ and $u \in L(0, +\infty; \mathbb{U})\}$. $\tag{2.6}$

In this problem,

- a tetrad $(0, y_0, t, u)$ is called admissible, if

$$t \in (0, +\infty), \quad u \in L(0, +\infty; \mathbb{U}) \quad \text{and} \quad y(t; 0, y_0, u) \in Y_E;$$

- a tetrad $(0, y_0, \widetilde{t}^*, \widetilde{u}^*)$ is called optimal if

$$\widetilde{t}^* \in (0, +\infty), \quad \widetilde{u}^* \in L(0, +\infty; \mathbb{U}) \quad \text{and} \quad y(\widetilde{t}^*; 0, y_0, \widetilde{u}^*) \in Y_E;$$

- when $(0, y_0, \widetilde{t}^*, \widetilde{u}^*)$ is an optimal tetrad, $\widetilde{t}^*$ and $\widetilde{u}^*$ are called the optimal time and an optimal control, respectively.

The following three facts can be checked easily:

- $t^* = \tilde{t}^*$;
- The infimum in (2.5) can be reached if and only if the infimum in (2.6) can be reached;
- The zero extension of each optimal control to $(P)_{min}$ over $(0, +\infty)$ is an optimal control to $(\tilde{P})_{min}$; the restriction of each optimal control to $(\tilde{P})_{min}$ over $(0, t^*)$ is an optimal control to $(P)_{min}$.

Because of the above three facts, we can treat $(P)_{min}$ and $(\tilde{P})_{min}$ as the same problem.

(ii) The second case is that t_E is fixed. We call the corresponding **Problem (TP)** a maximal time control problem. We use $(TP)_{max}^{Q_S, Q_E}$ to denote **Problem (TP)**. It covers the following standard maximal time control problem (see, for instance, [33] and [38]):

$$(P)_{max} \qquad t^* \triangleq \sup\{t \in [0, t_E) \mid y(t_E; t, y(t; 0, y_0, 0), u) \in \mathcal{O} \atop \text{and } u \in L(t, t_E; \mathbb{U})\}, \qquad (2.7)$$

where $t_E > 0$, $y_0 \in Y$, and $\mathcal{O} \subseteq Y$ are fixed. (The set $\mathcal{O}$ is called the target of this problem.) The above $(P)_{max}$ is indeed equivalent to the following problem: (We will explain the reasons later.)

$$(\tilde{P})_{max} \qquad \tilde{t}^* \triangleq \sup\{t \in [0, t_E) \mid y(t_E; 0, y_0, \chi_{(t,t_E)}u) \in \mathcal{O} \atop \text{and } u \in L(0, t_E; \mathbb{U})\}. \qquad (2.8)$$

In this problem,

- we call $(0, y_0, t_E, \chi_{(t,t_E)}u)$ an admissible tetrad, if

$$t \in [0, t_E), \quad u \in L(0, t_E; \mathbb{U}) \quad \text{and} \quad y(t_E; 0, y_0, \chi_{(t,t_E)}u) \in \mathcal{O};$$

- we call $(0, y_0, t_E, \chi_{(\tilde{t}^*, t_E)}\tilde{u}^*)$ an optimal tetrad, if

$$\tilde{t}^* \in [0, t_E), \quad \tilde{u}^* \in L(0, t_E; \mathbb{U}) \quad \text{and} \quad y(t_E, 0, y_0, \chi_{(\tilde{t}^*, t_E)}\tilde{u}^*) \in \mathcal{O};$$

- when $(0, y_0, t_E, \chi_{(\tilde{t}^*, t_E)}\tilde{u}^*)$ is an optimal tetrad, $t_E - \tilde{t}^*$ and $\tilde{u}^*$ are called the optimal time and an optimal control, respectively.

We can easily check the following three facts:

- $t^* = \tilde{t}^*$;
- The supreme in (2.7) can be reached if and only if the supreme in (2.8) can be reached;
- The zero extension of each optimal control to $(P)_{max}$ over $(0, t_E)$ is an optimal control to $(\tilde{P})_{max}$, while the restriction of each optimal control to $(\tilde{P})_{max}$ over (t^*, t_E) is an optimal control to $(P)_{max}$.

Based on the above-mentioned facts, we can treat $(P)_{max}$ and $(\tilde{P})_{max}$ as the same problem.

2.1.2 Different Ingredients in Time Optimal Control Problems

As mentioned in the above subsection, **Problem (TP)** consists of four ingredients. Different time optimal control problems have different ingredients. We now are going to introduce some special but important ingredients.

- For state equations, we will focus on what follows:

$$\dot{y}(t) = Ay(t) + f(t, y(t)) + B(t)u(t), \quad t \in (0, +\infty), \tag{2.9}$$

where $A : \mathscr{D}(A) \subseteq Y \mapsto Y$ generates a C_0 semigroup $\{e^{At}\}_{t \geq 0}$ on Y, $f : (0, +\infty) \times Y \mapsto Y$ is a given map, and $B(\cdot) \in L^\infty(0, +\infty; \mathscr{L}(U, Y))$. Equation (2.9) is a semilinear time-varying controlled equation where controls enter the equation linearly. It contains the following important cases:

- Semilinear time-invariant evolution equation:

$$\dot{y}(t) = Ay(t) + f(y(t)) + Bu(t), \quad t \in (0, +\infty). \tag{2.10}$$

- Linear time-varying evolution equation:

$$\dot{y}(t) = Ay(t) + D(t)y(t) + B(t)u(t), \quad t \in (0, +\infty), \tag{2.11}$$

where $D(\cdot) \in L^1_{loc}(0, +\infty; \mathscr{L}(Y))$.
- Linear time-invariant evolution equation:

$$\dot{y}(t) = Ay(t) + Bu(t), \quad t \in (0, +\infty). \tag{2.12}$$

- For the control constraint sets, we will focus on the following two important types:

- Ball type: $\mathbb{U}$ is the closed ball $B_r(0)$ in U, centered at 0 and of radius $r > 0$.
- Rectangle type: When $U = \mathbb{R}^m$, $\mathbb{U}$ is the following rectangle:

$$\mathbb{U} \triangleq \left\{ u \in \mathbb{R}^m \mid -\lambda_j \leq \langle u, e_j \rangle_{\mathbb{R}^m} \leq \lambda_j \text{ for any } j = 1, 2, \ldots, m \right\}, \tag{2.13}$$

where $\lambda_j, 1 \leq j \leq m$, are positive constants and $\{e_j\}_{j=1}^m$ forms the standard basis of $\mathbb{R}^m$;
When $\dim U = +\infty$, $\mathbb{U}$ is the following rectangle:

$$\mathbb{U} \triangleq \left\{ u \in U \mid -\lambda_j \leq \langle u, e_j \rangle_U \leq \lambda_j \text{ for any } j \geq 1 \right\}, \tag{2.14}$$

where $\lambda_j, j = 1, 2 \ldots$, are positive numbers and $\{e_j\}_{j \geq 1}$ is an orthonormal basis of U.

- For starting sets and ending sets, we mainly concern the following cases:

  - We take

$$Q_S \triangleq \{0\} \times Y_S \quad \text{and} \quad Q_E \subseteq (0, +\infty) \times Y, \qquad (2.15)$$

    where Y_S and Q_E are respectively nonempty subsets of Y and $(0, +\infty) \times Y$. In this case, $t_S = 0$.
  - We take

$$Q_S \triangleq \{(t_S, y(t_S; 0, y_0, 0)) : t_S \in [0, t_E)\} \quad \text{and} \quad Q_E \triangleq \{t_E\} \times B_r(z_d), \qquad (2.16)$$

    where $t_E > 0$, $r > 0$, $y_0 \in Y$ and $z_d \in Y$ are fixed so that $y(t_E; 0, y_0, 0) \notin B_r(z_d)$. In this case, $B_r(z_d)$ is called the target set (or the target).

Corresponding to (2.15), **Problem (TP)** is a minimal time control problem $(TP)_{min}^{Q_S, Q_E}$. Corresponding to (2.16), **Problem (TP)** is a maximal time control problem $(TP)_{max}^{Q_S, Q_E}$. In general, a minimal time control problem is to ask for a control (from a constraint set) driving the corresponding solution of a controlled equation from the initial state at the beginning to a target in the shortest time, while a maximal time control problem is to ask for a control (from a constraint set) driving the corresponding solution of a controlled equation from the initial state at the beginning to a given target at the fixed ending time so that the initiation of the control is delayed as late as possible.

We end this subsection with introducing a special time optimal control problem which is not under the above framework but is important. It is called a minimal blowup time control problem and can be stated as follows:

$$\inf\{\hat{t} \in (0, +\infty) \mid \lim_{t \to \hat{t}-} \|y(t; 0, y_0, u|_{(0,t)})\|_Y = +\infty \text{ and } u \in L(0, \hat{t}; \mathbb{U})\}, \qquad (2.17)$$

where $y_0 \in Y$. The problem (2.17) can be formally put into our framework by setting

$$Q_S \triangleq \{(0, y_0)\} \quad \text{and} \quad Q_E \triangleq (0, +\infty) \times \{\infty\}. \qquad (2.18)$$

By a convention, "$(t_E, y(t_E; 0, y_0, u)) \in Q_E$" means that the following conclusions are true:

(i) For each $t \in (0, t_E)$ and each $u \in L(0, t_E; \mathbb{U})$, the Equation (2.1), with the initial condition that $y(0) = y_0$, has a unique mild solution $y(\cdot; 0, y_0, u|_{(0,t)})$ on $[0, t]$;

(ii) It holds that

$$\lim_{t \to t_E-} \|y(t; 0, y_0, u|_{(0,t)})\|_Y = +\infty. \qquad (2.19)$$

The minimal blowup time control problem (2.17) differs essentially from the problem $(TP)_{min}^{Q_S,Q_E}$ defined by (2.4) with t_S fixed. The problem (2.17) cannot be really put into the framework of (2.4). When $Q_E = (0, +\infty) \times Y_E$, with Y_E a subset of Y, the target of $(TP)_{min}^{Q_S,Q_E}$ (defined by (2.4) with t_S fixed) is the subset Y_E in the state space Y, while the target of the problem (2.17) is outside the state space Y. The problem (2.17) is independently important. We explain it as follows: The differential systems whose solutions have the behavior of blowup replicate a large class of phenomena in applied science. In certain cases, the blowup of a solution is desirable. Thus, it is obviously appealing to minimize the blowup time by making use of controls in certain circumstances. For instance, the optimal process of starting a car engine can be thought of as minimizing the blowup time.

2.2 Connections of Minimal and Maximal Time Controls

First of all, we give an example. When a lady plans to drive her car from the city Shanghai to the city Wuhan along a fixed road, how can she arrive at Wuhan in the shortest time? This problem can be formulated as a time optimal control problem. When the departure time is fixed, it is a minimal time control problem. When the arrival time is fixed, it is a maximal time control problem. Under some conditions, these two problems are exchangeable. This hints us that some minimal and maximal time control problems are equivalent in a certain sense. The main purpose of this section is to present several cases, where a minimal time control problem is equivalent to a maximal time control problem in a certain sense, or a minimal time control problem is equivalent to another minimal time control problem in a certain sense.

We begin with studying the connection of time optimal control problems $(TP)_{min}^{Q_S,Q_E}$ and $(TP)_{max}^{\widehat{Q}_S,\widehat{Q}_E}$ in the following CASE ONE:

- The controlled system is (2.1), with $f(t, y, u) = f(y, u)$ (i.e., (2.1) is time-invariant);
- The control constraint set $\mathbb{U} \subseteq U$ is nonempty;
- The starting sets and the ending sets are as: $Q_S = \{0\} \times Y_1$, $Q_E = (0, +\infty) \times Y_2$; $\widehat{Q}_S = [0, \widehat{t}_E) \times Y_1$, and $\widehat{Q}_E = \{\widehat{t}_E\} \times Y_2$, where Y_1 and Y_2 are nonempty subsets of Y and $\widehat{t}_E > 0$.

In CASE ONE, the admissible sets for $(TP)_{min}^{Q_S,Q_E}$ and $(TP)_{max}^{\widehat{Q}_S,\widehat{Q}_E}$ are respectively as follows:

$$\mathscr{A}_{ad} = \{(0, y_0, t_E, u) \mid y_0 \in Y_1, t_E > 0, u \in L(0, t_E; \mathbb{U}), y(t_E; 0, y_0, u) \in Y_2\}$$

and

$$\widehat{\mathscr{A}}_{ad} = \{(\widehat{t}_S, \widehat{y}_0, \widehat{t}_E, \widehat{u}) \mid 0 \le \widehat{t}_S < \widehat{t}_E, \widehat{y}_0 \in Y_1,$$
$$\widehat{u} \in L(\widehat{t}_S, \widehat{t}_E; \mathbb{U}), y(\widehat{t}_E; \widehat{t}_S, \widehat{y}_0, \widehat{u}) \in Y_2\}.$$

Definition 2.1 In CASE ONE, the problems $(TP)_{min}^{Q_S,Q_E}$ and $(TP)_{max}^{\widehat{Q}_S,\widehat{Q}_E}$ are said to be equivalent if the following conditions are valid:

(i) If $(0, y_0^*, t_E^*, u^*)$ is an optimal tetrad to $(TP)_{min}^{Q_S,Q_E}$ and $\widehat{t}_E \geq t_E^*$, then $(\widehat{t}_E - t_E^*, y_0^*, \widehat{t}_E, v^*)$, with $v^*(\cdot) = u^*(\cdot - \widehat{t}_E + t_E^*)$, is an optimal tetrad to $(TP)_{max}^{\widehat{Q}_S,\widehat{Q}_E}$.

(ii) If $(\widehat{t}^*_S, \widehat{y}_0^*, \widehat{t}_E, \widehat{u}^*)$ is an optimal tetrad to $(TP)_{max}^{\widehat{Q}_S,\widehat{Q}_E}$, then $(0, \widehat{y}_0^*, \widehat{t}_E - \widehat{t}^*_S, \widehat{v}^*)$, with $\widehat{v}^*(\cdot) = \widehat{u}^*(\cdot + \widehat{t}^*_S)$, is an optimal tetrad to $(TP)_{min}^{Q_S,Q_E}$.

Theorem 2.1 *In CASE ONE, the problems* $(TP)_{min}^{Q_S,Q_E}$ *and* $(TP)_{max}^{\widehat{Q}_S,\widehat{Q}_E}$ *are equivalent.*

Proof We first prove (i) in Definition 2.1. Assume that $(0, y_0^*, t_E^*, u^*)$ is an optimal tetrad to $(TP)_{min}^{Q_S,Q_E}$ and $\widehat{t}_E \geq t_E^*$. In order to prove that $(\widehat{t}_E - t_E^*, y_0^*, \widehat{t}_E, v^*)$, with $v^*(\cdot) \triangleq u^*(\cdot - \widehat{t}_E + t_E^*)$, is an optimal tetrad to $(TP)_{max}^{\widehat{Q}_S,\widehat{Q}_E}$, it suffices to verify that

$$(\widehat{t}_E - t_E^*, y_0^*, \widehat{t}_E, v^*) \in \widehat{\mathscr{A}}_{ad} \tag{2.20}$$

and

$$t_E^* \leq \widehat{t}_E - \widehat{t}_S \quad \text{for all} \quad (\widehat{t}_S, \widehat{y}_0, \widehat{t}_E, \widehat{u}) \in \widehat{\mathscr{A}}_{ad}. \tag{2.21}$$

To prove (2.20), we notice that $(0, y_0^*, t_E^*, u^*) \in \mathscr{A}_{ad}$, which implies that

$$y_0^* \in Y_1, \quad t_E^* > 0, \quad u^* \in L(0, t_E^*; \mathbb{U}) \quad \text{and} \quad y(t_E^*; 0, y_0^*, u^*) \in Y_2. \tag{2.22}$$

Meanwhile, because (2.1) is time-invariant and $v^*(\cdot) = u^*(\cdot - \widehat{t}_E + t_E^*)$, we find that

$$y(t; \widehat{t}_E - t_E^*, y_0^*, v^*) = y(t - \widehat{t}_E + t_E^*; 0, y_0^*, u^*) \quad \text{for all} \quad t \in [\widehat{t}_E - t_E^*, \widehat{t}_E],$$

from which, it follows that

$$y(\widehat{t}_E; \widehat{t}_E - t_E^*, y_0^*, v^*) = y(t_E^*; 0, y_0^*, u^*). \tag{2.23}$$

This, combined with (2.22), indicates (2.20).

To show (2.21), we arbitrarily fix $(\widehat{t}_S, \widehat{y}_0, \widehat{t}_E, \widehat{u}) \in \widehat{\mathscr{A}}_{ad}$. Then we have that

$$0 \leq \widehat{t}_S < \widehat{t}_E, \quad \widehat{y}_0 \in Y_1, \quad \widehat{u} \in L(\widehat{t}_S, \widehat{t}_E; \mathbb{U}) \quad \text{and} \quad y(\widehat{t}_E; \widehat{t}_S, \widehat{y}_0, \widehat{u}) \in Y_2. \tag{2.24}$$

Let $\widehat{v}(\cdot) = \widehat{u}(\cdot + \widehat{t}_S)$. Then by the time-invariance of (2.1), we obtain that

$$y(t; 0, \widehat{y}_0, \widehat{v}) = y(t + \widehat{t}_S; \widehat{t}_S, \widehat{y}_0, \widehat{u}) \quad \text{for all} \quad t \in [0, \widehat{t}_E - \widehat{t}_S],$$

from which, it follows that

$$y(\widehat{t}_E - \widehat{t}_S; 0, \widehat{y}_0, \widehat{v}) = y(\widehat{t}_E; \widehat{t}_S, \widehat{y}_0, \widehat{u}). \tag{2.25}$$

This, combined with (2.24), implies that

$$(0, \widehat{y}_0, \widehat{t}_E - \widehat{t}_S, \widehat{v}) \in \mathscr{A}_{ad},$$

which, along with the optimality of t_E^* to $(TP)_{min}^{Q_S, Q_E}$, leads to (2.21).

We next show (ii) in Definition 2.1. Assume that $(\widehat{t}_S^*, \widehat{y}_0^*, \widehat{t}_E, \widehat{u}^*)$ is an optimal tetrad to $(TP)_{max}^{\widehat{Q}_S, \widehat{Q}_E}$. In order to prove that $(0, \widehat{y}_0^*, \widehat{t}_E - \widehat{t}_S^*, \widehat{v}^*)$, with $\widehat{v}^*(\cdot) = \widehat{u}^*(\cdot + \widehat{t}_S^*)$, is an optimal tetrad to $(TP)_{min}^{Q_S, Q_E}$, it suffices to verify that

$$(0, \widehat{y}_0^*, \widehat{t}_E - \widehat{t}_S^*, \widehat{v}^*) \in \mathscr{A}_{ad} \tag{2.26}$$

and

$$\widehat{t}_E - \widehat{t}_S^* \le t_E \quad \text{for all} \quad (0, y_0, t_E, u) \in \mathscr{A}_{ad}. \tag{2.27}$$

To prove (2.26), we note that $(\widehat{t}_S^*, \widehat{y}_0^*, \widehat{t}_E, \widehat{u}^*) \in \widehat{\mathscr{A}}_{ad}$, which implies that

$$0 \le \widehat{t}_S^* < \widehat{t}_E, \quad \widehat{y}_0^* \in Y_1, \quad \widehat{u}^* \in L(\widehat{t}_S^*, \widehat{t}_E; \mathbb{U}) \quad \text{and} \quad y(\widehat{t}_E; \widehat{t}_S^*, \widehat{y}_0^*, \widehat{u}^*) \in Y_2. \tag{2.28}$$

By a very similar way to that used to prove (2.25), we can verify that

$$y(\widehat{t}_E - \widehat{t}_S^*; 0, \widehat{y}_0^*, \widehat{v}^*) = y(\widehat{t}_E; \widehat{t}_S^*, \widehat{y}_0^*, \widehat{u}^*).$$

This, combined with (2.28), implies (2.26).

To show (2.27), we arbitrarily fix $(0, y_0, t_E, u) \in \mathscr{A}_{ad}$. Then it follows that

$$y_0 \in Y_1, \quad t_E > 0, \quad u \in L(0, t_E; \mathbb{U}) \quad \text{and} \quad y(t_E; 0, y_0, u) \in Y_2. \tag{2.29}$$

In the case that $t_E \ge \widehat{t}_E$, (2.27) is trivial. In the case that $t_E \in (0, \widehat{t}_E)$, by a similar way to that used to prove (2.23), we can verify that

$$y(\widehat{t}_E; \widehat{t}_E - t_E, y_0, v) = y(t_E; 0, y_0, u), \quad \text{with} \quad v(\cdot) = u(\cdot - \widehat{t}_E + t_E).$$

This, combined with (2.29), implies that

$$(\widehat{t}_E - t_E, y_0, \widehat{t}_E, v) \in \widehat{\mathscr{A}}_{ad},$$

which, together with the optimality of $(\widehat{t}_E - \widehat{t}_S^*)$ to $(TP)_{max}^{\widehat{Q}_S, \widehat{Q}_E}$, leads to that

$$\widehat{t}_E - \widehat{t}_S^* \le \widehat{t}_E - (\widehat{t}_E - t_E).$$

Hence, (2.27) is also true in the second case.

Thus, we complete the proof of Theorem 2.1. □

We next study the connection of time optimal control problems $(TP)_{min}^{Q_S,Q_E}$ and $(TP)_{max}^{\widehat{Q}_S,\widehat{Q}_E}$ in the following CASE TWO:

- The operator A in (2.1) generates a C_0 group on Y. Arbitrarily fix $T > 0$ and a nonempty subset $Y_1 \subseteq Y$;
- The control constraint set $\mathbb{U} \subseteq U$ is nonempty;
- The starting sets and the ending sets are as:

$$Q_S = \{0\} \times Y_1, \quad Q_E \subseteq (0, T] \times Y,$$

and

$$\widehat{Q}_S = \left\{ (\widehat{t}_S, \widehat{z}_0) \mid (T - \widehat{t}_S, \widehat{z}_0) \in Q_E \right\}, \quad \widehat{Q}_E = \{T\} \times Y_1;$$

- The controlled system for $(TP)_{min}^{Q_S,Q_E}$ is (2.1) (which is time-varying, in general), while the controlled system for $(TP)_{max}^{\widehat{Q}_S,\widehat{Q}_E}$ is as follows:

$$\dot{z}(t) = -Az(t) - f(T - t, z(t), \widehat{u}(t)), \quad t \in (0, T). \tag{2.30}$$

Write $z(\cdot; t_0, z_0, \widehat{u})$ for the solution of (2.30) with the initial data $(t_0, z_0) \in [0, T) \times Y$.

In CASE TWO, the sets of admissible tetrads for $(TP)_{min}^{Q_S,Q_E}$ and $(TP)_{max}^{\widehat{Q}_S,\widehat{Q}_E}$ are respectively as follows:

$$\mathscr{A}_{ad} = \{(0, y_0, t_E, u) \mid y_0 \in Y_1, u \in L(0, t_E; \mathbb{U}), (t_E, y(t_E; 0, y_0, u)) \in Q_E\} \tag{2.31}$$

and

$$\widehat{\mathscr{A}}_{ad} = \{(\widehat{t}_S, \widehat{z}_0, T, \widehat{u}) \mid (\widehat{t}_S, \widehat{z}_0) \in \widehat{Q}_S, \widehat{u} \in L(\widehat{t}_S, T; \mathbb{U}), z(T; \widehat{t}_S, \widehat{z}_0, \widehat{u}) \in Y_1\}. \tag{2.32}$$

One can easily check that

$$\widehat{\mathscr{A}}_{ad} = \{(\widehat{t}_S, \widehat{z}_0, T, \widehat{u}) \mid (T - \widehat{t}_S, \widehat{z}_0) \in Q_E, \widehat{u} \in L(\widehat{t}_S, T; \mathbb{U}), z(T; \widehat{t}_S, \widehat{z}_0, \widehat{u}) \in Y_1\}. \tag{2.33}$$

Definition 2.2 In CASE TWO, the problems $(TP)_{min}^{Q_S,Q_E}$ and $(TP)_{max}^{\widehat{Q}_S,\widehat{Q}_E}$ are said to be equivalent if the following conditions are true:

(i) If $(0, y_0^*, t_E^*, u^*)$ is an optimal tetrad to $(TP)_{min}^{Q_S,Q_E}$, then $(T - t_E^*, y(t_E^*; 0, y_0^*, u^*), T, v^*)$, with $v^*(\cdot) = u^*(T - \cdot)$, is an optimal tetrad to $(TP)_{max}^{\widehat{Q}_S,\widehat{Q}_E}$.

(ii) If $(\widehat{t}_S^*, \widehat{z}_0^*, T, \widehat{u}^*)$ is an optimal tetrad to $(TP)_{max}^{\widehat{Q}_S,\widehat{Q}_E}$, then $(0, z(T; \widehat{t}_S^*, \widehat{z}_0^*, \widehat{u}^*), T - \widehat{t}_S^*, v^*)$, with $\widehat{v}^*(\cdot) = \widehat{u}^*(T - \cdot)$ is an optimal tetrad to $(TP)_{min}^{Q_S,Q_E}$.

Theorem 2.2 *In CASE TWO, problems* $(TP)_{min}^{Q_S, Q_E}$ *and* $(TP)_{max}^{\widehat{Q}_S, \widehat{Q}_E}$ *are equivalent.*

Proof We begin with proving (i) in Definition 2.2. Assume that $(0, y_0^*, t_E^*, u^*)$ is an optimal tetrad to $(TP)_{min}^{Q_S, Q_E}$. In order to prove that $(T - t_E^*, y(t_E^*; 0, y_0^*, u^*), T, v^*)$, with $v^*(\cdot) = u^*(T - \cdot)$, is an optimal tetrad to $(TP)_{max}^{\widehat{Q}_S, \widehat{Q}_E}$, it suffices to verify that

$$(T - t_E^*, y(t_E^*; 0, y_0^*, u^*), T, v^*) \in \widehat{\mathscr{A}}_{ad} \tag{2.34}$$

and

$$t_E^* \leq T - \widehat{t}_S \quad \text{for all} \quad (\widehat{t}_S, \widehat{z}_0, T, \widehat{u}) \in \widehat{\mathscr{A}}_{ad}. \tag{2.35}$$

To show (2.34), we note that $(0, y_0^*, t_E^*, u^*) \in \mathscr{A}_{ad}$, which implies that

$$y_0^* \in Y_1, \quad u^* \in L(0, t_E^*; \mathbb{U}) \quad \text{and} \quad (t_E^*, y(t_E^*; 0, y_0^*, u^*)) \in Q_E. \tag{2.36}$$

Meanwhile, since the operator A (in (2.1)) generates a C_0 group, we find that

$$y(T - t; 0, y_0^*, u^*) = z(t; T - t_E^*, y(t_E^*; 0, y_0^*, u^*), v^*) \quad \text{for all} \quad t \in [T - t_E^*, T].$$

Especially, we have that

$$z(T; T - t_E^*, y(t_E^*; 0, y_0^*, u^*), v^*) = y_0^*.$$

This, combined with (2.36), implies (2.34).

To prove (2.35), we arbitrarily fix $(\widehat{t}_S, \widehat{z}_0, T, \widehat{u}) \in \widehat{\mathscr{A}}_{ad}$. Then we have that

$$(T - \widehat{t}_S, \widehat{z}_0) \in Q_E, \quad \widehat{u} \in L(\widehat{t}_S, T; \mathbb{U}) \quad \text{and} \quad z(T; \widehat{t}_S, \widehat{z}_0, \widehat{u}) \in Y_1. \tag{2.37}$$

Let $\widehat{v}(\cdot) = \widehat{u}(T - \cdot)$. Since the operator A (in (2.1)) generates a C_0 group, we see that

$$y(t; 0, z(T; \widehat{t}_S, \widehat{z}_0, \widehat{u}), \widehat{v}) = z(T - t; \widehat{t}_S, \widehat{z}_0, \widehat{u}) \quad \text{for all} \quad t \in [0, T - \widehat{t}_S].$$

This, along with (2.37), indicates that

$$(0, z(T; \widehat{t}_S, \widehat{z}_0, \widehat{u}), T - \widehat{t}_S, \widehat{v}) \in \mathscr{A}_{ad},$$

which, together with the optimality of t_E^* to $(TP)_{min}^{Q_S, Q_E}$, leads to (2.35).

We next show (ii) in Definition 2.2. Assume that $(\widehat{t}_S^*, \widehat{z}_0^*, T, \widehat{u}^*)$ is an optimal tetrad of $(TP)_{max}^{\widehat{Q}_S, \widehat{Q}_E}$. In order to prove that $(0, z(T; \widehat{t}_S^*, \widehat{z}_0^*, \widehat{u}^*), T - \widehat{t}_S^*, \widehat{v}^*)$, with $\widehat{v}^*(\cdot) = \widehat{u}^*(T - \cdot)$, is an optimal tetrad to $(TP)_{min}^{Q_S, Q_E}$, it suffices to show that

$$(0, z(T; \widehat{t}_S^*, \widehat{z}_0^*, \widehat{u}^*), T - \widehat{t}_S^*, \widehat{v}^*) \in \mathscr{A}_{ad} \tag{2.38}$$

and

$$T - \widehat{t}\,_S^* \leq t_E \quad \text{for all} \quad (0, y_0, t_E, u) \in \mathscr{A}_{ad}. \tag{2.39}$$

To prove (2.38), we notice that $(\widehat{t}\,_S^*, \widehat{z}_0^*, T, \widehat{u}^*) \in \widehat{\mathscr{A}}_{ad}$, which implies that

$$(T - \widehat{t}\,_S^*, \widehat{z}_0^*) \in Q_E, \widehat{u}^* \in L(\widehat{t}\,_S^*, T; \mathbb{U}), z(T; \widehat{t}\,_S^*, \widehat{z}_0^*, \widehat{u}^*) \in Y_1. \tag{2.40}$$

At the same time, since A in (2.1) generates a C_0 group, we obtain that

$$y(t; 0, z(T; \widehat{t}\,_S^*, \widehat{z}_0^*, \widehat{u}^*), \widehat{v}^*) = z(T - t; \widehat{t}\,_S^*, \widehat{z}_0^*, \widehat{u}^*) \quad \text{for all} \ \ t \in [0, T - \widehat{t}\,_S^*],$$

which, combined with (2.40), leads to (2.38).

To show (2.39), we let $(0, y_0, t_E, u) \in \mathscr{A}_{ad}$. Then we have that

$$y_0 \in Y_1, \quad u \in L(0, t_E; \mathbb{U}), \quad (t_E, y(t_E; 0, y_0, u)) \in Q_E. \tag{2.41}$$

Write $v(\cdot) = u(T - \cdot)$. Since A in (2.1) generates a C_0 group, we see that

$$y(T - t; 0, y_0, u) = z(t; T - t_E, y(t_E; 0, y_0, u), v) \quad \text{for all} \ \ t \in [T - t_E, T].$$

This, together with (2.41), implies that

$$(T - t_E, y(t_E; 0, y_0, u), T, v) \in \widehat{\mathscr{A}}_{ad},$$

which, along with the optimality of $(T - \widehat{t}\,_S^*)$ to $(TP)_{max}^{\widehat{Q}_S, \widehat{Q}_E}$, leads to (2.39).

Thus, we complete the proof of Theorem 2.2.   □

We finally study the connection of two different minimal time control problems $(TP)_{min}^{Q_S, Q_E}$ and $(TP)_{min}^{\widehat{Q}_S, \widehat{Q}_E}$ in the following CASE THREE:

- The controlled system is (2.1), with

$$f(t, y, u) = D(t)y + f_1(t, u), \quad t \in (0, +\infty), \tag{2.42}$$

where $D(\cdot) \in L^\infty(0, +\infty; \mathscr{L}(Y))$. Write $\Phi(\cdot, \cdot)$ for the evolution operator generated by $A + D(\cdot)$;
- The control constraint set $\mathbb{U} \subseteq U$ is nonempty;
- The starting set and the ending set are as: $Q_S = \{0\} \times Y_1$ and $Q_E \subseteq (0, +\infty) \times Y$, where Y_1 is a nonempty subset of Y. Define

$$T_E \triangleq \{t_E > 0 \mid \text{there exists} \ y_1 \in Y \ \text{so that} \ (t_E, y_1) \in Q_E\}, \tag{2.43}$$

and define a map $Y_E : T_E \mapsto 2^Y$ by

$$Y_E(t_E) \triangleq \{y_1 \in Y \mid (t_E, y_1) \in Q_E\} \quad \text{for each} \ t_E \in T_E; \tag{2.44}$$

- Arbitrarily fix $\widehat{y}_0 \in Y$. Take

$$\widehat{Q}_S \triangleq \widehat{Q}_S(\widehat{y}_0) = \{(0, \widehat{y}_0)\};$$

$$\widehat{Q}_E \triangleq \widehat{Q}_E(\widehat{y}_0)$$
$$= \left\{(t_E, z_1 - z_2) \mid t_E \in T_E, z_1 \in Y_E(t_E), z_2 \in \bigcup_{h \in Y_1} \Phi(t_E, 0)(h - \widehat{y}_0)\right\}.$$

Two notes are given in order:

- The set Y_1 of $(TP)_{min}^{Q_S, Q_E}$ is not a point in the state space, in general. Can we transfer this problem into a minimal time control problem, where the starting set is $\{(0, y_0)\}$ for some point y_0 in Y? The answer is positive. Indeed, we will see that the above $(TP)_{min}^{Q_S, Q_E}$ and $(TP)_{min}^{\widehat{Q}_S, \widehat{Q}_E}$ are equivalent.
- From (2.43) and (2.44), we see that T_E is the projection of Q_E into $(0, +\infty)$ and $Y_E(t_E)$ is the y-slice of Q_E at t_E.

One can easily check that the admissible set of $(TP)_{min}^{\widehat{Q}_S, \widehat{Q}_E}$ is as:

$$\widehat{\mathscr{A}}_{ad} \triangleq \widehat{\mathscr{A}}_{ad}(\widehat{y}_0)$$
$$= \{(0, \widehat{y}_0, \widehat{t}_E, \widehat{u}) \mid \widehat{t}_E > 0, \widehat{u} \in L(0, \widehat{t}_E; \mathbb{U}), (\widehat{t}_E, y(\widehat{t}_E; 0, \widehat{y}_0, \widehat{u})) \in \widehat{Q}_E\}.$$

Definition 2.3 In CASE THREE, the problems $(TP)_{min}^{Q_S, Q_E}$ and $(TP)_{min}^{\widehat{Q}_S(\widehat{y}_0), \widehat{Q}_E(\widehat{y}_0)}$, with $\widehat{y}_0 \in Y$ arbitrarily fixed, are said to be equivalent if the following conditions are true:

(i) If $(0, y_0^*, t_E^*, u^*)$ is an optimal tetrad to $(TP)_{min}^{Q_S, Q_E}$, then $(0, \widehat{y}_0, t_E^*, u^*)$ is an optimal tetrad to $(TP)_{min}^{\widehat{Q}_S(\widehat{y}_0), \widehat{Q}_E(\widehat{y}_0)}$.

(ii) If $(0, \widehat{y}_0, \widehat{t}_E^*, \widehat{u}^*)$ is an optimal tetrad to $(TP)_{min}^{\widehat{Q}_S(\widehat{y}_0), \widehat{Q}_E(\widehat{y}_0)}$, then $(0, h, \widehat{t}_E^*, \widehat{u}^*)$ is an optimal tetrad to $(TP)_{min}^{Q_S, Q_E}$. Here, $h \in Y_1$ satisfies that there is $z_1 \in Y_E(\widehat{t}_E^*)$ and $z_2 \in Y$ so that

$$y(\widehat{t}_E^*; 0, \widehat{y}_0, \widehat{u}^*) = z_1 - z_2 \text{ and } z_2 = \Phi(\widehat{t}_E^*, 0)(h - \widehat{y}_0).$$

Theorem 2.3 *In CASE THREE, the problems* $(TP)_{min}^{Q_S, Q_E}$ *and* $(TP)_{min}^{\widehat{Q}_S(\widehat{y}_0), \widehat{Q}_E(\widehat{y}_0)}$, *with* $\widehat{y}_0 \in Y$ *arbitrarily fixed, are equivalent.*

Proof Arbitrarily fix $\widehat{y}_0 \in Y$. We first show (i) in Definition 2.3. Assume that $(0, y_0^*, t_E^*, u^*)$ is an optimal tetrad to $(TP)_{min}^{Q_S, Q_E}$. In order to prove that $(0, \widehat{y}_0, t_E^*, u^*)$ is an optimal tetrad to $(TP)_{min}^{\widehat{Q}_S(\widehat{y}_0), \widehat{Q}_E(\widehat{y}_0)}$, it suffices to show that

$$(0, \widehat{y}_0, t_E^*, u^*) \in \widehat{\mathscr{A}}_{ad}(\widehat{y}_0) \tag{2.45}$$

and that

$$t_E^* \leq \widehat{t}_E \quad \text{for each} \quad (0, \widehat{y}_0, \widehat{t}_E, \widehat{u}) \in \widehat{\mathscr{A}_{ad}}(\widehat{y}_0). \tag{2.46}$$

To prove (2.45), we use the assumption that $(0, y_0^*, t_E^*, u^*) \in \mathscr{A}_{ad}$ to get that

$$y_0^* \in Y_1, \quad t_E^* > 0, \quad u^* \in L(0, t_E^*; \mathbb{U}), \quad t_E^* \in T_E \quad \text{and} \quad y(t_E^*; 0, y_0^*, u^*) \in Y_E(t_E^*). \tag{2.47}$$

Meanwhile, because of (2.42), we find that

$$\begin{aligned}
y(t_E^*; 0, \widehat{y}_0, u^*) &= y(t_E^*; 0, y_0^*, u^*) - y(t_E^*; 0, y_0^* - \widehat{y}_0, 0) \\
&= y(t_E^*; 0, y_0^*, u^*) - \Phi(t_E^*, 0)(y_0^* - \widehat{y}_0).
\end{aligned} \tag{2.48}$$

This, together with (2.47), implies (2.45).

To prove (2.46), we arbitrarily fix $(0, \widehat{y}_0, \widehat{t}_E, \widehat{u}) \in \widehat{\mathscr{A}_{ad}}(\widehat{y}_0)$. Then we have that

$$\widehat{t}_E > 0, \quad \widehat{u} \in L(0, \widehat{t}_E; \mathbb{U}), \quad \widehat{t}_E \in T_E, \tag{2.49}$$

and that

$$y(\widehat{t}_E; 0, \widehat{y}_0, \widehat{u}) = z_1 - z_2, \quad z_1 \in Y_E(\widehat{t}_E), \quad z_2 = \Phi(\widehat{t}_E, 0)(h - \widehat{y}_0) \quad \text{and} \quad h \in Y_1. \tag{2.50}$$

By the same way as that used to prove (2.48), we can verify that

$$y(\widehat{t}_E; 0, \widehat{y}_0, \widehat{u}) = y(\widehat{t}_E; 0, h, \widehat{u}) - \Phi(\widehat{t}_E, 0)(h - \widehat{y}_0).$$

This, along with (2.49) and (2.50), yields that

$$y(\widehat{t}_E; 0, h, \widehat{u}) = z_1 \in Y_E(\widehat{t}_E)$$

and

$$(0, h, \widehat{t}_E, \widehat{u}) \in \mathscr{A}_{ad}. \tag{2.51}$$

By (2.51) and the optimality of t_E^* to $(TP)_{min}^{Q_S, Q_E}$, (2.46) follows at once.

We next show (ii) in Definition 2.3. Assume that $(0, \widehat{y}_0, \widehat{t}_E^*, \widehat{u}^*)$ is an optimal tetrad to $(TP)_{min}^{\widehat{Q}_S(\widehat{y}_0), \widehat{Q}_E(\widehat{y}_0)}$. Let h, z_1 and z_2 be given in (ii) of Definition 2.3. Then we have that

$$\widehat{t}_E^* > 0 \quad \text{and} \quad \widehat{u}^* \in L(0, \widehat{t}_E^*; \mathbb{U}) \tag{2.52}$$

and that

$$y(\widehat{t_E^*}; 0, \widehat{y_0}, \widehat{u^*}) = z_1 - z_2, \quad z_1 \in Y_E(\widehat{t_E^*}), \quad z_2 = \varPhi(\widehat{t_E^*}, 0)(h - \widehat{y_0}) \text{ and } h \in Y_1.$$
(2.53)

In order to show that $(0, h, \widehat{t_E^*}, \widehat{u^*})$ is an optimal tetrad to the problem $(TP)_{min}^{Q_S, Q_E}$, it suffices to verify that

$$(0, h, \widehat{t_E^*}, \widehat{u^*}) \in \mathscr{A}_{ad}$$
(2.54)

and

$$\widehat{t_E^*} \leq t_E \text{ for each tetrad } (0, y_0, t_E, u) \in \mathscr{A}_{ad}.$$
(2.55)

We first show (2.54). By (2.53) and by the same way as that used to prove (2.48), we can easily verify that

$$z_1 = y(\widehat{t_E^*}; 0, \widehat{y_0}, \widehat{u^*}) + \varPhi(\widehat{t_E^*}, 0)(h - \widehat{y_0}) = y(\widehat{t_E^*}; 0, h, \widehat{u^*})$$

and

$$(\widehat{t_E^*}, y(\widehat{t_E^*}; 0, h, \widehat{u^*})) \in Q_E.$$
(2.56)

Since $h \in Y_1$, (2.54) follows from (2.52) and (2.56) at once.

We next show (2.55). Arbitrarily fix $(0, y_0, t_E, u) \in \mathscr{A}_{ad}$. Then we have that

$$y_0 \in Y_1, \quad t_E > 0, \quad u \in L(0, t_E; \mathbb{U}) \text{ and } (t_E, y(t_E; 0, y_0, u)) \in Q_E.$$
(2.57)

From (2.57), we see that

$$t_E \in T_E \text{ and } y(t_E; 0, y_0, u) \in Y_E(t_E).$$
(2.58)

Then, by a very similar way to that used to show (2.48), we can verify that

$$y(t_E; 0, \widehat{y_0}, u) = y(t_E; 0, y_0, u) - \varPhi(t_E, 0)(y_0 - \widehat{y_0}).$$
(2.59)

It follows from (2.57), (2.58), and (2.59) that

$$(0, \widehat{y_0}, t_E, u) \in \widehat{\mathscr{A}}_{ad}(\widehat{y_0}).$$

This, along with the optimality of $\widehat{t_E^*}$ to $(TP)_{min}^{\widehat{Q}_S(\widehat{y_0}), \widehat{Q}_E(\widehat{y_0})}$, leads to (2.55).

Hence, we end the proof of Theorem 2.3. □

2.3 Several Examples

In this section, we will present several examples of time optimal control problems.

Example 2.1 ([23]) How can we reach the top fastest by an elevator? In this problem, the controlled system is as:

$$\frac{d^2}{dt^2} y(t) = u(t) - g, \quad t \in (0, +\infty), \tag{2.60}$$

where $y(t)$ stands for the height of the mass at time t, $u(t)$ is the force of a unit mass given by an elevator at time t, and g is the acceleration of gravity. We can rewrite (2.60) as the linear equation:

$$\frac{d}{dt} \begin{pmatrix} y_1(t) \\ y_2(t) \end{pmatrix} = \begin{pmatrix} 0 & 1 \\ 0 & 0 \end{pmatrix} \begin{pmatrix} y_1(t) \\ y_2(t) \end{pmatrix} + \begin{pmatrix} 0 \\ 1 \end{pmatrix} u(t) + \begin{pmatrix} 0 \\ -g \end{pmatrix}, \quad t \in (0, +\infty). \tag{2.61}$$

Take the control constraint set $\mathbb{U}$ as $[-20, 20]$; the initial time as $t_S = 0$; the initial state as: $(y_1(0), y_2(0))^\top = (0, 0)^\top$; the terminal condition as $(y_1(t_E), y_2(t_E))^\top = (h, 0)^\top$, with a fixed $h > 0$. Then we can put this problem into our framework $(TP)_{min}^{Q_S, Q_E}$ in the following manner: The controlled system is (2.61);

$$\mathbb{U} = [-20, 20]; \quad Q_S = \{0\} \times \left\{ \begin{pmatrix} 0 \\ 0 \end{pmatrix} \right\}; \quad Q_E = (0, +\infty) \times \left\{ \begin{pmatrix} h \\ 0 \end{pmatrix} \right\}.$$

We now end Example 2.1.

Example 2.2 ([37]) Consider the undamped harmonic oscillator

$$\frac{d}{dt} \begin{pmatrix} y_1(t) \\ y_2(t) \end{pmatrix} = \begin{pmatrix} 0 & 1 \\ -1 & 0 \end{pmatrix} \begin{pmatrix} y_1(t) \\ y_2(t) \end{pmatrix} + \begin{pmatrix} 0 \\ 1 \end{pmatrix} u(t), \quad t \in (0, +\infty), \tag{2.62}$$

where y_1 is the displacement from its equilibrium position and y_2 is its velocity. Let $\mathbb{U} = [-1, 1]$. We are asked for a control (with the constraint $\mathbb{U}$) driving an initial state $(y_0, 0)^\top$ at time $t = 0$ to a target $(y_E, 0)^\top$ in the shortest time. This problem can be put into our framework $(TP)_{min}^{Q_S, Q_E}$ in the following manner: The controlled system is (2.62);

$$\mathbb{U} = [-1, 1]; \quad Q_S = \{0\} \times \left\{ \begin{pmatrix} y_0 \\ 0 \end{pmatrix} \right\}; \quad Q_E = (0, +\infty) \times \left\{ \begin{pmatrix} y_E \\ 0 \end{pmatrix} \right\}.$$

We end Example 2.2.

Example 2.3 ([19]) Observations of children playing on swings show that children who are good at this task all follow a similar strategy. Making the reasonable assumption that the objective is to get as high as possible and as quickly as possible, the strategy used by children is to crouch when the swing is at its highest point and to stand-up when the swing passes its lowest point. This is done both on the forward and return cycle.

To understand the physics behind this pumping, we shall make some simplified assumptions. The swing is ideally modeled as a pendulum. The rider is modeled as a point mass, and the variable distance of the center of mass of the rider from the fixed support is denoted by $l(t)$. The angle that the swing makes with the vertical is denoted by $\theta(t)$. Conservation of angular momentum for a point mass undergoing planar motion gives:

$$\frac{d}{dt}(l(t)^2\dot{\theta}(t)) = -gl(t)\sin(\theta(t)), \quad t \in (0, +\infty).$$

We rewrite the above equation as the following system of ordinary differential equations:

$$\begin{cases} \dot{z}_1 = z_2, \quad t \in (0, +\infty), \\ \dot{z}_2 = -\dfrac{2\dot{l}z_2}{l} - \dfrac{g\sin(z_1)}{l}, \quad t \in (0, +\infty). \end{cases}$$

Now by making the substitution $x_2 = z_2 l^2$ and rewriting z_1 by x_1, we have that

$$\begin{cases} \dot{x}_1 = \dfrac{x_2}{l^2}, \quad t \in (0, +\infty), \\ \dot{x}_2 = -gl\sin(x_1), \quad t \in (0, +\infty). \end{cases}$$

Let us now make the following approximation:

$$l(t) = L(1 + \varepsilon u(t)), \quad \text{with } L > 0, \quad 0 < \varepsilon << 1 \text{ and } |u(t)| = 1.$$

We then have after expanding and retaining only the first order terms:

$$\begin{cases} \dot{x}_1 = \dfrac{x_2}{L^2} - \dfrac{2\varepsilon x_2 u(t)}{L^2}, \\ \dot{x}_2 = -gL\sin(x_1) - \varepsilon gLu(t)\sin(x_1). \end{cases}$$

Now, the question how to make the objective to get as high as possible and as quickly as possible is a time optimal control problem. It can also be put into our framework. We omit the detail. Figure 2.1 will help us to understand this problem better. We now end Example 2.3.

Example 2.4 Assume that one person loans $10,000$ US dollar funds from a bank at the beginning of a year. He is required to return the funds at the end of that year. Let $y(t)$ denote his wealth at the time t. Then we have the following equation:

$$\begin{cases} \dot{y}(t) = u(t) - c(t), \quad t \in (0, +\infty), \\ y(0) = y_0, \end{cases} \tag{2.63}$$

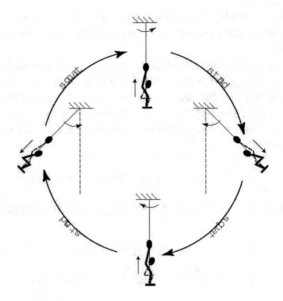

Fig. 2.1 Playing on swings

where $u(\cdot)$ is the earning rate and $c(\cdot)$ is the fixed consumption rate. Assume that for some $M > 0$, $u(t) \in [0, M]$ for all $t \in (0, +\infty)$. In order to pay off the loan on the schedule, when will he have to begin working? This problem can be put into the framework $(TP)_{max}^{Q_S, Q_E}$ in the following manner: The controlled system is (2.63);

$$\mathbb{U} = [0, M]; \quad Q_S = [0, 1) \times \{y_0\}; \quad Q_E = \{1\} \times [10000, +\infty).$$

Now we end Example 2.4.

Example 2.5 ([7, 44]) We consider a solid fuel model. The system is governed by canonical equations from combustion theory such as the nonstationary semilinear one-dimensional Frank-Kamenetskii equation:

$$\begin{cases} y_t = y_{xx} + e^{y(x,t)}, & (x, t) \in (0, 1) \times (0, +\infty), \\ y(0, t) = y(1, t) = 0, \, t \in (0, +\infty), \\ y(x, 0) = y_0(x), & x \in (0, 1), \end{cases} \tag{2.64}$$

where y stands for the temperature. A force enters the first equation of (2.64) in the following way:

$$y_t = y_{xx} + e^{y(x,t)} + \chi_{(1/3, 1/2)}(x)u(x, t), \quad (x, t) \in (0, 1) \times (0, +\infty),$$

where u satisfies that $\|u(t)\|_{L^2(0,1)} \leq 1$ for a.e. $t > 0$. The problem now is to find a control, with the aforementioned constraint, so that the corresponding solution explodes as soon as possible. This problem can be put into our framework (2.17). We end Example 2.5.

2.4   Main Subjects on Time Optimal Control Problems

This section presents main subjects (on time optimal control problems) which
will be studied in this monograph. They are the existence of admissible controls
and optimal controls; the Pontryagin Maximum Principle for optimal controls; the
bang-bang property of optimal controls; and connections of time optimal control
problems with other kinds of optimal control problems, such as minimal norm
control problems and optimal target problems.

We begin with introducing the existence of admissible controls and optimal
controls. **Problem (TP)** can be treated as an optimization problem, where the
admissible set $\mathscr{A}_{ad}$ is the set of available alternatives. When $\mathscr{A}_{ad} = \emptyset$, this
problem does not make sense. Hence, one of the main purposes in the studies of
Problem (TP) is to prove that $\mathscr{A}_{ad} \neq \emptyset$, which is the existence of admissible
controls. It connects with some controllability of controlled equations. Both are
to ask for a control driving the corresponding solution of a controlled equation
from an initial set to a target set. Their main differences are as follows: First, in
general, the controllability is to ask for a control (without any constraint) driving
the corresponding solution of a controlled system from an initial set to a target in a
fixed time interval, while the existence of admissible controls is to ask for a control
(in a constraint set), which drives the corresponding solution of a controlled system
from an initial set to a target in an unfixed time interval. Second, the aim of most
controllability problems is to drive the solution of a controlled system from any
initial state $y_0 \in Y$ to a given target, with the aid of controls, while in most time
optimal control problems, the starting set is fixed.

The next example shows that for some time optimal control problems, admissible
sets can be empty.

Example 2.6 Consider the time optimal control problem $(TP)_{min}^{Q_S,Q_E}$, where the
controlled system is as:

$$\dot{y}(t) = y(t) + u(t), \quad t \in (0, +\infty),$$

and where

$$\mathbb{U} = [-1, 1]; \quad Q_S = \{(0, 1)\}; \quad Q_E = (0, +\infty) \times \{0\}.$$

One can easily check that for any $u \in L(0, +\infty; \mathbb{U})$,

$$y(t; 0, 1, u|_{(0,t)}) \geq 1 \quad \text{for all } t \in (0, +\infty).$$

(Here, $u|_{(0,t)}$ denotes the restriction of u over $(0, t)$.) Hence, the corresponding
admissible set $\mathscr{A}_{ad} = \emptyset$. We now end Example 2.6.

In many cases, the existence of optimal controls can be implied by the existence
of admissible controls, as well as some properties on the controlled equations. This

will be explained in the next chapter. A natural question is as follows: When $\mathscr{A}_{ad} \neq \emptyset$, is it necessary that the corresponding time optimal control problem has an optimal control? The answer is negative. This can be seen from the following example:

Example 2.7 Consider the time optimal control problem $(TP)_{min}^{Q_S, Q_E}$, where the controlled system is as:

$$\dot{y}(t) = u(t), \quad t \in (0, +\infty),$$

and where

$$U = \mathbb{R}; \quad Q_S = \{(0, 1)\}; \quad Q_E = (0, +\infty) \times \{0\}.$$

For any $\varepsilon > 0$, we define a control u^ε over $(0, \varepsilon)$ by

$$u^\varepsilon(t) = -\frac{1}{\varepsilon}, \quad t \in (0, \varepsilon).$$

Then one can directly check that for each $\varepsilon > 0$, $(0, 1, \varepsilon, u^\varepsilon)$ is an admissible tetrad for the above problem $(TP)_{min}^{Q_S, Q_E}$, i.e.,

$$(0, 1, \varepsilon, u^\varepsilon) \in \mathscr{A}_{ad}, \tag{2.65}$$

which leads to that $\mathscr{A}_{ad} \neq \emptyset$. On the other hand, from (2.65), we see that the infimum in the corresponding time optimal control problem cannot be reached. Hence, the above problem $(TP)_{min}^{Q_S, Q_E}$ has no optimal control. We end Example 2.7.

We now turn to the Pontryagin Maximum Principle (or Pontryagin's maximum principle). In general, the standard Pontryagin Maximum Principle for **Problem (TP)** is stated as follows: If u^* is an optimal control (associated with an optimal tetrade $(t_S^*, y_0^*, t_E^*, u^*)$) to **Problem (TP)**, then there is a vector $h \in Y$ with $h \neq 0$ so that

$$\langle \varphi(t), f(t, y^*(t), u^*(t)) \rangle = \max_{u \in U} \langle \varphi(t), f(t, y^*(t), u) \rangle \quad \text{for a.e. } t \in (t_S^*, t_E^*), \tag{2.66}$$

where $y^*(\cdot)$ is the optimal state given by

$$y^*(t) = y(t; t_S^*, y_0^*, u^*), \quad t \in (t_S^*, t_E^*),$$

and where $\varphi(\cdot)$ is the solution to the following adjoint equation (or co-state equation):

$$\begin{cases} \dot{\varphi}(t) = -A^* \varphi(t) - f_y(t, y^*(t), u^*(t))^* \varphi(t), \quad t \in (t_S^*, t_E^*), \\ \varphi(t_E^*) = h. \end{cases} \tag{2.67}$$

Several notes on the standard Pontryagin Maximum Principle are given in order:

- This principle may not hold for some time optimal control problems, in particular, in infinitely dimensional settings. We will provide such a counterexample in Chapter 4. We also refer readers to [24] and [11] for such counterexamples.
- This principle has a weaker version, where h may not be a vector in the state space Y, but the Equation (2.67) is still well posed in some sense and (2.66) still holds (see, for instance, [11, 41] and [42]).
- In general, there are two ways to derive Pontryagin's maximum principle for time optimal control problems. One is a geometric way which will be introduced in Chapter 4 of this monograph. (We also refer the readers to [35] and [11] for this way.) Another is an analytical way. We will not introduce it in our monograph. We refer readers to books [24] and [3] for this method.

We next explain the bang-bang property. An optimal control u^* (associated with an optimal tetrad $(t_S^*, y_0^*, t_E^*, u^*)$) of **Problem (TP)** is said to have the bang-bang property if

$$u^*(t) \in \partial \mathbb{U} \quad \text{for a.e. } t \in (t_S^*, t_E^*). \tag{2.68}$$

From geometric point of view, it says that the optimal control u^* takes its value on the boundary of the constraint set $\mathbb{U}$ at almost every time in the domain of u^*. **Problem (TP)** is said to have the bang-bang property if any optimal control has the bang-bang property. So the bang-bang property is indeed a property of optimal controls. This property is not only important from perspective of applications, but also very interesting from perspective of mathematics. In some cases, from the bang-bang property of **Problem (TP)**, one can easily get the uniqueness of the optimal control to this problem. The bang-bang property may help us to do better numerical analyses and algorithm on optimal controls in some cases (see, for instance, [20, 21] and [31]). We now are going to explain how the bang-bang property implies the uniqueness of the optimal control.

Definition 2.4 A convex subset Z of a Banach space X is said to be strictly convex, provided that for any two points z_1 and z_2 on ∂Z (the boundary of Z in X), the whole line segment connecting z_1 and z_2 meets ∂Z only at z_1 and z_2.

Theorem 2.4 *Let $\mathbb{U}$ be strictly convex and closed. Let Q_S and Q_E be convex and closed subsets of $(0, +\infty) \times Y$. Suppose that the minimal time control problem $(TP)_{min}^{Q_S, Q_E}$ (or the maximal time control problem $(TP)_{max}^{Q_S, Q_E}$), associated with the controlled equation (2.11), has the bang-bang property. Then the optimal control to $(TP)_{min}^{Q_S, Q_E}$ (or $(TP)_{max}^{Q_S, Q_E}$) is unique.*

Proof We only prove the uniqueness of the optimal control to $(TP)_{min}^{Q_S, Q_E}$. The uniqueness of the optimal control to $(TP)_{max}^{Q_S, Q_E}$ can be verified by the same way. Let (t_S, y_0^*, t_E^*, u^*) and $(t_S, \widehat{y_0^*}, \widehat{t_E^*}, \widehat{u^*})$ be two optimal tetrads to $(TP)_{min}^{Q_S, Q_E}$. It is obvious that

$$t_S < t_E^* = \widehat{t}_E^*, \quad u^* \in L(t_S, t_E^*; \mathbb{U}), \quad \widehat{u}^* \in L(t_S, t_E^*; \mathbb{U}), \tag{2.69}$$

$$(t_S, y_0^*), \quad (t_S, \widehat{y}_0^*) \in Q_S, \tag{2.70}$$

and that

$$(t_E^*, y(t_E^*; t_S, y_0^*, u^*)) \in Q_E, \quad (t_E^*, y(t_E^*; t_S, \widehat{y}_0^*, \widehat{u}^*)) \in Q_E. \tag{2.71}$$

We claim that

$$\left(t_S, \frac{y_0^* + \widehat{y}_0^*}{2}, t_E^*, \frac{u^* + \widehat{u}^*}{2} \right) \in \mathscr{A}_{ad}. \tag{2.72}$$

Indeed, it follows from the convexity of Q_S and $\mathbb{U}$, (2.69) and (2.70) that

$$\left(t_S, \frac{y_0^* + \widehat{y}_0^*}{2} \right) \in Q_S \quad \text{and} \quad \frac{u^* + \widehat{u}^*}{2} \in L(t_S, t_E^*; \mathbb{U}). \tag{2.73}$$

Meanwhile, the linearity of the Equation (2.11) implies that

$$y \left(t_E^*; t_S, \frac{y_0^* + \widehat{y}_0^*}{2}, \frac{u^* + \widehat{u}^*}{2} \right) = \frac{1}{2} y(t_E^*; t_S, y_0^*, u^*) + \frac{1}{2} y(t_E^*; t_S, \widehat{y}_0^*, \widehat{u}^*),$$

which, combined with the convexity of Q_E and (2.71), indicates that

$$\left(t_E^*, y \left(t_E^*; t_S, \frac{y_0^* + \widehat{y}_0^*}{2}, \frac{u^* + \widehat{u}^*}{2} \right) \right) \in Q_E. \tag{2.74}$$

Since $t_S < t_E^*$, (2.72) follows from (2.73) and (2.74) at once.

Because $(t_E^* - t_S)$ is the optimal time of $(TP)_{min}^{Q_S, Q_E}$, we obtain from (2.72) that the following tetrad:

$$(t_S, (y_0^* + \widehat{y}_0^*)/2, t_E^*, (u^* + \widehat{u}^*)/2)$$

is also an optimal tetrad of the problem $(TP)_{min}^{Q_S, Q_E}$. Then, by the bang-bang property of this problem, we have that

$$u^*(t) \in \partial \mathbb{U}, \quad \widehat{u}^*(t) \in \partial \mathbb{U} \quad \text{and} \quad (u^* + \widehat{u}^*)/2 \in \partial \mathbb{U} \quad \text{for a.e. } t \in (t_S, t_E^*).$$

From this and the strict convexity of $\mathbb{U}$, we can easily verify that

$$u^*(t) = \widehat{u}^*(t) \quad \text{for a.e. } t \in (t_S, t_E^*).$$

This completes the proof of Theorem 2.4.   □

The next example presents a time optimal control problem, which holds the Pontryagin Maximum Principle, but does not hold the bang-bang property.

Example 2.8 Consider the time optimal control problem $(TP)_{min}^{Q_S, Q_E}$, where the controlled equation is as:

$$
\frac{d}{dt}\begin{pmatrix} y_1(t) \\ y_2(t) \end{pmatrix} = \begin{cases} \begin{pmatrix} 1 \\ 0 \end{pmatrix} u(t), & t \in [0, 1), \\[2ex] \begin{pmatrix} 0 \\ 1 \end{pmatrix} u(t), & t \in [1, +\infty), \end{cases} \tag{2.75}
$$

and where

$$
\mathbb{U} = [-1, 1], \quad Q_S = \{0\} \times \left\{ \begin{pmatrix} 0 \\ 1 \end{pmatrix} \right\} \quad \text{and} \quad Q_E = (0, +\infty) \times \left\{ \begin{pmatrix} 0 \\ 0 \end{pmatrix} \right\}. \tag{2.76}
$$

By the second equality in (2.76) and by (2.75), after some computations, we get that

$$
y_1(t) = \begin{cases} \displaystyle\int_0^t u(s)\,ds & \text{for all } t \in [0, 1), \\[2ex] \displaystyle\int_0^1 u(s)\,ds & \text{for all } t \geq 1 \end{cases} \tag{2.77}
$$

and

$$
y_2(t) = \begin{cases} 1 & \text{for all } t \in [0, 1), \\[2ex] 1 + \displaystyle\int_1^t u(s)\,ds & \text{for all } t \geq 1. \end{cases} \tag{2.78}
$$

We first show that this problem holds the Pontryagin Maximum Principle. To this end, we let $(t_S^*, y_0^*, t_E^*, u^*)$ be an optimal tetrad to this problem. Then, by (2.76), (2.77), and (2.78), we obtain that

$$
t_S^* = 0; \quad y_0^* = \begin{pmatrix} 0 \\ 1 \end{pmatrix}; \quad t_E^* > 1, \quad \int_0^1 u^*(t)\,dt = 0 \quad \text{and} \quad 1 + \int_1^{t_E^*} u^*(t)\,dt = 0. \tag{2.79}
$$

From the first equality in (2.76) and by (2.79), we see that

$$
t_E^* = 2, \quad \int_0^1 u^*(t)\,dt = 0, \quad \text{and} \quad u^*(t) = -1 \quad \text{for a.e. } t \in (1, 2). \tag{2.80}
$$

Hence, we have that

$$
(t_S^*, y_0^*, t_E^*, u^*) = \left(0, \begin{pmatrix} 0 \\ 1 \end{pmatrix}, 2, u^* \right).
$$

Consider the equation:

$$\begin{cases} \dfrac{d}{dt}\begin{pmatrix} \varphi_1(t) \\ \varphi_2(t) \end{pmatrix} = \begin{pmatrix} 0 \\ 0 \end{pmatrix}, & t \in (0,2) \\ \begin{pmatrix} \varphi_1(2) \\ \varphi_2(2) \end{pmatrix} = \begin{pmatrix} 0 \\ -1 \end{pmatrix}. \end{cases} \qquad (2.81)$$

From (2.75), one can easily check that (2.81) is the dual equation (2.67), with $h = \begin{pmatrix} 0 \\ -1 \end{pmatrix}$, corresponding to the above optimal tetrad. By solving (2.81), we obtain that

$$\varphi_1(t) = 0, \quad \varphi_2(t) = -1 \text{ for all } t \in [0,2]. \qquad (2.82)$$

Then, by the third equality in (2.80) and by (2.82), we can easily check that

$$\left\langle \begin{pmatrix} \varphi_1(t) \\ \varphi_2(t) \end{pmatrix}, \begin{pmatrix} u^*(t) \\ 0 \end{pmatrix} \right\rangle = \max_{u \in [-1,1]} \left\langle \begin{pmatrix} \varphi_1(t) \\ \varphi_2(t) \end{pmatrix}, \begin{pmatrix} u \\ 0 \end{pmatrix} \right\rangle \quad \text{for a.e. } t \in (0,1),$$

and

$$\left\langle \begin{pmatrix} \varphi_1(t) \\ \varphi_2(t) \end{pmatrix}, \begin{pmatrix} 0 \\ u^*(t) \end{pmatrix} \right\rangle = \max_{u \in [-1,1]} \left\langle \begin{pmatrix} \varphi_1(t) \\ \varphi_2(t) \end{pmatrix}, \begin{pmatrix} 0 \\ u \end{pmatrix} \right\rangle \quad \text{for a.e. } t \in (1,2).$$

From (2.81), (2.75), and (2.76), we see that the above two equalities are exactly the equality (2.66) corresponding to the above optimal tetrad $(t_S^*, y_0^*, t_E^*, u^*)$ and the dual equation (2.81). Since $(t_S^*, y_0^*, t_E^*, u^*)$ was assumed to be an arbitrary optimal tetrad, we see that this time optimal control problem $(TP)_{min}^{Q_S, Q_E}$ holds the Pontryagin Maximum Principle.

We next show that this problem has no bang-bang property. To this end, we take

$$v^*(t) = \begin{cases} 0, & t \in [0,1), \\ -1, & t \in [1,2]. \end{cases}$$

One can directly check that $\left(0, \begin{pmatrix} 0 \\ 1 \end{pmatrix}, 2, v^*\right)$ is an optimal tetrad to the problem $(TP)_{min}^{Q_S, Q_E}$. It is clear that v^* does not hold the bang-bang property. Hence, the problem $(TP)_{min}^{Q_S, Q_E}$ does not have the bang-bang property.

Now we end Example 2.8.

We end this section with introducing briefly connections between time optimal control problems with other kind of optimal control problems. Among all kinds of optimal control problems, the time optimal control problem is one of the most difficult problems. In some cases, Problem $(TP)_{min}^{Q_S, Q_E}$ (or Problem $(TP)_{max}^{Q_S, Q_E}$) is equivalent to a norm optimal control problem (or a target optimal problem, as well as a norm optimal control problem). The aforementioned norm optimal control problems and target optimal control problems will be introduced in Chapter 5. In many cases, the studies of the above-mentioned norm optimal control problems (or target optimal control problems) are easier than those of time optimal control problems. Thus, if we can build up connections between time optimal control problems with norm optimal control problems (or target optimal control problems), then we could get desired information on time optimal control problems through studies of norm optimal control problems (or target optimal control problems).

Miscellaneous Notes

The studies on time optimal control problem for ordinary differential equations can be dated back to 1950s (see, for instance, [5, 6] and [35]). Then such studies extended to infinitely dimensional cases in 1960s (see, for instance, [9] and [10]). In [35], the problem is stated in the following manner: Given two points x_0 and x_1 in the state space, a control constraint set $\mathbb{U}$, a controlled ordinary differential equation (which is called a controlled system) and a functional f^0 of states and controls, we ask for an admissible control u (i.e., it drives the corresponding solution of the system from $x(t_0) = x_0$ to $x(t_1) = x_1$ for some $t_0 < t_1$) so that it minimizes the functional:

$$J = \int_{t_0}^{t_1} f^0(x(t), u(t))\mathrm{d}t.$$

Notice that the above t_0 and t_1 are not fixed and depend on admissible controls. When $f^0 \equiv 1$, the aforementioned problem can be put into our framework. From this point of view, our monograph focuses on a special kind of time optimal control problems. But this kind of problems is extremely important. It contains the minimal time control problems and the maximal time control problems. There have been many literatures studying the minimal time control problems (see, for instance, [2, 3, 8, 12–15, 17, 20, 27, 32, 36] and [39]). About studies on the maximal time control problems, we refer the readers to [33, 34] and [38]. Though many years have passed, the study on the time optimal controls, in particular, for infinite dimensional cases, is not very mature. There are still many interesting open problems.

The differential equations whose solutions have the behavior of blowup replicate a large class of phenomena in applied science (see [16]). In [28], an optimal control problem governed by the following differential equation was studied:

$$\partial_t y - \Delta y = y^\lambda + u \quad \text{in} \ \ \Omega \times (0, T), \tag{2.83}$$

where Ω is a bounded domain in $\mathbb{R}^d$, $T > 0$ is a fixed constant, and $\lambda = 2, 3$. For a given control u, solutions to the above equation may blow up. To deal with such kind of optimal control problem, the author introduced the concept of admissible pair: the pair (u, y) is called as an admissible one if y exists globally.

The restriction of global existence of y precludes the behavior of blowup. The same method was used in [1]. For the controlled equation (2.83), an interesting problem is as: how to find a control, which is supported on a small ball of Ω, to eliminate the blowup? On the other hand, the blowup of a solution is desirable in certain cases. For example, one may desire to end a chemical reaction as fast as possible with the aid of some catalyst. The blowup of the solution might be thought of the end of a chemical reaction, while the control might play the role of a catalyst. The minimal blowup time control problem introduced at the end of Section 2.1 was first studied for controlled ODEs in [25], where the existence of optimal controls and Pontryagin's maximum principle were obtained. Then it was studied for PDEs in [26], where the existence of optimal controls was obtained. However, the corresponding Pontryagin's maximum principle was unsolved. This might be an interesting problem. Before [25], a maximal blowup time control problem was studied in [4], where a necessary condition for an optimal control was derived by the dynamic programming method. To our surprise, the published papers on minimal (or maximal) blowup time control problems are very limited (see, for instance, [29] and [30]). We think that this direction has bright development prospect.

The numerical analysis on time optimal controls and optimal time is an interesting topic in the field of time optimal control problems. However, we do not touch it in this monograph since it is little bit beyond the scope of this book. We would like to refer readers to [13, 18, 20–22, 31, 40], and [43] for this subject. An interesting problem is what happens for bang-bang control from the numerical perspective.

Several issues on time optimal control problems deserve to be studied:

- Numerical analysis on time optimal control problems.
- Time optimal control problems with noise in some data.
- Stability of time optimal controls with respect to perturbations of systems.
- Properties of the Bellman function, i.e., regarding the optimal time as a function of the initial data (in the case that the starting set is a singleton set).
- Time optimal control problems for some concrete controlled systems, such as affine controlled ODEs, differential equations with delay, impulsive differential equations, differential equations with sampled-data controls or impulse controls, wave equation, and wave-like equations.

References

1. H. Amann, P. Quittner, Optimal control problems with final observation governed by explosive parabolic equations. SIAM J. Control Optim. **44**, 1215–1238 (2005)
2. V. Barbu, The dynamic programming equation for the time-optimal control problem in infinite dimensions. SIAM J. Control Optim. **29**, 445–456 (1991)
3. V. Barbu, *Analysis and Control of Nonlinear Infinite Dimensional Systems*. Mathematics in Science and Engineering, vol. 190 (Academic, Boston, MA, 1993)
4. E.N. Barron, W.X. Liu, Optimal control of the blowup time. SIAM J. Control Optim. **34**, 102–123 (1996)

5. R. Bellman, I. Glicksberg, O. Gross, On the "bang-bang" control problem. Q. Appl. Math. **14**, 11–18 (1956)
6. V.G. Boltyanskiĭ, R.V. Gamkrelidz, L.S. Pontryagin, On the theory of optimal processes (Russian). Dokl. Akad. Nauk. SSSR **110**, 7–10 (1956)
7. C.J. Budd, V.A. Galaktionov, J.F. Williams, Self-similar blow-up in higher-order semilinear parabolic equations. SIAM J. Appl. Math. **64**, 1775–1809 (2004)
8. O. Cârjă, The minimal time function in infinite dimensions. SIAM J. Control Optim. **31**, 1103–1114 (1993)
9. J.V. Egorov, Optimal control in Banach spaces (Russian). Dokl. Akad. Nauk SSSR **150**, 241–244 (1963)
10. H.O. Fattorini, Time-optimal control of solution of operational differential equations. J. SIAM Control **2**, 54–59 (1964)
11. H.O. Fattorini, Vanishing of the costate in Pontryagin's maximum principle and singular time optimal controls. J. Evol. Equ. **4**, 99–123 (2004)
12. H.O. Fattorini, *Infinite Dimensional Linear Control Systems, the Time Optimal and Norm Optimal Problems*. North-Holland Mathematics Studies, vol. 201 (Elsevier Science B.V., Amsterdam, 2005)
13. W. Gong, N. Yan, Finite element method and its error estimates for the time optimal controls of heat equation. Int. J. Numer. Anal. Model. **13**, 265–279 (2016)
14. F. Gozzi, P. Loreti, Regularity of the minimum time function and minimum energy problems: the linear case. SIAM J. Control Optim. **37**, 1195–1221 (1999)
15. B. Guo, D. Yang, Optimal actuator location for time and norm optimal control of null controllable heat equation. Math. Control Signals Syst. **27**, 23–48 (2015)
16. B. Hu, *Blow-Up Theories for Semilinear Parabolic Equations*. Lecture Notes in Mathematics (Springer, Heidelberg, 2011)
17. K. Ito, K. Kunisch, Semismooth newton methods for time-optimal control for a class of ODEs. SIAM J. Control Optim. **48**, 3997–4013 (2010)
18. G. Knowles, Finite element approximation of parabolic time optimal control problem. SIAM J. Control Optim. **20**, 414–427 (1982)
19. J.E. Kulkarni, Time-optimal control of a swing. https://courses.cit.cornell.edu/ee476/ideas/swing.pdf
20. K. Kunisch, D. Wachsmuth, On time optimal control of the wave equation, its regularization and optimality system. ESAIM Control Optim. Calc. Var. **19**, 317–336 (2013)
21. K. Kunisch, D. Wachsmuth, On time optimal control of the wave equation and its numerical realization as parametric optimization problem. SIAM J. Control Optim. **51**, 1232–1262 (2013)
22. I. Lasiecka, Ritz-Galerkin approximation of the time optimal boundary control problem for parabolic systems with Dirichlet boundary conditions. SIAM J. Control Optim. **22**, 477–500 (1984)
23. D. Levy, M. Yadin, A. Alexandrovitz, Optimal control of elevators. Int. J. Syst. Sci. **8**, 301–320 (1977)
24. X. Li, J. Yong, *Optimal Control Theory for Infinite-Dimensional Systems*. Systems & Control: Foundations & Applications (Birkhäuser Boston, Boston, MA, 1995)
25. P. Lin, G. Wang, Blowup time optimal control for ordinary differential equations. SIAM J. Control Optim. **49**, 73–105 (2011)
26. P. Lin, G. Wang, Some properties for blowup parabolic equations and their application. J. Math. Pures Appl. **101**, 223–255 (2014)
27. J.-L. Lions, Optimisation pour certaines classes d'équations d'évolution non linéaires (French). Ann. Mat. Pura Appl. **72**, 275–293 (1966)
28. J.-L. Lions, Contrôle des systèmes distribués singuliers. Méthodes Mathématiques de l'Informatique, vol. 13 (Gauthier-Villars, Montrouge, 1983)
29. H. Lou, W. Wang, Optimal blowup/quenching time for controlled autonomous ordinary differential equation. Math. Control Relat. Fields **5**, 517–527 (2015)
30. H. Lou, W. Wang, Optimal blowup time for controlled ordinary differential equations. ESAIM Control Optim. Calc. Var. **21**, 815–834 (2015)

31. X. Lü, L. Wang, Q. Yan, Computation of time optimal control problems governed by linear ordinary differential equations. J. Sci. Comput. **73**, 1–25 (2017)
32. S. Micu, I. Roventa, M. Tucsnak, Time optimal boundary controls for the heat equation. J. Funct. Anal. **263**, 25–49 (2012)
33. V.J. Mizel, T.I. Seidman, An abstract bang-bang principle and time optimal boundary control of the heat equation. SIAM J. Control Optim. **35**, 1204–1216 (1997)
34. K.D. Phung, G. Wang, An observability estimate for parabolic equations from a measurable set in time and its application. J. Eur. Math. Soc. **15**, 681–703 (2013)
35. L.S. Pontryagin, V.G. Boltyanskii, R.V. Gamkrelidze, E.F. Mischenko, *The Mathematical Theory of Optimal Processes*, Translated from the Russian by K. N. Trirogoff, ed. by L.W. Neustadt (Interscience Publishers Wiley, New York, London, 1962)
36. J.P. Raymond, H. Zidani, Time optimal problems with boundary controls. Differ. Integral Equ. **13**, 1039–1072 (2000)
37. E.D. Sontag, *Mathematical Control Theory*. Deterministic Finite-Dimensional Systems, 2nd edn. Texts in Applied Mathematics, vol. 6 (Springer, New York, 1998)
38. G. Wang, Y. Xu, Equivalence of three different kinds of optimal control problems for heat equations and its applications. SIAM J. Control Optim. **51**, 848–880 (2013)
39. G. Wang, Y. Zhang, Decompositions and bang-bang problems. Math. Control Relat. Fields **7**, 73–170 (2017)
40. G. Wang, G. Zheng, An approach to the optimal time for a time optimal control problem of an internally controlled heat equation. SIAM J. Control Optim. **50**, 601–628 (2012)
41. G. Wang, E. Zuazua, On the equivalence of minimal time and minimal norm controls for the internally controlled heat equations. SIAM J. Control Optim. **50**, 2938–2958 (2012)
42. G. Wang, Y. Xu, Y. Zhang, Attainable subspaces and the bang-bang property of time optimal controls for heat equations. SIAM J. Control Optim. **53**, 592–621 (2015)
43. G. Wang, D. Yang, Y. Zhang, Time optimal sampled-data controls for the heat equation. C. R. Math. Acad. Sci. Paris **355**, 1252–1290 (2017)
44. Y.B. Zel'dovich, G.I. Barenblatt, V.B. Librovich, G.M. Makhviladze, The mathematical theory of combustion and explosions, Translated from the Russian by Donald H. McNeill (Consultants Bureau [Plenum], New York, 1985)

Chapter 3
Existence of Admissible Controls and Optimal Controls

In this chapter, we will study the existence of admissible controls and optimal controls for **Problem (TP)** (given by (2.4) in Chapter 1) for some special cases. Since minimal time control problems and maximal time control problems are two kinds of most important time optimal control problems and because these two kinds of problems can be mutually transformed in many cases (see Theorem 2.1, Theorem 2.2, and Theorem 2.3), we will only study minimal time control problems throughout this chapter. More precisely, we will focus on the problem $(TP)_{min}^{Q_S,Q_E}$ under the following framework $(\mathscr{A}_{TP})$:

  (i)   The state space Y and the control space U are real separable Hilbert spaces.
 (ii)   The controlled system (or the state system) is the Equation (2.1), i.e.,

$$\dot{y}(t) = Ay(t) + f(t, y(t), u(t)), \quad t \in (0, +\infty),$$

where $A : \mathscr{D}(A) \subseteq Y \mapsto Y$ generates a C_0 semigroup $\{e^{At}\}_{t \geq 0}$ on Y and $f : (0, +\infty) \times Y \times U \mapsto Y$ is a given map. Moreover, for any $0 \leq a < b < +\infty$, $y_0 \in Y$ and $u(\cdot) \in L(a, b; U)$, this equation, with the initial condition $y(a) = y_0$, has a unique mild solution $y(\cdot; a, y_0, u)$ on $[a, b]$.
(iii)   The starting set is as: $Q_S = \{0\} \times Y_S$ with Y_S a nonempty subset of Y.
 (iv)   The ending set Q_E is a nonempty subset of $(0, +\infty) \times Y$.
  (v)   The control constraint set $\mathbb{U}$ is a nonempty subset of U.

The main part of this chapter is devoted to the existence of admissible controls for the problem $(TP)_{min}^{Q_S,Q_E}$. We will study the above existence from three different viewpoints: the controllability, the minimal norm, and reachable sets. At the last section of this chapter, we introduce ways deriving the existence of optimal controls from the existence of admissible controls. In each section of this chapter (except for the last one), we will first present our main theorem in an abstract setting which can be embedded into the framework of this book (see the beginning of Chapter 1), then apply the main theorem to some examples.

© Springer International Publishing AG, part of Springer Nature 2018
G. Wang et al., *Time Optimal Control of Evolution Equations*, Progress in Nonlinear Differential Equations and Their Applications 92, https://doi.org/10.1007/978-3-319-95363-2_3

3.1   Admissible Controls and Controllability

In this section, we will discuss the existence of admissible controls of $(TP)_{min}^{Q_S, Q_E}$ from the perspective of controllability.

3.1.1   General Results in Abstract Setting

With the aid of $(\mathscr{A}_{TP})$ (given at the beginning of Chapter 3), we can define the following two sets:

$$Y_C^{Q_E, \mathbb{U}}(t) \triangleq \left\{ y_0 \in Y \mid \exists\, u \in L(0, t; \mathbb{U}) \text{ s.t. } (t, y(t; 0, y_0, u)) \in Q_E \right\}, \ t \in T_E; \tag{3.1}$$

$$Y_C^{Q_E, \mathbb{U}} \triangleq \bigcup_{t \in T_E} Y_C^{Q_E, \mathbb{U}}(t). \tag{3.2}$$

Here, T_E was defined by (2.43). We call respectively $Y_C^{Q_E, \mathbb{U}}(t)$ and $Y_C^{Q_E, \mathbb{U}}$ the controllable set at time t and the controllable set for the Equation (2.1), associated with Q_E and $\mathbb{U}$. The main result of this subsection is as:

Theorem 3.1 *The problem* $(TP)_{min}^{Q_S, Q_E}$ *has an admissible control if and only if* $Y_S \cap Y_C^{Q_E, \mathbb{U}} \neq \emptyset.$

Proof Suppose that $(TP)_{min}^{Q_S, Q_E}$ has an admissible control u (associated with an admissible tetrad (t_S, y_0, t_E, u)). Then we have that

$$(t_S, y_0, t_E, u) \in \mathscr{A}_{ad},$$

where $\mathscr{A}_{ad}$ is defined by (2.3). From (2.3), (iii) of $(\mathscr{A}_{TP})$ and (2.43), we have that

$$t_S = 0, \quad y_0 \in Y_S, \quad t_E \in T_E, \quad u \in L(0, t_E; \mathbb{U}) \quad \text{and} \quad (t_E, y(t_E; 0, y_0, u)) \in Q_E.$$

These, along with (3.1) and (3.2), yield that

$$y_0 \in Y_S \cap Y_C^{Q_E, \mathbb{U}}(t_E) \subseteq Y_S \cap Y_C^{Q_E, \mathbb{U}}.$$

Conversely, we assume that $y_0 \in Y_S \cap Y_C^{Q_E, \mathbb{U}}$. Then by (3.2), there is $t_E \in T_E$ so that

$$y_0 \in Y_C^{Q_E, \mathbb{U}}(t_E) \quad \text{and} \quad y_0 \in Y_S.$$

These, along with (iii) of $(\mathscr{A}_{TP})$ and (3.1), yield that

$$(0, y_0) \in Q_S; \quad \text{and} \quad (t_E, y(t_E; 0, y_0, u)) \in Q_E \quad \text{for some} \quad u \in L(0, t_E; \mathbb{U}).$$

From these and (2.3), we find that $(0, y_0, t_E, u) \in \mathscr{A}_{ad}$. Hence, u is an admissible control of $(TP)_{min}^{Q_S, Q_E}$. This ends the proof of Theorem 3.1.   □

Two facts deserve to be mentioned: First, both $Y_C^{Q_E, \mathbb{U}}(t)$ (with $t \in T_E$) and $Y_C^{Q_E, \mathbb{U}}$ are independent of Q_S. Second, it is hard to characterize controllable sets $Y_C^{Q_E, \mathbb{U}}(t)$ and $Y_C^{Q_E, \mathbb{U}}$, in general. In the next subsection, we will discuss the controllable sets for some linear systems.

3.1.2 Linear ODEs with Ball-Type Control Constraints

In this subsection, we will use Theorem 3.1 to study the existence of some minimal time control problems, where the controlled system is as:

$$\dot{y}(t) = A(t)y(t) + B(t)u(t), \quad t \in (0, +\infty), \tag{3.3}$$

with $A(\cdot) \in L_{loc}^\infty(0, +\infty; \mathbb{R}^{n \times n})$ and $B(\cdot) \in L_{loc}^\infty(0, +\infty; \mathbb{R}^{n \times m})$. Equation (3.3) can be put into the framework of the Equation (2.1), in the following manner:

- Set $Y \triangleq \mathbb{R}^n$; $U \triangleq \mathbb{R}^m$; $\mathbb{U} \triangleq B_r(0)$ (the closed ball in $\mathbb{R}^m$, centered at the origin and of radius $r > 0$);
- Take $A = 0$ and $f(t, y, u) = A(t)y + B(t)u$.

The minimal time control problem studied in this subsection is $(TP)_{min}^{Q_S, Q_E}$, where the controlled system is (3.3); $Q_S \triangleq \{0\} \times Y_S$ with $Y_S \triangleq \{\widehat{y_0}\} \subseteq \mathbb{R}^n \setminus \{0\}$; $Q_E \triangleq (0, +\infty) \times \{0\}$; and $\mathbb{U} \triangleq B_r(0)$. It is clear that $(TP)_{min}^{Q_S, Q_E}$ can be put into the framework $(\mathscr{A}_{TP})$ in this case. *Throughout this subsection, we simply write* $(TP)_r^{\widehat{y_0}}$ *for* $(TP)_{min}^{Q_S, Q_E}$; *write* $y(\cdot; 0, y_0, u)$ *for the solution of (3.3), with the initial condition that* $y(0) = y_0$.

We now define a special kind of null controllability for (3.3) (see [24]).

Definition 3.1 Several definitions are given in order.

(i) The Equation (3.3) is r-ball-null controllable at $y_0 \in \mathbb{R}^n$ if there exists $u \in L(0, +\infty; B_r(0))$ so that $y(\cdot) \triangleq y(\cdot; 0, y_0, u)$ satisfies that $y(T) = 0$ for some $T \in (0, +\infty)$.

(ii) The Equation (3.3) is globally r-ball-null controllable if it is r-ball-null controllable at each $y_0 \in \mathbb{R}^n$.

By Definition 3.1, we can easily get the following result.

Theorem 3.2 *Let $(\widehat{y}_0, r) \in \mathbb{R}^n \times (0, +\infty)$. Then the Equation (3.3) is r-ball-null controllable at $\widehat{y}_0$ if and only if the problem $(TP)_r^{\widehat{y}_0}$ has an admissible control.*

Several notes are given in order:

- The controllable sets $Y_C^{Q_E, \mathbb{U}}$ and $Y_C^{Q_E, \mathbb{U}}(t)$ for the Equation (3.3), associated with $Q_E \triangleq (0, +\infty) \times \{0\}$ and $\mathbb{U} \triangleq B_r(0)$, depend only on the Equation (3.3) and r (see (3.1) and (3.2)). *To stress such dependence, we denote them by Y_C^r and $Y_C^r(t)$ respectively, and call them respectively the controllable set and the controllable set at t for (3.3), associated with r.*

- It follows by the definition of T_E (see (2.43)) that for the problem $(TP)_r^{\widehat{y}_0}$, $T_E = (0, +\infty)$. Thus, we have that

$$Y_C^r = \bigcup_{t \in (0, +\infty)} Y_C^r(t) \quad \text{for each } r > 0. \tag{3.4}$$

- Let $\Phi(\cdot, \cdot)$ be the evolution operator generated by $A(\cdot)$. Then, by (3.1), we can easily check that for each $r > 0$ and $t \in (0, +\infty)$,

$$Y_C^r(t) = \left\{ \int_0^t \Phi(s, 0)^{-1} B(s) u(s) ds \mid u(\cdot) \in L(0, t; B_r(0)) \right\}. \tag{3.5}$$

Notice that $Y_S = \{\widehat{y}_0\}$ in the problem $(TP)_r^{\widehat{y}_0}$. Thus, we can easily obtain the following consequence of Theorem 3.1:

Corollary 3.1 *Let $(\widehat{y}_0, r) \in \mathbb{R}^n \times (0, +\infty)$. Then the problem $(TP)_r^{\widehat{y}_0}$ has an admissible control if and only if $\widehat{y}_0 \in Y_C^r$.*

We now give an example where the controllable set can be precisely expressed.

Example 3.1 Consider the time optimal control problem $(TP)_1^{\widehat{y}_0}$, where the controlled system is as:

$$\dot{y}(t) = y(t) + u(t), \quad t \in (0, +\infty),$$

with $y(\cdot)$ and $u(\cdot)$ taking values on $\mathbb{R}$ and $[-1, 1]$, respectively. Then we see that

$$Y_C^1 = (-1, 1). \tag{3.6}$$

Indeed, given $\alpha \in Y_C^1$, it follows by (3.5) and (3.4) that there is $t_0 \in (0, +\infty)$ and $u \in L(0, t_0; B_1(0))$ so that

$$\alpha = \int_0^{t_0} e^{-s} u(s) ds,$$

which indicates that

$$|\alpha| \leq \int_0^{t_0} e^{-s} ds = 1 - e^{-t_0} < 1.$$

Hence

$$Y_C^1 \subseteq (-1, 1). \tag{3.7}$$

To show the reverse of (3.7), we arbitrarily fix $\beta \in (-1, 1)$. Then there exists $t_0 \in (0, +\infty)$ so that $|\beta| < 1 - e^{-t_0}$. We define a control $\widehat{u}$ by

$$\widehat{u}(s) \triangleq \frac{\beta}{1 - e^{-t_0}}, \quad s \in (0, t_0).$$

Then we have that

$$\widehat{u} \in L(0, t_0; B_1(0)) \quad \text{and} \quad \beta = \int_0^{t_0} e^{-s} \widehat{u}(s) \mathrm{d}s.$$

These, together with (3.5) and (3.4), yield that $\beta \in Y_C^1(t_0) \subseteq Y_C^1$. Hence,

$$(-1, 1) \subseteq Y_C^1. \tag{3.8}$$

Finally, (3.6) follows from (3.7) and (3.8) at once. we now end Example 3.1.

The next lemma is concerned with some properties on controllable sets of (3.3).

Lemma 3.1 *Let $r > 0$. Then the following conclusions are true:*

(i) *For each $t > 0$, $Y_C^r(t)$ is convex, closed and symmetric with respect to 0 (i.e., $y_0 \in Y_C^r(t)$ if and only if $-y_0 \in Y_C^r(t)$).*
(ii) *When $0 < t_1 \leq t_2$, $Y_C^r(t_1) \subseteq Y_C^r(t_2)$.*
(iii) *The set Y_C^r is convex and symmetric with respect to 0.*

Proof The conclusions (i)–(iii) can be directly verified from (3.5) and (3.4). This ends the proof of Lemma 3.1. □

Theorem 3.3 *The following statements are equivalent:*

(i) *For any $r > 0$, the Equation (3.3) is globally r-ball-null controllable.*
(ii) *The pair $[A(\cdot), B(\cdot)]$ satisfies the following Conti's condition (see [24] and [5]):*

$$\int_0^{+\infty} \|B(s)^* \varphi(s)\|_{\mathbb{R}^m} \mathrm{d}s = +\infty \tag{3.9}$$

for all nonzero solutions φ to the adjoint equation:

$$\dot{\varphi}(t) = -A(t)^* \varphi(t), \quad t \in (0, +\infty). \tag{3.10}$$

(iii) *For any $r > 0$, the controllable set Y_C^r of (3.3), associated with r, is $\mathbb{R}^n$.*
(iv) *For any $r > 0$ and $\widehat{y}_0 \neq 0$, $(TP)_r^{\widehat{y}_0}$ has an admissible control.*

Proof (i)⇒(ii). Suppose that the conclusion (i) is true. Arbitrarily fix $\varphi_0 \in \mathbb{R}^n \setminus \{0\}$. Let φ be the solution of (3.10) with the initial condition $\varphi(0) = \varphi_0$. Arbitrarily fix $r > 0$. Since (3.3) is globally r-ball-null controllable, we see that (3.3) is r-ball-null controllable at each $-k\varphi_0$, with $k \in \mathbb{N}^+$. Arbitrarily fix $k \in \mathbb{N}^+$. Then, there is $t_k \in (0, +\infty)$ and $u_k \in L(0, t_k; B_r(0))$ so that

$$\Phi(t_k, 0)(-k\varphi_0) + \int_0^{t_k} \Phi(t_k, s)B(s)u_k(s)ds = 0.$$

This implies that

$$k\varphi_0 = \int_0^{t_k} \Phi(s, 0)^{-1} B(s)u_k(s)ds. \tag{3.11}$$

Since $\|u_k(t)\|_{\mathbb{R}^m} \le r$ for a.e. $t \in (0, t_k)$, we see that

$$\int_0^{+\infty} \|B(s)^*\varphi(s)\|_{\mathbb{R}^m}\, ds \ge \int_0^{t_k} \|B(s)^*\varphi(s)\|_{\mathbb{R}^m}\, ds$$

$$\ge \frac{1}{r} \int_0^{t_k} \langle B(s)^*\varphi(s), u_k(s)\rangle_{\mathbb{R}^m}\, ds.$$

Since $k \in \mathbb{N}^+$ was arbitrarily fixed, the above, along with (3.11), yields that for all $k \in \mathbb{N}^+$,

$$\int_0^{+\infty} \|B(s)^*\varphi(s)\|_{\mathbb{R}^m}\, ds \ge \frac{1}{r} \int_0^{t_k} \langle \varphi_0, \Phi(s, 0)^{-1} B(s)u_k(s)\rangle_{\mathbb{R}^n}\, ds = \frac{k}{r}\|\varphi_0\|_{\mathbb{R}^n}^2.$$

Sending $k \to +\infty$ in the above leads to (3.9). Hence, the conclusion (ii) is true.

(ii)⇒(iii). By contradiction, suppose that the conclusion (ii) was true, but the conclusion (iii) did not hold. Then there would be $r_0 > 0$ so that $Y_C^{r_0} \ne \mathbb{R}^n$. Thus, we have that $\partial Y_C^{r_0} \ne \emptyset$. This, together with the convexity of $Y_C^{r_0}$ (see Lemma 3.1), yields that for an arbitrarily fixed $z \in \partial Y_C^{r_0}$, there is $\lambda \in \mathbb{R}^n \setminus \{0\}$ so that

$$\langle \lambda, y\rangle_{\mathbb{R}^n} \le \langle \lambda, z\rangle_{\mathbb{R}^n} \quad \text{for all } y \in Y_C^{r_0}. \tag{3.12}$$

Meanwhile, we arbitrarily fix $t \in (0, +\infty)$, and then define a control u by

$$u(s) \triangleq \begin{cases} r_0 \dfrac{B(s)^*(\Phi(s, 0)^{-1})^*\lambda}{\|B(s)^*(\Phi(s, 0)^{-1})^*\lambda\|_{\mathbb{R}^m}}, & \text{if } B(s)^*(\Phi(s, 0)^{-1})^*\lambda \ne 0, \\ 0, & \text{if } B(s)^*(\Phi(s, 0)^{-1})^*\lambda = 0, \end{cases} \quad \text{for a.e. } s \in (0, t).$$

Then one can easily check that

$$\left\langle \lambda, \int_0^t \Phi(s, 0)^{-1} B(s)u(s)ds \right\rangle_{\mathbb{R}^n} = r_0 \int_0^t \|B(s)^*\varphi(s)\|_{\mathbb{R}^m}\, ds, \tag{3.13}$$

where $\varphi(\cdot)$ is the solution of (3.10) with the initial condition that $\varphi(0) = \lambda$.

Now it follows from (3.12) and (3.13) that

$$\int_0^t \|B(s)^*\varphi(s)\|_{\mathbb{R}^m}\, ds \leq \frac{1}{r_0}\langle \lambda,\, z\rangle_{\mathbb{R}^n},$$

which indicates that

$$\int_0^{+\infty} \|B(s)^*\varphi(s)\|_{\mathbb{R}^m}\, ds \leq \frac{1}{r_0}\langle \lambda,\, z\rangle_{\mathbb{R}^n}.$$

This contradicts (ii).

(iii)$\Rightarrow$(iv). It follows from Corollary 3.1.

(iv)$\Rightarrow$(i). It follows from Definition 3.1.

Hence, we end the proof of Theorem 3.3. $\square$

Corollary 3.2 *For the time-invariant system:*

$$\dot{y}(t) = Ay(t) + Bu(t), \quad t \in (0, +\infty) \tag{3.14}$$

(where A and B are $n \times n$ and $n \times m$ constant matrices), the condition (3.9) is equivalent to the combination of the following two conditions:

$$\mathrm{rank}(B, AB, A^2 B, \ldots, A^{n-1} B) = n$$

$$\textit{(Kalman controllability rank condition, see [12])}; \tag{3.15}$$

$$\sigma(A) \subseteq \mathbb{C}_- \triangleq \{c \in \mathbb{C} \mid Re(c) \leq 0\} \quad \textit{(The spectral condition, see [25])}. \tag{3.16}$$

Proof We show it by two steps as follows:

Step 1. We prove that (3.9)$\Rightarrow$ (3.15) and (3.16).

By contradiction, suppose that (3.9) was true, but either (3.15) or (3.16) did not hold.

In the case that (3.15) is not true, there is $\eta \neq 0$ so that

$$\eta^\top A^k B = 0, \quad k = 0, 1, \ldots, n-1. \tag{3.17}$$

This, along with the Hamilton-Cayley Theorem, yields that $\eta^\top A^k B = 0$ for all $k \geq 0$. From this, one can easily check that

$$\eta^\top e^{tA} B = 0 \quad \text{for all } t \in \mathbb{R}. \tag{3.18}$$

Let $\widehat{\varphi}$ be the solution to

$$\begin{cases} \dot{\varphi}(t) = -A^*\varphi(t), \quad t \in (0, +\infty), \\ \varphi(0) = \eta. \end{cases}$$

Then it follows by (3.18) that

$$B^*\widehat{\varphi}(t) = 0 \quad \text{for all} \ t \in (0, +\infty),$$

which contradicts (3.9).

We now turn to the case when (3.16) is not true. Since $\sigma(A) = \sigma(A^*)$, there is $\alpha + i\beta \in \mathbb{C}$ (where $(\alpha, \beta) \in (0, +\infty) \times \mathbb{R}$) and $\xi_1 + i\xi_2$ (where $\|\xi_1\|_{\mathbb{R}^n}^2 + \|\xi_2\|_{\mathbb{R}^n}^2 \neq 0$) so that

$$A^*(\xi_1 + i\xi_2) = (\alpha + i\beta)(\xi_1 + i\xi_2). \tag{3.19}$$

If $\xi_1 = 0$, then by (3.19), we have that

$$A^*\xi_2 = \alpha\xi_2 \quad \text{and} \ \beta = 0. \tag{3.20}$$

Since

$$\|B^* e^{-A^* t}\xi_2\|_{\mathbb{R}^m} \leq \|B\|_{\mathscr{L}(\mathbb{R}^m, \mathbb{R}^n)} \left\|e^{-A^* t}\xi_2\right\|_{\mathbb{R}^n},$$

it follows by (3.20) that

$$\|B^* e^{-A^* t}\xi_2\|_{\mathbb{R}^m} \leq e^{-\alpha t} \|B\|_{\mathscr{L}(\mathbb{R}^m, \mathbb{R}^n)} \|\xi_2\|_{\mathbb{R}^n} \quad \text{for all} \ t \in (0, +\infty).$$

This contradicts (3.9). If $\xi_1 \neq 0$, then by (3.19), we can use a very similar way to the above to obtain that

$$\|B^* e^{-A^* t}\xi_1\|_{\mathbb{R}^m} \leq \|B^* e^{-A^* t}(\xi_1 + i\xi_2)\|_{\mathbb{R}^m} \leq \|B\|_{\mathscr{L}(\mathbb{R}^m, \mathbb{R}^n)} \|e^{-(\alpha+i\beta)t}(\xi_1 + i\xi_2)\|_{\mathbb{R}^n}$$

$$\leq e^{-\alpha t} \|B\|_{\mathscr{L}(\mathbb{R}^m, \mathbb{R}^n)} \sqrt{\|\xi_1\|_{\mathbb{R}^n}^2 + \|\xi_2\|_{\mathbb{R}^n}^2} \quad \text{for all} \ t \in (0, +\infty),$$

which contradicts (3.9). Hence, we have finished Step 1.

Step 2. We show that (3.15) and (3.16) imply (3.9).

Assume that (3.15) and (3.16) hold. According to Theorem 3.3, it suffices to show that for any $r > 0$, the controllable set Y_C^r of (3.14), associated with r, is $\mathbb{R}^n$.

By contradiction, suppose that it was not true. Then there would be $r_0 > 0$ so that $Y_C^{r_0} \neq \mathbb{R}^n$. Thus, by the convexity of $Y_C^{r_0}$ (see Lemma 3.1), there exists $y_0 \in \mathbb{R}^n \setminus Y_C^{r_0}$ and $\lambda_0 \in \mathbb{R}^n \setminus \{0\}$ so that

$$\langle \lambda_0, y \rangle_{\mathbb{R}^n} \leq \langle \lambda_0, y_0 \rangle_{\mathbb{R}^n} \quad \text{for all} \ y \in Y_C^{r_0}. \tag{3.21}$$

Define

$$v(t) \triangleq B^* e^{-t A^*} \lambda_0, \quad t \geq 0. \tag{3.22}$$

By (3.15), we have that $v(\cdot) \neq 0$ (i.e., v is not a zero function).

We next claim that

$$\int_0^{+\infty} \|v(t)\|_{\mathbb{R}^m}\, dt = +\infty. \tag{3.23}$$

For this purpose, we set

$$g(t) \triangleq \int_t^{+\infty} v(s)\, ds, \quad t > 0. \tag{3.24}$$

By contradiction, suppose that (3.23) was not true. Then we would have that

$$\int_0^{+\infty} \|v(t)\|_{\mathbb{R}^m}\, dt < +\infty.$$

This, along with (3.24), yields that

$$|g(t)| \le \int_t^{+\infty} \|v(s)\|_{\mathbb{R}^m}\, ds \to 0 \text{ as } t \to +\infty. \tag{3.25}$$

Let $P(\cdot)$ be the characteristic polynomial of A. According to the Hamilton-Cayley Theorem, $P(A) = 0$. Then by (3.22), we get that

$$P\left(-\frac{d}{dt}\right) v(t) = B^*\left(P\left(-\frac{d}{dt}\right) e^{-tA^*}\right) \lambda_0$$

$$= B^*\left(P(A^*) e^{-tA^*}\right) \lambda_0 = 0 \quad \text{for each } t > 0.$$

This, together with (3.24), yields that

$$-\frac{d}{dt} P\left(-\frac{d}{dt}\right) g(t) = P\left(-\frac{d}{dt}\right)\left(-\frac{dg(t)}{dt}\right) = P\left(-\frac{d}{dt}\right) v(t) = 0,$$

i.e., $g(\cdot)$ is a solution to the following $(n+1)$-order ordinary differential equation:

$$\frac{d}{dt} P\left(-\frac{d}{dt}\right) z(t) = 0. \tag{3.26}$$

In the case that $g(\cdot) \equiv 0$, it follows by (3.24) that $v(\cdot) \equiv 0$, which leads to a contradiction (since $v(\cdot) \neq 0$).

In the case when $g(\cdot) \neq 0$, let $\mu_1, \mu_2, \ldots, \mu_{n+1}$ be solutions to the equation: $\mu P(-\mu) = 0$, which is the characteristic equation of (3.26). Without loss of generality, we can assume that $\mu_{n+1} = 0$ and $\mu_k = -\lambda_k (1 \le k \le n)$, where $\lambda_1, \lambda_2, \ldots, \lambda_n$ are eigenvalues of the matrix A. On one hand, it follows by (3.16) that

$$Re\mu_k \ge 0, \quad \text{for each } 1 \le k \le n. \tag{3.27}$$

On the other hand, since

$$g(t) = \sum_{k=1}^{n+1} P_k(t) e^{\mu_k t},$$

where $P_k(t)(1 \le k \le n)$ is a certain polynomial, we can easily get a contradiction from (3.27) and (3.25).

Finally, by (3.23), we can choose $t_0 > 0$ so that

$$r_0 \int_0^{t_0} \|v(t)\|_{\mathbb{R}^m} dt > \langle \lambda_0, y_0 \rangle_{\mathbb{R}^n}. \tag{3.28}$$

We set

$$u(t) \triangleq \begin{cases} r_0 v(t)/\|v(t)\|_{\mathbb{R}^m}, & \text{if } v(t) \ne 0, \\ 0, & \text{if } v(t) = 0, \end{cases} \quad t \in (0, t_0). \tag{3.29}$$

It is obvious that $u(\cdot) \in L(0, t_0; B_{r_0}(0))$. This, together with (3.5) and (3.4), yields that

$$\int_0^{t_0} e^{-tA} Bu(t) dt \in Y_C^{r_0},$$

which, combined with (3.21), (3.22), and (3.29), indicates that

$$r_0 \int_0^{t_0} \|v(t)\|_{\mathbb{R}^m} dt \le \langle \lambda_0, y_0 \rangle_{\mathbb{R}^n}.$$

This leads to a contradiction with (3.28).

Hence, we end the proof of Corollary 3.2.   □

From Corollary 3.1, we see that $\widehat{y}_0 \in Y_C^r$ if and only if $(TP)_r^{\widehat{y}_0}$ has an admissible control. Thus, for the problem $(TP)_r^{\widehat{y}_0}$, it is important and interesting to give some necessary and sufficient conditions on $(\widehat{y}_0, r)$ so that $\widehat{y}_0 \in Y_C^r$. We next provide several such criteria for the case when the controlled system (3.3) is time-invariant, i.e., $A(t) \equiv A$ and $B(t) \equiv B$. (In other words, we now turn to consider the Equation (3.14).) For this purpose, we define, for each $A \in \mathbb{R}^{n \times n}$ and each $B \in \mathbb{R}^{n \times m}$, the following three subsets of $\mathbb{R}^n$:

$$V_{A,B} \triangleq \left\{ y_0 \in \mathbb{R}^n \ \middle| \ \exists \, T \in (0, +\infty) \text{ and } v \in L^2(0, T; \mathbb{R}^m) \right. \\ \left. \text{so that } y_0 = -\int_0^T e^{-sA} Bv(s) ds \right\}, \tag{3.30}$$

$$N_{A,B} \triangleq \left\{ \xi \in \mathbb{R}^n \ \middle| \ B^* e^{-A^* s} \xi = 0 \text{ for all } s > 0 \right\} \tag{3.31}$$

and

$$D_{A,B} \triangleq N_{A,B}^{\perp} \bigcap \partial B_1(0). \tag{3.32}$$

Theorem 3.4 Let $A \in \mathbb{R}^{n \times n}$ and $B \in \mathbb{R}^{n \times m}$. Let $(\widehat{y}_0, r) \in \mathbb{R}^n \times (0, +\infty)$. Then the following conclusions are equivalent:

(i) The pair $(\widehat{y}_0, r)$ satisfies that $\widehat{y}_0 \in Y_C^r$. Here, Y_C^r is the controllable set of (3.14), associated with r.
(ii) The pair $(\widehat{y}_0, r)$ satisfies that $\widehat{y}_0 \in V_{A,B}$ and that

$$\langle \widehat{y}_0, \xi \rangle_{\mathbb{R}^n} < r \int_0^{+\infty} \left\| B^* e^{-sA^*} \xi \right\|_{\mathbb{R}^m} ds \ \text{for all} \ \xi \in D_{A,B}. \tag{3.33}$$

(iii) The pair $(\widehat{y}_0, r)$ satisfies that $\widehat{y}_0 \in V_{A,B}$ and that

$$\langle \widehat{y}_0, \xi \rangle_{\mathbb{R}^n} < r \int_0^{+\infty} \left\| B^* e^{-sA^*} \xi \right\|_{\mathbb{R}^m} ds \ \text{for all} \ \xi \notin N_{A,B}. \tag{3.34}$$

In order to prove Theorem 3.4, we need some preliminary results.

Lemma 3.2 Let $A \in \mathbb{R}^{n \times n}$ and $B \in \mathbb{R}^{n \times m}$. Then $V_{A,B} = N_{A,B}^{\perp}$.

Proof We first show that $V_{A,B} \subseteq N_{A,B}^{\perp}$. To this end, we let $y_0 \in V_{A,B}$. Then by (3.30), we can find $T \in (0, +\infty)$ and $v \in L^2(0, T; \mathbb{R}^m)$ so that

$$y_0 = -\int_0^T e^{-sA} Bv(s)ds.$$

From the above, it follows that

$$\langle \xi, y_0 \rangle_{\mathbb{R}^n} = -\int_0^T \langle B^* e^{-sA^*} \xi, v(s) \rangle_{\mathbb{R}^m} ds = 0 \ \text{for all} \ \xi \in N_{A,B}.$$

This implies that $y_0 \in N_{A,B}^{\perp}$. Hence, $V_{A,B} \subseteq N_{A,B}^{\perp}$.

We next show that $N_{A,B}^{\perp} \subseteq V_{A,B}$. By contradiction, suppose that there was $\eta \in \mathbb{R}^n$ so that $\eta \in N_{A,B}^{\perp} \setminus V_{A,B}$. Then by the convexity and the closedness of $V_{A,B}$ (in $\mathbb{R}^n$), we could use Theorem 1.11 to find a vector $\zeta \in \mathbb{R}^n$ so that

$$\langle \zeta, \eta \rangle_{\mathbb{R}^n} > \langle \zeta, w \rangle_{\mathbb{R}^n} \ \text{for all} \ w \in V_{A,B}. \tag{3.35}$$

By (3.30), one can easily check that $0 \in V_{A,B}$. This, along with (3.35), yields that $\langle \zeta, \eta \rangle_{\mathbb{R}^n} > 0$. Since $\eta \in N_{A,B}^{\perp}$, the above implies that $\zeta \notin N_{A,B}$, i.e., $B^* e^{-A^* s_0} \zeta \neq 0$ for some $s_0 > 0$. Hence, there is $\delta_0 \in (0, s_0)$ so that

$$B^* e^{-A^* s} \zeta \neq 0 \ \text{for each} \ s \in (s_0 - \delta_0, s_0 + \delta_0). \tag{3.36}$$

Meanwhile, for each $k > 0$, we set

$$u_k(s) \triangleq k\chi_{(0,s_0+\delta_0)}(s)B^*e^{-A^*s}\zeta, \quad s > 0; \quad w_k \triangleq \int_0^{s_0+\delta_0} e^{-sA}Bu_k(s)ds.$$

It is obvious that $w_k \in V_{A,B}$ for all $k \geq 1$. Thus, by (3.35), we have that

$$\langle \zeta, \eta \rangle_{\mathbb{R}^n} > \langle \zeta, w_k \rangle_{\mathbb{R}^n} = k\int_0^{s_0+\delta_0} \|B^*e^{-sA^*}\zeta\|_{\mathbb{R}^m}^2 ds \quad \text{for all } k \geq 1. \tag{3.37}$$

Now, it follows from (3.36) and (3.37) that $\langle \zeta, \eta \rangle_{\mathbb{R}^n} = +\infty$, which leads to a contradiction. This ends the proof of Lemma 3.2. $\qquad\square$

The next lemma is quoted from [11].

Lemma 3.3 ([11]) *Let Z be a closed and convex set in $\mathbb{R}^n$. Then*

$$Z = \bigcap_{\xi \in \partial B_1(0)} (CHS)_{Z,\xi},$$

where

$$(CHS)_{Z,\xi} \triangleq \left\{ \eta \in \mathbb{R}^n \mid \langle \eta, \xi \rangle_{\mathbb{R}^n} \leq \sup_{z \in Z}\langle z, \xi \rangle_{\mathbb{R}^n} \right\} \quad \text{for each } \xi \in \partial B_1(0).$$

We now are on the position to show Theorem 3.4.

Proof (Proof of Theorem 3.4) To show that (i)$\Rightarrow$(ii), we arbitrarily fix a pair $(\widehat{y}_0, r)$ with $\widehat{y}_0 \in Y_C^r$. Then, by (3.4) and (3.5), we see that there is $T > 0$ so that

$$\widehat{y}_0 = -\int_0^T e^{-sA}Bu(s)ds \quad \text{for some } u \in L(0, T; B_r(0)) \subseteq L^2(0, T; \mathbb{R}^m),$$

which, along with (3.30), implies that $\widehat{y}_0 \in V_{A,B}$. Moreover, for any $\xi \in \mathbb{R}^n$, we have that

$$\langle \widehat{y}_0, \xi \rangle_{\mathbb{R}^n} = -\int_0^T \langle B^*e^{-sA^*}\xi, u(s) \rangle_{\mathbb{R}^m} ds \leq r\int_0^T \|B^*e^{-sA^*}\xi\|_{\mathbb{R}^m} ds. \tag{3.38}$$

Since $s \to B^*e^{-A^*s}\xi, s \geq 0$, is an analytic function, we see that

$$\int_0^T \|B^*e^{-A^*s}\xi\|_{\mathbb{R}^m} ds < \int_0^{+\infty} \|B^*e^{-A^*s}\xi\|_{\mathbb{R}^m} ds \quad \text{for all } \xi \in D_{A,B}. \tag{3.39}$$

From (3.38) and (3.39), (3.33) follows at once. So the conclusion (ii) is true.

To show that (ii)$\Rightarrow$(iii), we let $(\widehat{y}_0, r)$ satisfy $\widehat{y}_0 \in V_{A,B}$ and (3.33). In order to prove (iii), it suffices to verify that $(\widehat{y}_0, r)$ satisfies (3.34). To this end, we let

$\xi \notin N_{A,B}$. Then there are two vectors $\xi_1 \in N_{A,B}$ and $\xi_2 \in N_{A,B}^{\perp} \setminus \{0\}$ so that

$$\xi = \xi_1 + \xi_2. \tag{3.40}$$

Since $\widehat{y}_0 \in V_{A,B}$ and $\xi_1 \in N_{A,B}$, it follows by Lemma 3.2 that

$$\langle \widehat{y}_0, \xi_1 \rangle_{\mathbb{R}^n} = 0. \tag{3.41}$$

From (3.40), (3.41), (3.32), (3.33), and (3.31), we find that

$$\langle \widehat{y}_0, \xi \rangle_{\mathbb{R}^n} = \langle \widehat{y}_0, \xi_2 \rangle_{\mathbb{R}^n} < r \int_0^{+\infty} \|B^* e^{-sA^*} \xi_2\|_{\mathbb{R}^m} ds = r \int_0^{+\infty} \|B^* e^{-sA^*} \xi\|_{\mathbb{R}^m} ds,$$

which leads to (3.34). Hence, (iii) is true.

The proof of (iii)$\Rightarrow$(ii) follows from (3.32) immediately.

To show (ii)$\Rightarrow$(i), we let $(\widehat{y}_0, r)$ satisfy $\widehat{y}_0 \in V_{A,B}$ and (3.33). We first claim that there is $T \in (0, +\infty)$ so that

$$\langle \widehat{y}_0, \xi \rangle_{\mathbb{R}^n} \leq r \int_0^T \|B^* e^{-sA^*} \xi\|_{\mathbb{R}^m} ds \quad \text{for all } \xi \in D_{A,B}. \tag{3.42}$$

By contradiction, suppose that it was not true. Then there would be a sequence $\{\xi_\ell\}_{\ell \geq 1}$ in $D_{A,B}$ so that

$$\langle \widehat{y}_0, \xi_\ell \rangle_{\mathbb{R}^n} > r \int_0^\ell \|B^* e^{-sA^*} \xi_\ell\|_{\mathbb{R}^m} ds \quad \text{for all } \ell \geq 1. \tag{3.43}$$

By (3.32), there is a subsequence of $\{\xi_\ell\}_{\ell \geq 1}$, denoted in the same manner, so that

$$\lim_{\ell \to +\infty} \xi_\ell = \xi^* \quad \text{for some } \xi^* \in D_{A,B}.$$

This, along with (3.43), yields that for each $j \geq 1$,

$$\langle \widehat{y}_0, \xi^* \rangle_{\mathbb{R}^n} = \lim_{\ell \to +\infty} \langle \widehat{y}_0, \xi_\ell \rangle_{\mathbb{R}^n}$$

$$\geq r \lim_{\ell \to +\infty} \int_0^j \|B^* e^{-sA^*} \xi_\ell\|_{\mathbb{R}^m} ds = r \int_0^j \|B^* e^{-sA^*} \xi^*\|_{\mathbb{R}^m} ds,$$

which implies that

$$\langle \widehat{y}_0, \xi^* \rangle_{\mathbb{R}^n} \geq r \int_0^{+\infty} \|B^* e^{-sA^*} \xi^*\|_{\mathbb{R}^m} ds.$$

This contradicts (3.33). So (3.42) is true.

Next, for each $\zeta \in \partial B_1(0)$, there are two vectors $\zeta_1 \in N_{A,B}$ and $\zeta_2 \in N_{A,B}^\perp$ so that

$$\zeta = \zeta_1 + \zeta_2. \tag{3.44}$$

Since $\widehat{y}_0 \in V_{A,B}$ and $\zeta_1 \in N_{A,B}$, it follows by Lemma 3.2 that

$$\langle \widehat{y}_0, \zeta_1 \rangle_{\mathbb{R}^n} = 0. \tag{3.45}$$

From (3.45), (3.44), (3.32), (3.42), and (3.31), we can easily check that

$$\langle \widehat{y}_0, \zeta \rangle_{\mathbb{R}^n} = \langle \widehat{y}_0, \zeta_2 \rangle_{\mathbb{R}^n} \leq r \int_0^T \|B^* e^{-sA^*} \zeta_2\|_{\mathbb{R}^m} ds = r \int_0^T \|B^* e^{-sA^*} \zeta\|_{\mathbb{R}^m} ds. \tag{3.46}$$

Since $Y_C^r(T)$ is a convex and closed subset in $\mathbb{R}^n$ (see Lemma 3.1), it follows by Lemma 3.3 and (3.5) that

$$Y_C^r(T) = \left\{ \eta \in \mathbb{R}^n \mid \langle \eta, \xi \rangle_{\mathbb{R}^n} \leq r \int_0^T \|B^* e^{-sA^*} \xi\|_{\mathbb{R}^m} ds \text{ for all } \xi \in \mathbb{R}^n \right\}.$$

This, together with (3.46) and (3.4), implies that $\widehat{y}_0 \in Y_C^r(T) \subseteq Y_C^r$.

In summary, we end the proof of Theorem 3.4.   □

We denote

$$\frac{0}{0} \triangleq 0, \quad \frac{1}{0} \triangleq +\infty, \quad \frac{-1}{0} \triangleq -\infty. \tag{3.47}$$

For each $A \in \mathbb{R}^{n \times n}$ and each $B \in \mathbb{R}^{n \times m}$, we define a function $N(\cdot) : \mathbb{R}^n \mapsto [0, +\infty]$ by

$$N(z) \triangleq \sup_{\xi \in \mathbb{R}^n} \langle z, \xi \rangle_{\mathbb{R}^n} \Big/ \int_0^{+\infty} \|B^* e^{-sA^*} \xi\|_{\mathbb{R}^m} ds, \quad z \in \mathbb{R}^n. \tag{3.48}$$

We next present the following consequence of Theorem 3.4:

Corollary 3.3 *Let $A \in \mathbb{R}^{n \times n}$, $B \in \mathbb{R}^{n \times m}$, and $(\widehat{y}_0, r) \in \mathbb{R}^n \times (0, +\infty)$. Let Y_C^r be the controllable set of (3.14), associated with r, and let $N(\cdot)$ be given by (3.48). Then $\widehat{y}_0 \in Y_C^r$ if and only if $N(\widehat{y}_0) < r$.*

Proof We first show that $\widehat{y}_0 \in Y_C^r$ implies $N(\widehat{y}_0) < r$. To this end, we arbitrarily fix $(\widehat{y}_0, r)$ with $\widehat{y}_0 \in Y_C^r$. By Theorem 3.4, we have that

$$\widehat{y}_0 \in V_{A,B} \text{ and } \langle \widehat{y}_0, \xi \rangle_{\mathbb{R}^n} < r \int_0^{+\infty} \|B^* e^{-sA^*} \xi\|_{\mathbb{R}^m} ds \text{ for all } \xi \notin N_{A,B}. \tag{3.49}$$

Meanwhile, by the first conclusion in (3.49), Lemma 3.2, and (3.31), we obtain that

$$\langle \widehat{y}_0, \xi \rangle_{\mathbb{R}^n} = r \int_0^{+\infty} \|B^* e^{-sA^*} \xi\|_{\mathbb{R}^m} ds = 0 \quad \text{for all} \ \xi \in N_{A,B}. \tag{3.50}$$

From (3.48), (3.47), (3.49), and (3.50), it follows that

$$N(\widehat{y}_0) = \sup_{\xi \notin N_{A,B}} \langle \widehat{y}_0, \xi \rangle_{\mathbb{R}^n} \Big/ \int_0^{+\infty} \|B^* e^{-sA^*} \xi\|_{\mathbb{R}^m} ds \leq r. \tag{3.51}$$

Next, we claim that

$$\sup_{\xi \notin N_{A,B}} \langle \widehat{y}_0, \xi \rangle_{\mathbb{R}^n} \Big/ \int_0^{+\infty} \|B^* e^{-sA^*} \xi\|_{\mathbb{R}^m} ds \neq r. \tag{3.52}$$

By contradiction, suppose that (3.52) was not true. Then there would be a sequence $\{\xi_\ell\}_{\ell \geq 1}$, with $\xi_\ell \notin N_{A,B}$, so that

$$\langle \widehat{y}_0, \xi_\ell \rangle_{\mathbb{R}^n} \Big/ \int_0^{+\infty} \|B^* e^{-sA^*} \xi_\ell\|_{\mathbb{R}^m} ds \to r. \tag{3.53}$$

Since $\xi_\ell \notin N_{A,B}$, we can write

$$\xi_\ell = \xi_\ell^{(1)} + \xi_\ell^{(2)}, \quad \text{with} \ \xi_\ell^{(1)} \in N_{A,B} \ \text{and} \ \xi_\ell^{(2)} \in N_{A,B}^\perp \setminus \{0\}.$$

Then, by the first conclusion in (3.49), Lemma 3.2, (3.31), and (3.53), we find that

$$\langle \widehat{y}_0, \xi_\ell^{(2)} \rangle_{\mathbb{R}^n} \Big/ \int_0^{+\infty} \|B^* e^{-sA^*} \xi_\ell^{(2)}\|_{\mathbb{R}^m} ds \to r. \tag{3.54}$$

Set

$$\widetilde{\xi}_\ell \triangleq \xi_\ell^{(2)} / \|\xi_\ell^{(2)}\|_{\mathbb{R}^n}.$$

It is obvious that $\widetilde{\xi}_\ell \in D_{A,B}$. Thus, there exists a subsequence of $\{\widetilde{\xi}_\ell\}_{\ell \geq 1}$, denoted in the same manner, so that

$$\widetilde{\xi}_\ell \to \widetilde{\xi} \quad \text{for some} \ \widetilde{\xi} \in D_{A,B}. \tag{3.55}$$

Meanwhile, by (3.54), we see that

$$\langle \widehat{y}_0, \widetilde{\xi}_\ell \rangle_{\mathbb{R}^n} \Big/ \int_0^{+\infty} \|B^* e^{-sA^*} \widetilde{\xi}_\ell\|_{\mathbb{R}^m} ds \to r. \tag{3.56}$$

Now, it follows from (3.55) and (3.56) that

$$\langle \widehat{y}_0, \widetilde{\xi} \rangle_{\mathbb{R}^n} \geq r \int_0^{+\infty} \|B^* e^{-sA^*} \widetilde{\xi}\|_{\mathbb{R}^m} ds,$$

which contradicts (3.49). So we obtain (3.52), which, along with (3.51), leads to that $N(\widehat{y}_0) < r$.

We next show that $N(\widehat{y}_0) < r$ implies $\widehat{y}_0 \in Y_C^r$. For this purpose, we arbitrarily fix $(\widehat{y}_0, r)$ with $N(\widehat{y}_0) < r$. Then, by (3.48), we have that

$$\langle \widehat{y}_0, \xi \rangle_{\mathbb{R}^n} \Big/ \int_0^{+\infty} \|B^* e^{-sA^*} \xi\|_{\mathbb{R}^m} ds < r \quad \text{for all } \xi \in \mathbb{R}^n. \tag{3.57}$$

It follows from (3.57) and (3.47) that

$$\langle \widehat{y}_0, \xi \rangle_{\mathbb{R}^n} \leq 0 \quad \text{for all } \xi \in N_{A,B},$$

which implies that

$$\langle \widehat{y}_0, \xi \rangle_{\mathbb{R}^n} = 0 \quad \text{for all } \xi \in N_{A,B}.$$

This, together with Lemma 3.2, indicates that

$$\widehat{y}_0 \in N_{A,B}^\perp = V_{A,B}. \tag{3.58}$$

Meanwhile, it follows by (3.57) that

$$\langle \widehat{y}_0, \xi \rangle_{\mathbb{R}^n} < r \int_0^{+\infty} \|B^* e^{-sA^*} \xi\|_{\mathbb{R}^m} ds \quad \text{for all } \xi \notin N_{A,B}.$$

From this and (3.58), we can apply Theorem 3.4 to see that $\widehat{y}_0 \in Y_C^r$.

Thus, we complete the proof of Corollary 3.3.   □

3.1.3 Heat Equations with Ball-Type Control Constraints

In this subsection, we let $\Omega \subseteq \mathbb{R}^d, d \geq 1$, be a bounded domain, with a C^2 boundary $\partial\Omega$. Let $\omega \subseteq \Omega$ be an open and nonempty subset, with its characteristic function χ_ω. Consider the following controlled semilinear heat equation:

$$\begin{cases} \partial_t y - \Delta y + f(y) = \chi_\omega u & \text{in } \Omega \times (0, +\infty), \\ y = 0 & \text{on } \partial\Omega \times (0, +\infty), \\ y(0) = \widehat{y}_0 & \text{in } \Omega. \end{cases} \tag{3.59}$$

Here, $\widehat{y}_0 \in L^2(\Omega)$ is a nonzero function; control u is taken from the constraint set:

$$U_\rho \triangleq \{u : (0, +\infty) \mapsto L^2(\Omega) \text{ is measurable} \mid \|u(t)\|_{L^2(\Omega)} \leq \rho$$

$$\text{for a.e. } t \in (0, +\infty)\},$$

where $\rho > 0$ is a given constant; $f : \mathbb{R} \mapsto \mathbb{R}$ satisfies the following hypothesis:

(A_1) f is globally Lipschitz and continuously differentiable;
(A_2) $f(r)r \geq 0$ for all $r \in \mathbb{R}$.

The Equation (3.59) can be put into our framework (2.1) (with the initial condition that $y(0) = \widehat{y}_0$) in the following manner:

- Let $Y \triangleq L^2(\Omega); U \triangleq L^2(\Omega)$;
- Let $A \triangleq \Delta$, with its domain $\mathscr{D}(A) \triangleq H^2(\Omega) \cap H_0^1(\Omega)$. It generates an analytic semigroup on Y;
- The function χ_ω is treated as a linear and bounded operator on U;
- The function f provides a nonlinear operator $\widehat{f} : Y \mapsto Y$ via

$$\widehat{f}(y)(x) \triangleq f(y(x)), \quad x \in \Omega.$$

Thus, the Equation (3.59) can be rewritten as

$$\begin{cases} \dot{y} - Ay + \widehat{f}(y) = \chi_\omega u, \ t \in (0, +\infty), \\ y(0) = \widehat{y}_0. \end{cases} \tag{3.60}$$

In this subsection, we shall study the problem $(TP)_{min}^{Q_S, Q_E}$, where the controlled system is (3.60), and where

$$Q_S \triangleq \{0\} \times \{\widehat{y}_0\}, \quad Q_E \triangleq (0, +\infty) \times \{0\} \text{ and } \mathbb{U} \triangleq \{u \in U \mid \|u\|_{L^2(\Omega)} \le \rho\}.$$

One can easily check that the problem $(TP)_{min}^{Q_S, Q_E}$ can be put into the framework $(\mathscr{A}_{TP})$ (given at the beginning of Chapter 2). We will simply write $(TP)_\rho^{\widehat{y}_0}$ for the above $(TP)_{min}^{Q_S, Q_E}$. We denote by Y_C^ρ the controllable set $Y_C^{Q_E, \mathbb{U}}$ of the Equation (3.60), associated with Q_E and $\mathbb{U}$. The same can be said about $Y_C^\rho(t)$. The main result of this subsection is as follows:

Theorem 3.5 *For each $\rho > 0$, it holds that $Y_C^\rho = Y$.*

Remark 3.1 By Theorem 3.5 and Theorem 3.1, we see that for any $\widehat{y}_0 \in L^2(\Omega) \backslash \{0\}$ and $\rho > 0$, the problem $(TP)_\rho^{\widehat{y}_0}$ has an admissible control.

To prove Theorem 3.5, we need the next Proposition 3.1.

Proposition 3.1 *There exists a positive constant κ so that for any $z_0 \in L^2(\Omega)$, there is a control $u \in L^\infty(0, 1; L^2(\Omega))$, with*

$$\|u\|_{L^\infty(0,1;L^2(\Omega))} \le \kappa \|z_0\|_{L^2(\Omega)},$$

so that the solution $z \in C([0, 1]; L^2(\Omega))$ to the equation:

$$\begin{cases} \partial_t z - \Delta z + f(z) = \chi_\omega u \text{ in } \Omega \times (0, 1), \\ z = 0 \qquad\qquad\qquad\quad \text{on } \partial\Omega \times (0, 1), \\ z(0) = z_0 \qquad\qquad\qquad \text{in } \Omega \end{cases}$$

satisfies that $z(1) = 0$ over Ω.

Proof We will use the Kakutani-Fan-Glicksberg Theorem (see Theorem 1.14) to prove this proposition. To this end, we define

$$a(r) \triangleq \begin{cases} \dfrac{f(r)}{r} & \text{if } r \neq 0, \\ f'(0) & \text{if } r = 0. \end{cases}$$

Set

$$\mathcal{K} \triangleq \{\xi \in L^2(0, 1; L^2(\Omega)) \mid \|\xi\|_{L^2(0,1;H_0^1(\Omega))} + \|\xi\|_{W^{1,2}(0,1;H^{-1}(\Omega))} \leq c_0\},$$

where $c_0 > 0$ will be determined later. By assumptions (A_1) and (A_2) (given at the beginning of Section 3.1.3), we have that

$$|a(r)| \leq L, \quad \text{for all } r \in \mathbb{R}, \tag{3.61}$$

where $L > 0$ is the Lipschitz constant of the function f. For each $\xi \in \mathcal{K}$, we consider the following linear equation:

$$\begin{cases} z_t - \Delta z + a(\xi(x, t))z = \chi_\omega u & \text{in } \Omega \times (0, 1), \\ z = 0 & \text{on } \partial\Omega \times (0, 1), \\ z(0) = z_0 & \text{in } \Omega. \end{cases} \tag{3.62}$$

Simply write $z(\cdot)$ for the solution of (3.62). According to Theorems 1.22 and 1.21, there is a positive constant κ (independent of z_0 and ξ) and a control u so that

$$\|u\|_{L^\infty(0,1;L^2(\Omega))} \leq \kappa \|z_0\|_{L^2(\Omega)} \quad \text{and} \quad z(1) = 0. \tag{3.63}$$

Define a set-valued mapping

$$\Phi : \mathcal{K} \to 2^{L^2(0,1;L^2(\Omega))}$$

by setting

$$\Phi(\xi) \triangleq \{z \mid \text{there exists a control } u \in L^\infty(0, 1; L^2(\Omega))$$
$$\text{so that (3.62) and (3.63) hold}\}, \quad \xi \in \mathcal{K}.$$

One can easily check that $\Phi(\xi) \neq \emptyset$ for each $\xi \in \mathcal{K}$.

The rest of the proof will be organized by several steps as follows:

Step 1. We show that $\mathcal{K}$ is compact and convex in $L^2(0, 1; L^2(\Omega))$, and that each $\Phi(\xi)$ is convex in $L^2(0, 1; L^2(\Omega))$.

These can be directly checked.

Step 2. We show that $\Phi(\mathcal{K}) \subseteq \mathcal{K}$.

Given $\xi \in \mathcal{K}$, there is a control u satisfying (3.62) and (3.63). By (i) of Theorem 1.12 and (3.61)–(3.63), we can easily check that

$$\|z\|_{L^2(0,1;H_0^1(\Omega))} + \|z\|_{W^{1,2}(0,1;H^{-1}(\Omega))} \leq c_1 \|z_0\|_{L^2(\Omega)},$$

for a positive constant c_1 (independent of z_0 and ξ). Hence,

$$z \in \mathcal{K} \text{ if } c_0 = c_1 \|z_0\|_{L^2(\Omega)}.$$

Step 3. We show that Graph(Φ) is closed.
It suffices to show that $z \in \Phi(\xi)$, provided that

$$\xi_\ell \in \mathcal{K} \to \xi \text{ strongly in } L^2(0, 1; L^2(\Omega))$$

and

$$z_\ell \in \Phi(\xi_\ell) \to z \text{ strongly in } L^2(0, 1; L^2(\Omega)).$$

To this end, we first observe that $\xi \in \mathcal{K}$, since $\mathcal{K}$ is convex and closed. Next we claim that there exists a subsequence of $\{\ell\}_{\ell \geq 1}$, denoted in the same manner, so that

$$a(\xi_\ell)z_\ell \to a(\xi)z \text{ strongly in } L^2(0, 1; L^2(\Omega)). \tag{3.64}$$

Indeed, since

$$\xi_\ell \to \xi \text{ strongly in } L^2(0, 1; L^2(\Omega)),$$

we have a subsequence of $\{\ell\}_{\ell \geq 1}$, denoted in the same manner, so that

$$\xi_\ell(x, t) \to \xi(x, t) \text{ for a.e. } (x, t) \in \Omega \times (0, 1).$$

Then, by the definition of the function a, we conclude that

$$a(\xi_\ell(x, t)) \to a(\xi(x, t)) \text{ for a.e. } (x, t) \in \Omega \times (0, 1).$$

By this and (3.61), we can apply the Lebesgue Dominated Convergence Theorem to obtain that

$$\|a(\xi_\ell)z_\ell - a(\xi)z\|_{L^2(0,1;L^2(\Omega))}^2$$
$$\leq 2\|a(\xi_\ell)(z_\ell - z)\|_{L^2(0,1;L^2(\Omega))}^2 + 2\|(a(\xi_\ell) - a(\xi))z\|_{L^2(0,1;L^2(\Omega))}^2$$
$$\leq 2L^2\|z_\ell - z\|_{L^2(0,1;L^2(\Omega))}^2 + 2\|(a(\xi_\ell) - a(\xi))z\|_{L^2(0,1;L^2(\Omega))}^2$$
$$\to 0.$$

This leads to (3.64).

Finally, for each $\ell \geq 1$, since $z_\ell \in \Phi(\xi_\ell) \subseteq \mathcal{K}$, there is $u_\ell \in L^\infty(0, 1; L^2(\Omega))$ with

$$\|u_\ell\|_{L^\infty(0,1;L^2(\Omega))} \leq \kappa \|z_0\|_{L^2(\Omega)}, \tag{3.65}$$

so that

$$\begin{cases} \partial_t z_\ell - \Delta z_\ell + a(\xi_\ell(x,t))z_\ell = \chi_\omega u_\ell & \text{in } \Omega \times (0,1), \\ z_\ell = 0 & \text{on } \partial\Omega \times (0,1), \\ z_\ell(0) = z_0 & \text{in } \Omega, \\ z_\ell(1) = 0 & \text{in } \Omega \end{cases} \tag{3.66}$$

and

$$\|z_\ell\|_{L^2(0,1;H_0^1(\Omega))} + \|z_\ell\|_{W^{1,2}(0,1;H^{-1}(\Omega))} \leq c_0. \tag{3.67}$$

By (3.65) and (3.67), there is a control u and a subsequence of $\{\ell\}_{\ell \geq 1}$, denoted in the same manner, so that

$$u_\ell \to u \text{ weakly star in } L^\infty(0, 1; L^2(\Omega)), \tag{3.68}$$

$$z_\ell \to z \text{ weakly in } L^2(0, 1; H_0^1(\Omega)) \cap W^{1,2}(0, 1; H^{-1}(\Omega)), \tag{3.69}$$

and

$$z_\ell(1) \to z(1) \text{ strongly in } L^2(\Omega). \tag{3.70}$$

Passing to the limit for $\ell \to +\infty$ in (3.65) and (3.66), making use of (3.64) and (3.68)–(3.70), we obtain that $z \in \Phi(\xi)$.

Step 4. We apply the Kakutani-Fan-Glicksberg Theorem.

From the conclusions in the above three steps, we find that the map Φ satisfies conditions of the Kakutani-Fan-Glicksberg Theorem. Thus we can apply this theorem to obtain such z that $z \in \Phi(z)$. Then, by the definition of Φ and using the fact that

$$a(z(x,t))z(x,t) = f(z(x,t)),$$

one can easily get the desired results of this proposition.

Hence, we end the proof of Proposition 3.1. □

We now give the proof of Theorem 3.5.

Proof (Proof of Theorem 3.5) First, for any $y_0 \in L^2(\Omega) \setminus \{0\}$, we consider the following equation:

$$\begin{cases} \partial_t y - \Delta y + f(y) = 0 & \text{in } \Omega \times (0, t_0), \\ y = 0 & \text{on } \partial\Omega \times (0, t_0), \\ y(0) = y_0 & \text{in } \Omega, \end{cases} \tag{3.71}$$

where $t_0 > 0$ will be determined later. Write $y(\cdot)$ for the solution of (3.71). By (ii) of Theorem 1.12 and the assumption (A_2), we have that

$$\|y(t_0)\|_{L^2(\Omega)} \le e^{-\lambda_1 t_0}\|y_0\|_{L^2(\Omega)}, \tag{3.72}$$

where $\lambda_1 > 0$ is the first eigenvalue of $-\Delta$ with the zero Dirichlet boundary condition.

Next, by Proposition 3.1, there is $u \in L^\infty(0, 1; L^2(\Omega))$ and $z \in C([0, 1]; L^2(\Omega))$ so that

$$\|u\|_{L^\infty(0,1;L^2(\Omega))} \le \kappa \|y(t_0)\|_{L^2(\Omega)} \tag{3.73}$$

and

$$
\begin{cases}
\partial_t z - \Delta z + f(z) = \chi_\omega u & \text{in } \Omega \times (0, 1), \\
z = 0 & \text{on } \partial\Omega \times (0, 1), \\
z(0) = y(t_0) & \text{in } \Omega, \\
z(1) = 0 & \text{in } \Omega.
\end{cases}
\tag{3.74}
$$

Finally, we set

$$t_0 \triangleq \frac{1}{\lambda_1}\left(\left|\ln\frac{\kappa\|y_0\|_{L^2(\Omega)}}{\rho}\right| + 1\right) \quad \text{and} \quad \widehat{u}(t) \triangleq \begin{cases} 0, & t \in (0, t_0), \\ u(t - t_0), & t \in (t_0, t_0 + 1). \end{cases}$$

Then, by (3.71)–(3.74), we can easily check that

$$y_0 \in Y_C^\rho(t_0 + 1) \subseteq Y_C^\rho.$$

This completes the proof of Theorem 3.5. □

3.2 Admissible Controls and Minimal Norm Problems

In this section, we will study the existence of admissible controls of $(TP)_{min}^{Q_S, Q_E}$ from the perspective of minimal norm problems. Throughout this section, we focus on the problem $(TP)_{min}^{Q_S, Q_E}$ under the following framework $(\mathscr{H}_{TP})$:

(i) The state space Y and the control space U are real separable Hilbert spaces.
(ii) The controlled system (or the state system) is the Equation (2.1), i.e.,

$$\dot{y}(t) = Ay(t) + f(t, y(t), u(t)), \quad t \in (0, +\infty),$$

where $A : \mathscr{D}(A) \subseteq Y \mapsto Y$ generates a C_0 semigroup $\{e^{At}\}_{t \ge 0}$ on Y and $f : (0, +\infty) \times Y \times U \mapsto Y$ is some given map. Moreover, for any $0 \le a < b < +\infty$, $y_0 \in Y$ and $u(\cdot) \in L(a, b; \mathbb{U})$, this equation, with the initial condition $y(a) = y_0$, has a unique mild solution $y(\cdot; a, y_0, u)$ on $[a, b]$.

(iii) The starting set is as: $Q_S = \{0\} \times Y_S$, with Y_S a nonempty subset of Y.

(iv) The ending set is as: $Q_E = (0, +\infty) \times Y_E$, with Y_E a nonempty subset of Y.

(v) The control constraint set $\mathbb{U} \subseteq U$ is a nonempty, bounded, and convex closed subset, which is symmetric with respect to 0 (i.e., $u \in \mathbb{U} \Leftrightarrow -u \in \mathbb{U}$) and has a nonzero in-radius. The in-radius of $\mathbb{U}$ is defined to be

$$\inf_{u \in \mathbb{U} \setminus \{0\}} \sup_{\lambda > 0, \lambda u \in \mathbb{U}} \|\lambda u\|_U. \qquad (3.75)$$

Two facts deserve to be mentioned: First, $B_r(0) \subseteq U$ (with $r > 0$) is an example of such $\mathbb{U}$ that has properties in the (v) of $(\mathscr{H}_{TP})$. Second, $(\mathscr{H}_{TP})$ implies $(\mathscr{A}_{TP})$ (given at the beginning of Chapter 2). Indeed, (i)–(iii) in $(\mathscr{H}_{TP})$ are exactly (i)–(iii) in $(\mathscr{A}_{TP})$, while (iv) and (v) in $(\mathscr{H}_{TP})$ are special cases of (iv) and (v) in $(\mathscr{A}_{TP})$, respectively.

3.2.1 General Results in Abstract Setting

To study the problem $(TP)_{min}^{Q_S, Q_E}$, we need to introduce a norm optimal control problem. For this purpose, we will present several lemmas, as well as one corollary. The proofs of these results rely on (v) of $(\mathscr{H}_{TP})$ (given at the beginning of Section 3.2).

Let $\lambda \in [0, +\infty)$ and let

$$\lambda \mathbb{U} \triangleq \{\lambda u \mid u \in \mathbb{U}\} \subseteq U.$$

Since $\mathbb{U}$ is convex and symmetric with respect to 0, we have that $\lambda \mathbb{U} \subseteq \mathbb{U}$ for each $\lambda \in (0, 1)$. Set

$$V \triangleq \bigcup_{\lambda \geq 0} \lambda \mathbb{U} \subseteq U \qquad (3.76)$$

and define the Minkowski functional $p : V \mapsto [0, +\infty)$ by

$$p(v) \triangleq \min \{\lambda \in [0, +\infty) \mid v \in \lambda \mathbb{U}\}, \quad v \in V. \qquad (3.77)$$

Lemma 3.4 *The following conclusions are true: (i) $V = span\mathbb{U}$. (ii) $(V, \|\cdot\|_p)$ is a normed linear space, where $\|v\|_p \triangleq p(v)$ for each $v \in V$.*

Proof We prove the conclusions one by one.

(i) It is obvious that $V \subseteq span\mathbb{U}$. To show the reverse, we arbitrarily fix u in $span\mathbb{U}$. Then, there are two sequences $\{u_j\}_{j=1}^k \subseteq \mathbb{U}$ and $\{\ell_j\}_{j=1}^k \subseteq \mathbb{R} \setminus \{0\}$ so that $u = \sum_{j=1}^k \ell_j u_j$. By setting $\ell \triangleq \sum_{j=1}^k |\ell_j|$, we see that

$$u = \ell \sum_{j=1}^{k} \frac{|\ell_j|}{\ell} \left(\frac{\ell_j}{|\ell_j|} u_j \right).$$

This, along with the symmetry and the convexity of $\mathbb{U}$, yields that $u \in V$. Thus, we obtain that $\mathrm{span} \mathbb{U} \subseteq V$. Hence, the conclusion (i) is true.

(ii) It suffices to show that $p(\cdot)$ defines a norm on V. To this end, we first observe from (3.77) that

$$p(v) = 0 \text{ if and only if } v = 0, \tag{3.78}$$

and that

$$p(\mu v) = |\mu| p(v) \text{ for all } v \in V \text{ and } \mu \in \mathbb{R}. \tag{3.79}$$

We next claim that

$$\|v_1 + v_2\|_p \leq \|v_1\|_p + \|v_2\|_p \text{ for all } v_1, v_2 \in V \setminus \{0\}. \tag{3.80}$$

Indeed, given $v_1, v_2 \in V \setminus \{0\}$, we let

$$c_1 \triangleq \|v_1\|_p > 0 \text{ and } c_2 \triangleq \|v_2\|_p > 0. \tag{3.81}$$

By (3.81) and (3.77), we find that

$$v_1/c_1, \ v_2/c_2 \in \mathbb{U}. \tag{3.82}$$

Since $\mathbb{U}$ is a convex subset of U, it follows by (3.82) that

$$\frac{v_1 + v_2}{c_1 + c_2} = \frac{c_1}{c_1 + c_2} \cdot \frac{v_1}{c_1} + \frac{c_2}{c_1 + c_2} \cdot \frac{v_2}{c_2} \in \mathbb{U}.$$

Hence, we have that

$$\|v_1 + v_2\|_p = p(v_1 + v_2) \leq c_1 + c_2,$$

which, together with (3.81), leads to (3.80).

Finally, we see from (3.78), (3.79), and (3.80) that $p(\cdot)$ defines a norm on V. This completes the proof of Lemma 3.4. $\square$

The in-radius of $\mathbb{U}$ (see (3.75)) has more intuitive expression. Indeed, one can easily check that

$$\inf_{u \in \mathbb{U} \setminus \{0\}} \sup_{\lambda > 0, \lambda u \in \mathbb{U}} \|\lambda u\|_U = \inf_{w \in \partial B_1(0) \cap V} \sup_{r w \in \mathbb{U}} r, \tag{3.83}$$

where $B_1(0)$ is the closed unit ball in U. We now define

$$C_1 \triangleq \sup_{u \in U} \|u\|_U \quad \text{and} \quad C_2 \triangleq \inf_{w \in \partial B_1(0) \cap V} \sup_{r w \in U} r. \tag{3.84}$$

The above C_1 is the radius of U, while C_2 is the in-radius of U (see (3.75) and (3.83)). Several lemmas on properties of V and $p(\cdot)$ are given in order.

Lemma 3.5 *Let C_1 and C_2 be given by (3.84). Then $C_2\|v\|_p \leq \|v\|_U \leq C_1\|v\|_p$ for each $v \in V$.*

Proof It suffices to show that

$$C_2\|v\|_p \leq \|v\|_U \leq C_1\|v\|_p \quad \text{for all} \quad v \in V \setminus \{0\}. \tag{3.85}$$

To this end, we arbitrarily fix $v \in V \setminus \{0\}$. On one hand, by (3.77), there is $u \in U$ so that $v = \|v\|_p u$. From which it follows that

$$\|v\|_U = \|v\|_p \cdot \|u\|_U \leq C_1\|v\|_p. \tag{3.86}$$

On the other hand, it is obvious that

$$\frac{v}{\|v\|_U} \in \partial B_1(0) \cap V.$$

Since U is closed and convex, the above, along with (3.84) and the fact that $0 \in U$, yields that $(C_2 v)/\|v\|_U \in U$, which implies that $(C_2\|v\|_p)/\|v\|_U \leq 1$. This, along with (3.86), leads to (3.85). Thus, we end the proof of Lemma 3.5. $\qquad\square$

Define two normed spaces as follows:

$$V_U \triangleq (V, \|\cdot\|_U) \quad \text{and} \quad V_p \triangleq (V, \|\cdot\|_p).$$

It deserves mentioning that when $U = B_r(0)$ (with $r > 0$), $V_p = (U, r^{-1}\|\cdot\|_U)$.

Lemma 3.6 *The following conclusions are true:*

(i) *V_U is a Hilbert space with the inner product inherited from U.*
(ii) *V_p is a reflexive Banach space.*
(iii) *The identity operator on V, denoted by I_p, is an isomorphism from V_U to V_p.*

Proof We prove the conclusions one by one.

(i) It suffices to show that V is a closed subset of U. To this end, we let $\{v_\ell\}_{\ell \geq 1} \subseteq V$ satisfy that

$$v_\ell \to \bar{v} \quad \text{strongly in} \quad U. \tag{3.87}$$

By (3.77) and (3.76), there is a sequence $\{u_\ell\}_{\ell \geq 1} \subseteq U$ so that

$$v_\ell = \|v_\ell\|_p u_\ell \quad \text{for all} \quad \ell \geq 1. \tag{3.88}$$

Since $\mathbb{U}$ is a bounded, convex, and closed subset of U, there is a subsequence of $\{u_\ell\}_{\ell \geq 1}$, denoted in the same way, so that

$$u_\ell \to \bar{u} \quad \text{weakly in } U \text{ for some } \bar{u} \in \mathbb{U}. \tag{3.89}$$

Meanwhile, by (3.87) and Lemma 3.5, there is a subsequence of $\{\ell\}_{\ell \geq 1}$, denoted in the same manner, so that

$$\|v_\ell\|_p \to a_0 \quad \text{for some } a_0 \geq 0. \tag{3.90}$$

Now, by (3.87), (3.89), and (3.90), we can pass to the limit for $\ell \to +\infty$ in (3.88) to obtain that $\bar{v} = a_0 \bar{u}$. This, along with (3.76), yields that $\bar{v} \in V$. Hence, V is a closed subspace of U. So V_U is a Hilbert space.

(ii) By Lemma 3.4, V_p is a normed space. We now show that V_p is a Banach space. To this end, we let $\{v_\ell\}_{\ell \geq 1}$ be a Cauchy sequence in V_p. Then by Lemma 3.5, we find that it is a Cauchy sequence in V_U. Since V_U is a Hilbert space, we have that

$$v_\ell \to \bar{v} \quad \text{strongly in } V_U \text{ for some } \bar{v} \in V.$$

By this and by making use of Lemma 3.5 again, we find that V_p is a Banach space.

We next show that V_p is reflexive. To this end, we let f be a linear and continuous functional on V_p. It follows from Lemma 3.5 that f is also a linear and continuous functional on V_U. Since V_U is a Hilbert space, it follows by the Riesz Representation Theorem (see Theorem 1.3) that there is $u_f \in V$ so that

$$f(v) = \langle u_f, v \rangle_U \quad \text{for all } v \in V. \tag{3.91}$$

Arbitrarily fix $\widehat{f} \in V_p^*$ and consider the following problem: $\sup\limits_{\|v\|_p \leq 1} \widehat{f}(v)$. Then by (3.91), there exists a sequence $\{v_\ell\}_{\ell \geq 1} \subseteq V_p$ with $\|v_\ell\|_p \leq 1$ so that

$$\sup_{\|v\|_p \leq 1} \widehat{f}(v) = \lim_{\ell \to +\infty} \widehat{f}(v_\ell) = \lim_{\ell \to +\infty} \langle u_{\widehat{f}}, v_\ell \rangle_U. \tag{3.92}$$

Since $\|v_\ell\|_p \leq 1$, it follows from Lemma 3.5 that $\{v_\ell\}_{\ell \geq 1}$ is bounded in V_U. Hence, there is a subsequence of $\{v_\ell\}_{\ell \geq 1}$, denoted in the same way, so that

$$v_\ell \to \widehat{v} \quad \text{weakly in } V_U \text{ for some } \widehat{v} \in V. \tag{3.93}$$

Since $\|v_\ell\|_p \leq 1$, it follows by (3.91) and (3.93) that

$$\|\widehat{v}\|_p = \sup_{\|f\|_{V_p^*} \leq 1} f(\widehat{v}) = \sup_{\|f\|_{V_p^*} \leq 1} \langle u_f, \widehat{v} \rangle_U = \sup_{\|f\|_{V_p^*} \leq 1} \lim_{\ell \to +\infty} \langle u_f, v_\ell \rangle_U$$
$$= \sup_{\|f\|_{V_p^*} \leq 1} \lim_{\ell \to +\infty} f(v_\ell) \leq \sup_{\|f\|_{V_p^*} \leq 1} \overline{\lim_{\ell \to +\infty}} \|f\|_{V_p^*} \|v_\ell\|_p$$
$$\leq 1.$$
$$\tag{3.94}$$

Meanwhile, it follows from (3.92), (3.93), and (3.91) that

$$\sup_{\|v\|_p \leq 1} \widehat{f}(v) = \langle u_{\widehat{f}}, \widehat{v} \rangle_U = \widehat{f}(\widehat{v}).$$

This, together with (3.94), implies that

$$\sup_{\|v\|_p \leq 1} \widehat{f}(v) = \max_{\|v\|_p \leq 1} \widehat{f}(v). \tag{3.95}$$

Since we have proved that V_p is a Banach space, it follows from (3.95) and the James Theorem (see Theorem 1.6) that V_p is a reflexive Banach space.

(iii) It follows from Lemma 3.5 that I_p is an isomorphism from V_U to V_p.
       Thus, we end the proof of Lemma 3.6.   □

Based on Lemma 3.5 and Lemma 3.6, we can easily verify the following result:

Corollary 3.4 *It holds that* $L^\infty(0, T; V_p) = L^\infty(0, T; V_U)$.

To introduce the desired norm optimal control, we also need to make the following additional assumption $(\widehat{\mathscr{H}_{TP}})$:

- For any $y_0 \in Y_S$, $t > 0$ and $v \in L^\infty(0, t; V_p)$, the controlled system (2.1) has a unique mild solution on $[0, t]$, denoted by $y(\cdot; 0, y_0, v)$.

Now, given $y_0 \in Y_S$, $t \in (0, +\infty)$ and $Y_E \subseteq Y$, we define the following norm optimal control problem $(NP)_{y_0}^{t, Y_E}$:

$$N(t, y_0; Y_E) \triangleq \inf \left\{ \|v\|_{L^\infty(0, t; V_p)} \mid y(t; 0, y_0, v) \in Y_E \right\}. \tag{3.96}$$

Several notes related to the problem $(NP)_{y_0}^{t, Y_E}$ are given in order:

(a) When the set in the right-hand side of (3.96) is empty, we agree that $N(t, y_0; Y_E) = +\infty$.

(b) It is a norm optimal control problem corresponding to the problem $(TP)_{min}^{Q_S, Q_E}$.

(c) In the studies of this problem, Corollary 3.4 will be used.

The next theorem gives connections between the problem $(NP)_{y_0}^{t, Y_E}$ and admissible controls of $(TP)_{min}^{Q_S, Q_E}$.

Theorem 3.6 *Suppose that* $(\widehat{\mathscr{H}_{TP}})$ *holds. Then the problem* $(TP)_{min}^{Q_S, Q_E}$ *has an admissible control if and only if either of the following two conditions is true:*

(i) $1 > \inf_{y_0 \in Y_S, t > 0} N(t, y_0; Y_E)$.

(ii) $1 = \inf_{y_0 \in Y_S, t > 0} N(t, y_0; Y_E)$; *there exists* $\widehat{y}_0 \in Y_S$ *and* $\widehat{t} \in (0, +\infty)$ *satisfying that* $N(\widehat{t}, \widehat{y}_0; Y_E) = \inf_{y_0 \in Y_S, t > 0} N(t, y_0; Y_E)$; *and the problem* $(NP)_{\widehat{y}_0}^{\widehat{t}, Y_E}$ *has a solution.*

Proof Let $(\widehat{\mathscr{H}_{TP}})$ hold. Then the problem $(NP)_{y_0}^{t,Y_E}$ is well defined. We first show the necessity. Assume that the problem $(TP)_{min}^{Q_S,Q_E}$ has an admissible control. Then by $(\mathscr{H}_{TP})$ (given at the beginning of Section 3.2), we see that there exists $\widehat{y}_0 \in Y_S$, $\widehat{t} \in (0, +\infty)$, and $\widehat{u} \in L(0, \widehat{t}; \mathbb{U})$ so that $y(\widehat{t}; 0, \widehat{y}_0, \widehat{u}) \in Y_E$. This implies that

$$\inf_{y_0 \in Y_S, t > 0} N(t, y_0; Y_E) \leq N(\widehat{t}, \widehat{y}_0; Y_E) \leq \|\widehat{u}\|_{L^\infty(0, \widehat{t}; V_p)}. \tag{3.97}$$

Meanwhile, since $\widehat{u}(t) \in \mathbb{U}$ for a.e. $t \in (0, \widehat{t})$, it follows by (3.77) that $p(\widehat{u}(t)) \leq 1$ for a.e. $t \in (0, \widehat{t})$. This, along with Lemma 3.4, yields that $\|\widehat{u}\|_{L^\infty(0, \widehat{t}; V_p)} \leq 1$, from which and (3.97), it follows that

$$\inf_{y_0 \in Y_S, t > 0} N(t, y_0; Y_E) \leq N(\widehat{t}, \widehat{y}_0; Y_E) \leq \|\widehat{u}\|_{L^\infty(0, \widehat{t}; V_p)} \leq 1. \tag{3.98}$$

When $\inf_{y_0 \in Y_S, t > 0} N(t, y_0; Y_E) = 1$, we see from (3.98) that

$$\inf_{y_0 \in Y_S, t > 0} N(t, y_0; Y_E) = N(\widehat{t}, \widehat{y}_0; Y_E) = \|\widehat{u}\|_{L^\infty(0, \widehat{t}; V_p)}.$$

From these, one can easily check that (ii) is true. From this and (3.98), we find that either (i) or (ii) is true.

We now show the sufficiency. We first prove that (i) implies the existence of admissible controls for the problem $(TP)_{min}^{Q_S, Q_E}$. To this end, we suppose that (i) holds. Then we can find $\varepsilon > 0$ so that

$$2\varepsilon < 1 - \inf_{y_0 \in Y_S, t > 0} N(t, y_0; Y_E). \tag{3.99}$$

Thus, there exists $y_0^\varepsilon \in Y_S$ and $t_\varepsilon \in (0, +\infty)$ so that

$$\inf_{y_0 \in Y_S, t > 0} N(t, y_0; Y_E) \leq N(t_\varepsilon, y_0^\varepsilon; Y_E) < \inf_{y_0 \in Y_S, t > 0} N(t, y_0; Y_E) + \varepsilon. \tag{3.100}$$

By the definition of $N(t_\varepsilon, y_0^\varepsilon; Y_E)$, there exists a control v_ε, with

$$v_\varepsilon \in L^\infty(0, t_\varepsilon; V_p) \quad \text{and} \quad y(t_\varepsilon; 0, y_0^\varepsilon, v_\varepsilon) \in Y_E, \tag{3.101}$$

so that

$$N(t_\varepsilon, y_0^\varepsilon; Y_E) \leq \|v_\varepsilon\|_{L^\infty(0, t_\varepsilon; V_p)} < N(t_\varepsilon, y_0^\varepsilon; Y_E) + \varepsilon. \tag{3.102}$$

It follows from (3.100) and (3.102) that

$$\inf_{y_0 \in Y_S, t > 0} N(t, y_0; Y_E) \leq \|v_\varepsilon\|_{L^\infty(0, t_\varepsilon; V_p)} < \inf_{y_0 \in Y_S, t > 0} N(t, y_0; Y_E) + 2\varepsilon.$$

This implies that

$$\|v_\varepsilon(t)\|_{V_p} < \inf_{y_0 \in Y_S, t>0} N(t, y_0; Y_E) + 2\varepsilon \quad \text{for a.e. } t \in (0, t_\varepsilon),$$

which, together with the definition of V_p and (3.99), indicates that

$$p(v_\varepsilon(t)) < \inf_{y_0 \in Y_S, t>0} N(t, y_0; Y_E) + 2\varepsilon < 1 \quad \text{for a.e. } t \in (0, t_\varepsilon). \tag{3.103}$$

Meanwhile, since $v_\varepsilon(t) \in V$ for a.e. $t \in (0, t_\varepsilon)$, it follows from (3.77) that

$$v_\varepsilon(t) \in p(v_\varepsilon(t))\mathbb{U} \quad \text{for a.e. } t \in (0, t_\varepsilon).$$

Since $\mathbb{U}$ is convex and $0 \in \mathbb{U}$, it follows by the above and (3.103) that

$$v_\varepsilon(t) \in \mathbb{U} \quad \text{for a.e. } t \in (0, t_\varepsilon). \tag{3.104}$$

By the first conclusion in (3.101) and by Corollary 3.4, we see that $v_\varepsilon \in L^\infty(0, t_\varepsilon; U)$, which, along with (3.104), shows that $v_\varepsilon \in L(0, t_\varepsilon; \mathbb{U})$. From this and the second conclusion in (3.101), we can easily verify that $(0, y_0^\varepsilon, t_\varepsilon, v_\varepsilon)$ is an admissible tetrad of the problem $(TP)_{min}^{Q_S, Q_E}$. Hence, v_ε is an admissible control to this problem.

We next show that (ii) implies the existence of admissible controls for the problem $(TP)_{min}^{Q_S, Q_E}$. To this end, we suppose that (ii) holds. Then there is $\widehat{y}_0 \in Y_S$, $\widehat{t} > 0$ and a control $\widehat{v}$, with

$$\widehat{v} \in L^\infty(0, \widehat{t}; V_p) \quad \text{and} \quad y(\widehat{t}; 0, \widehat{y}_0, \widehat{v}) \in Y_E, \tag{3.105}$$

so that

$$1 = \|\widehat{v}\|_{L^\infty(0,\widehat{t}; V_p)} = N(\widehat{t}, \widehat{y}_0; Y_E).$$

From the latter, it follows that

$$\|\widehat{v}(t)\|_{V_p} \leq 1 \quad \text{for a.e. } t \in (0, \widehat{t}).$$

This implies that

$$0 \leq p(\widehat{v}(t)) \leq 1 \quad \text{for a.e. } t \in (0, \widehat{t}). \tag{3.106}$$

Since $\mathbb{U}$ is convex, $0 \in \mathbb{U}$ and $\widehat{v}(t) \in p(\widehat{v}(t))\mathbb{U}$ for a.e. $t \in (0, \widehat{t})$, by (3.106), we obtain that

$$\widehat{v}(t) \in \mathbb{U} \quad \text{for a.e. } t \in (0, \widehat{t}),$$

which, combined with (3.105) and Corollary 3.4, indicates that $(0, \widehat{y}_0, \widehat{t}, \widehat{v})$ is an admissible tetrad of the problem $(TP)_{min}^{Q_S, Q_E}$. Hence, $\widehat{v}$ is an admissible control to this problem.

In summary, we end the proof of Theorem 3.6. $\square$

3.2.2 Heat Equations with Ball-Type Control Constraints

In this subsection, we will apply Theorem 3.6 to a minimal time control problem for heat equations. Throughout this subsection, $\Omega \subseteq \mathbb{R}^d, d \geq 1$, is a bounded domain with a C^2 boundary $\partial\Omega$, and $\omega \subseteq \Omega$ is a nonempty open subset with its characteristic function χ_ω. Consider the following controlled heat equation:

$$\begin{cases} \partial_t y - \Delta y + a(x,t)y = \chi_\omega u & \text{in } \Omega \times (0, +\infty), \\ y = 0 & \text{on } \partial\Omega \times (0, +\infty), \end{cases} \tag{3.107}$$

where $a \in L^\infty(\Omega \times (0, +\infty))$ and $u \in L^\infty(0, +\infty; L^2(\Omega))$.

The Equation (3.107) can be put into our framework (2.1) in the same manner as that used to deal with the Equation (3.59) in Section 3.1.3, except for the term $a(x,t)y$. To deal with it, we notice that since $a \in L^\infty(\Omega \times (0, +\infty))$, there is a measurable set $E \subseteq \Omega \times (0, +\infty)$ with $|E| = 0$ and a positive constant c_0 so that

$$|a(x,t)| \leq c_0 \text{ for each } (x,t) \in (\Omega \times (0, +\infty)) \setminus E.$$

Define

$$\widehat{a}(x,t) \triangleq \begin{cases} a(x,t), & \text{for all } (x,t) \in (\Omega \times (0, +\infty)) \setminus E, \\ 0, & \text{for all } (x,t) \in E. \end{cases}$$

It is obvious that

$$\widehat{a} \in L^\infty(\Omega \times (0, +\infty)); \quad \widehat{a}(x,t) = a(x,t) \text{ for a.e. } (x,t) \in \Omega \times (0, +\infty), \tag{3.108}$$

and

$$|\widehat{a}(x,t)| \leq c_0 \text{ for all } (x,t) \in \Omega \times (0, +\infty). \tag{3.109}$$

By (3.109), we can define, for each $t \in (0, +\infty)$, a linear and bounded operator $D(t) : Y \mapsto Y$ in the following manner:

$$(D(t)y)(x) = \widehat{a}(x,t)y(x) \text{ for each } x \in \Omega. \tag{3.110}$$

With the aid of (3.108) and (3.110), the Equation (3.107) can be put into our framework (2.1) in the following manner:

$$\dot{y} - \Delta y + D(t)y = \chi_\omega u, \quad t \in (0, +\infty). \tag{3.111}$$

In this subsection, we will use Theorem 3.6 to study $(TP)_{min}^{Q_S, Q_E}$, where the controlled system is (3.111) (or equivalently (3.107)); and where

$$Q_S \triangleq \{0\} \times Y_S, \quad Q_E \triangleq (0, +\infty) \times \{0\}, \quad \mathbb{U} \triangleq \{u \in U \mid \|u\|_{L^2(\Omega)} \leq \rho\},$$

with $\rho > 0$, Y_S a nonempty subset of $Y \setminus \{0\}$. For simplicity, we will denote $(TP)_{min}^{Qs,QE}$ by $(TP)_\rho^{Qs}$. We aim to give a sufficient and necessary condition for the existence of admissible controls of $(TP)_\rho^{Qs}$. We would like to mention what follows:

- Both $(\mathcal{H}_{TP})$ (given at the beginning of Section 3.2) and $(\widehat{\mathcal{H}_{TP}})$ (given after Corollary 3.4) hold for this case.

The condition $(\mathcal{H}_{TP})$ can be easily verified in this case. Let us explain why $(\widehat{\mathcal{H}_{TP}})$ holds in the current case. By (3.76), (3.77), and Lemma 3.4, we can easily check that

$$V = L^2(\Omega); \quad p(v) = \frac{\|v\|_{L^2(\Omega)}}{\rho} \text{ for each } v \in L^2(\Omega),$$

and

$$\|v\|_{L^\infty(0,t;V_p)} = \frac{1}{\rho}\|v\|_{L^\infty(0,t;L^2(\Omega))} \text{ for each } v \in L^\infty(0,t;V_p).$$

From these, we see that for any $y_0 \in Y_S$, $t \in (0,+\infty)$, and $v \in L^\infty(0,t;V_p)$, the Equation (3.111), with the initial condition that $y(0) = y_0$, has a unique mild solution on $[0,t]$.

The corresponding norm optimal control problem $(NP)_{y_0}^{t,Y_E}$ (see (3.96)), with $y_0 \in Y_S$ and $t \in (0,+\infty)$, is as follows:

$$(NP)_{y_0}^{t,\{0\}} \qquad N(t,y_0;\{0\}) \triangleq \inf\left\{\|v\|_{L^\infty(0,t;V_p)} \mid y(t;0,y_0,v) = 0\right\}. \tag{3.112}$$

Notice that (3.112) can be rewritten as:

$$(NP)_{y_0}^{t,\{0\}} \qquad N(t,y_0;\{0\}) = \frac{1}{\rho}\inf\left\{\|v\|_{L^\infty(0,t;L^2(\Omega))} \mid y(t;0,y_0,v) = 0\right\}. \tag{3.113}$$

In order to use Theorem 3.6 to study the existence of admissible controls for the minimal time control problem of this subsection, we introduce another norm optimal control problem as follows:

$$(NP)_{y_0}^t \qquad N(t,y_0) \triangleq \inf\left\{\|v\|_{L^\infty(0,t;L^2(\Omega))} \mid y(t;0,y_0,v) = 0\right\}. \tag{3.114}$$

From (3.113) and (3.114), we see that

$$N(t,y_0;\{0\}) = \frac{1}{\rho}N(t,y_0) \text{ for all } y_0 \in Y_S \text{ and } t \in (0,+\infty). \tag{3.115}$$

In the problem $(NP)_{y_0}^t$, $N(t,y_0)$ is called the optimal norm; a function v^* is called an optimal control if

$$y(t;0,y_0,v^*) = 0 \text{ and } \|v^*\|_{L^\infty(0,t;L^2(\Omega))} = N(t,y_0);$$

a function $\widehat{v}$ is called an admissible control if

$$y(t; 0, y_0, \widehat{v}) = 0 \quad \text{and} \quad \widehat{v} \in L^\infty(0, t; L^2(\Omega)).$$

We say that the problem $(NP)_{y_0}^t$ has the bang-bang property if any optimal control v^* to this problem verifies that

$$\left\| v^*(s) \right\|_{L^2(\Omega)} = N(t, y_0) \quad \text{for a.e. } s \in (0, t).$$

Lemma 3.7 *Let $y_0 \in L^2(\Omega) \backslash \{0\}$ and $t \in (0, +\infty)$. Then the following conclusions are true:*

(i) *The problem $(NP)_{y_0}^t$ has the bang-bang property.*
(ii) *The problem $(NP)_{y_0}^t$ has a unique optimal control.*

Proof We prove the conclusions one by one.

(i) By contradiction, suppose that $(NP)_{y_0}^t$ did not hold the bang-bang property. Since $N(t, y_0) \neq 0$ (see (i) of Remark 1.5 after Theorem 1.22), there would be an optimal control v^* (to $(NP)_{y_0}^t$), $\varepsilon \in (0, N(t, y_0))$ and a subset $E \subseteq (0, t)$ of positive measure so that

$$\left\| v^*(s) \right\|_{L^2(\Omega)} \leq N(t, y_0) - \varepsilon \quad \text{for all } s \in E. \tag{3.116}$$

By Theorems 1.22 and 1.21, we can find a $\delta > 0$ so that for each initial state z, with $\|z\|_{L^2(\Omega)} \leq \delta$, there is a control v^z verifying that

$$y\left(t; 0, z, \chi_E v^z\right) = 0 \quad \text{and} \quad \left\| v^z \right\|_{L^\infty(0,t;L^2(\Omega))} \leq \varepsilon.$$

Here χ_E is the characteristic function of E. By taking $z = \delta_1 y_0$, where $\delta_1 > 0$ verifies $\|\delta_1 y_0\|_{L^2(\Omega)} \leq \delta$, in the above, we find that

$$y\left(t; 0, \delta_1 y_0, \chi_E v^{\delta_1 y_0}\right) = 0. \tag{3.117}$$

Since v^* is an optimal control to $(NP)_{y_0}^t$, we see that

$$y\left(t; 0, y_0, v^*\right) = 0 \quad \text{and} \quad \left\| v^* \right\|_{L^\infty(0,t;L^2(\Omega))} = N(t, y_0). \tag{3.118}$$

From (3.117) and the first equality in (3.118), it follows that

$$y\left(t; 0, \delta_1 y_0 + y_0, v^* + \chi_E v^{\delta_1 y_0}\right) = 0,$$

which leads to

$$y\left(t; 0, y_0, (1 + \delta_1)^{-1} \left(v^* + \chi_E v^{\delta_1 y_0}\right)\right) = 0. \tag{3.119}$$

Meanwhile, since

$$\left\| v^{\delta_1 y_0} \right\|_{L^\infty(0,t;L^2(\Omega))} \leq \varepsilon,$$

it follows from the second equality in (3.118) and (3.116) that

$$\left\| (1+\delta_1)^{-1} \left(v^* + \chi_E v^{\delta_1 y_0} \right) \right\|_{L^\infty(0,t;L^2(\Omega))} \leq \frac{1}{1+\delta_1} N(t, y_0).$$

This, along with (3.119), contradicts the optimality of $N(t, y_0)$. Hence, the problem $(NP)_{y_0}^t$ has the bang-bang property.

(ii) The existence of admissible controls follows from Theorems 1.22 and 1.21.

To prove the existence of optimal controls, we let $\{v_\ell\}_{\ell \geq 1} \subseteq L^\infty(0, t; L^2(\Omega))$ be a minimization sequence of $(NP)_{y_0}^t$. Then

$$y(t; 0, y_0, v_\ell) = 0 \quad \text{and} \quad \lim_{\ell \to +\infty} \|v_\ell\|_{L^\infty(0,t;L^2(\Omega))} = N(t, y_0). \qquad (3.120)$$

From the second equality of (3.120), there is a subsequence of $\{v_\ell\}_{\ell \geq 1}$, still denoted in the same way, so that

$$v_\ell \to v^* \quad \text{weakly star in} \quad L^\infty\left(0, t; L^2(\Omega)\right) \qquad (3.121)$$

and

$$y(t; 0, y_0, v_\ell) \to y\left(t; 0, y_0, v^*\right) \quad \text{strongly in} \quad L^2(\Omega).$$

This, together with (3.120) and (3.121), implies that

$$y\left(t; 0, y_0, v^*\right) = 0 \quad \text{and} \quad \|v^*\|_{L^\infty(0,t;L^2(\Omega))} \leq N(t, y_0).$$

These show that v^* is an optimal control of the problem $(NP)_{y_0}^t$.

To show the uniqueness, we let v_1 and v_2 be two optimal controls of the problem $(NP)_{y_0}^t$. Then one can easily check that $(v_1 + v_2)/2$ is also an optimal control to the problem $(NP)_{y_0}^t$. By (i), we find that

$$\|v_1(s)\|_{L^2(\Omega)} = \|v_2(s)\|_{L^2(\Omega)}$$
$$= \|(v_1 + v_2)(s)/2\|_{L^2(\Omega)} = N(t, y_0) \quad \text{for a.e. } s \in (0, t).$$

From the above and from the parallelogram law in $L^2(\Omega)$, we can easily verify that

$$v_1(s) = v_2(s) \quad \text{for a.e. } s \in (0, t).$$

Thus, we complete the proof of Lemma 3.7.   □

Lemma 3.8 *For each $y_0 \in L^2(\Omega) \setminus \{0\}$, the function $N(\cdot, y_0)$ is strictly monotonically decreasing over $(0, +\infty)$. Furthermore,*

$$\lim_{t \to 0+} N(t, y_0) = +\infty \tag{3.122}$$

and

$$\lim_{t \to +\infty} N(t, y_0) \triangleq \widehat{N}(y_0) \in [0, +\infty). \tag{3.123}$$

Proof Let $0 < t_1 < t_2$. By (ii) of Lemma 3.7, we can let v_1 be the optimal control to the problem $(NP)_{y_0}^{t_1}$. Then we have that

$$\|v_1\|_{L^\infty(0,t_1;L^2(\Omega))} = N(t_1, y_0) \quad \text{and} \quad y(t_1; 0, y_0, v_1) = 0. \tag{3.124}$$

Construct a new control v_2 by setting

$$v_2(s) \triangleq \begin{cases} v_1(s) & \text{for a.e. } s \in (0, t_1), \\ 0 & \text{for a.e. } s \in (t_1, t_2). \end{cases} \tag{3.125}$$

By the first equality of (3.124) and (3.125), we see that

$$\|v_2\|_{L^\infty(0,t_2;L^2(\Omega))} = N(t_1, y_0). \tag{3.126}$$

From the second equality of (3.124) and (3.125), we find that

$$y(t_2; 0, y_0, v_2) = 0. \tag{3.127}$$

This yields that v_2 is an admissible control to $(NP)_{y_0}^{t_2}$. Then by (3.126) and the optimality of $N(t_2, y_0)$, it follows that

$$N(t_1, y_0) = \|v_2\|_{L^\infty(0,t_2;L^2(\Omega))} \geq N(t_2, y_0).$$

Namely, $N(\cdot, y_0)$ is monotonically decreasing.

We next show that

$$N(t_1, y_0) > N(t_2, y_0).$$

By contradiction, suppose that it was not true. Then by the above monotonicity of $N(\cdot, y_0)$, we would have that

$$N(t_1, y_0) = N(t_2, y_0),$$

which, along with (3.126), shows that

$$\|v_2\|_{L^\infty(0,t_2;L^2(\Omega))} = \|v_1\|_{L^\infty(0,t_1;L^2(\Omega))} = N(t_1, y_0) = N(t_2, y_0),$$

where v_1 is the optimal control to the problem $(NP)_{y_0}^{t_1}$ and v_2 is given by (3.125). This, together with (3.127), shows that v_2 is the optimal control to the problem $(NP)_{y_0}^{t_2}$. From (3.125), we see that v_2 is not a bang-bang control. This contradicts the bang-bang property of the problem $(NP)_{y_0}^{t_2}$ (see Lemma 3.7).

The conclusion (3.123) follows from the monotonicity of $N(\cdot, y_0)$ at once. Finally, we will show (3.122). By contradiction, we suppose that (3.122) was not true. Then there would be a sequence $\{t_\ell\}_{\ell \geq 1} \subseteq (0, 1)$ so that

$$t_\ell \searrow 0 \quad \text{and} \quad N(t_\ell, y_0) \nearrow \tilde{N} \in (0, +\infty).$$

Let u_ℓ be the optimal control to $(NP)_{y_0}^{t_\ell}$. Then we have that

$$\|u_\ell\|_{L^\infty(0,t_\ell;L^2(\Omega))} = N(t_\ell, y_0) \leq \tilde{N} \quad \text{for all } \ell \geq 1, \tag{3.128}$$

and

$$0 = y(t_\ell; 0, y_0, u_\ell) = y(t_\ell; 0, y_0, 0) + y(t_\ell; 0, 0, u_\ell). \tag{3.129}$$

Since $t_\ell \searrow 0$ and because of (3.128), we can easily verify that $y(t_\ell; 0, 0, u_\ell) \to 0$, through multiplying (3.107) (with $y(0) = 0$ and $u = u_\ell$) by y, and then integrating over $\Omega \times (0, t), t \in (0, t_\ell)$, and finally using Gronwall's inequality. Meanwhile, since $t_\ell \searrow 0$, we can use the continuity of the solution $y(\cdot; 0, y_0, 0)$ at time $t = 0$ to get that $y(t_\ell; 0, y_0, 0) \to y_0$. These, along with (3.129), yield that $0 = y_0$, which leads to a contradiction (since we assumed that $y_0 \neq 0$). Hence (3.122) holds. This completes the proof of Lemma 3.8. □

Theorem 3.7 *The problem $(TP)_\rho^{Qs}$ has an admissible control if and only if $\rho > \hat{N}$. Here,*

$$\hat{N} \triangleq \inf_{y_0 \in Y_S} \hat{N}(y_0), \quad \text{with } \hat{N}(y_0) \text{ given by } (3.123). \tag{3.130}$$

Proof We first claim that

$$\inf_{y_0 \in Y_S, t > 0} N(t, y_0; \{0\}) = \frac{1}{\rho} \hat{N}. \tag{3.131}$$

Indeed, by (3.115), we get that

$$\inf_{y_0 \in Y_S, t > 0} N(t, y_0; \{0\}) = \frac{1}{\rho} \inf_{y_0 \in Y_S, t > 0} N(t, y_0) = \frac{1}{\rho} \inf_{y_0 \in Y_S} \inf_{t > 0} N(t, y_0). \tag{3.132}$$

Meanwhile, by the strict monotonicity of the function $N(\cdot, y_0)$ (see Lemma 3.8) and by (3.123), we find that

$$\inf_{t > 0} N(t, y_0) = \hat{N}(y_0); \tag{3.133}$$

$$N(t, y_0) > \hat{N}(y_0) \quad \text{for all } t > 0 \text{ and } y_0 \in L^2(\Omega) \setminus \{0\}. \tag{3.134}$$

Hence, (3.131) follows from (3.132), (3.133), and (3.130).

We next claim that for any $\widehat{y}_0 \in Y_S$ and $\widehat{t} \in (0, +\infty)$,

$$\inf_{y_0 \in Y_S, t > 0} N(t, y_0; \{0\}) < N(\widehat{t}, \widehat{y}_0; \{0\}). \tag{3.135}$$

By contradiction, suppose that it was not true. Then there would exist $\widehat{y}_0 \in Y_S$ and $\widehat{t} \in (0, +\infty)$ so that

$$\inf_{y_0 \in Y_S, t > 0} N(t, y_0; \{0\}) = N(\widehat{t}, \widehat{y}_0; \{0\}).$$

This, together with (3.115), (3.134), and (3.130), implies that

$$\inf_{y_0 \in Y_S, t > 0} N(t, y_0; \{0\}) = \frac{1}{\rho} N(\widehat{t}, \widehat{y}_0) > \frac{1}{\rho} \widehat{N}(\widehat{y}_0) \geq \frac{1}{\rho} \widehat{N},$$

which contradicts (3.131). Hence, (3.135) is true.

Finally, from (3.131), (3.135), and Theorem 3.6, we can easily get the desired result of this theorem. This completes the proof of Theorem 3.7.   □

3.3   Admissible Controls and Reachable Sets

In this section, we will study the existence of admissible controls of $(TP)_{min}^{Q_S, Q_E}$ from perspective of reachable sets.

3.3.1   General Results in Abstract Setting

With the aid of $(\mathscr{A}_{TP})$ (given at the beginning of Chapter 3), we can define the following three sets:

$$Q_R \triangleq \{(t, y(t; 0, y_0, u)) \in (0, +\infty) \times Y \mid y_0 \in Y_S$$

$$\text{and } u \in L(0, t; \mathbb{U})\}; \tag{3.136}$$

$$Y_R(t) \triangleq \{y \in Y \mid (t, y) \in Q_R\}, \quad \text{for each } t \in (0, +\infty); \tag{3.137}$$

$$Y_R \triangleq \bigcup_{t > 0} Y_R(t). \tag{3.138}$$

These sets depend on Y_S and $\mathbb{U}$. We call respectively Q_R, $Y_R(t)$, and Y_R the reachable horizon, the reachable set at the time t and the reachable set of the Equation (2.1), associated with Q_S and $\mathbb{U}$. *In most time, we simply call them the reachable horizon, the reachable set at the time t and the reachable set of the Equation (2.1), if there is no risk causing any confusion.* For any $t \in T_E$ (Recall (2.43) for the definition of T_E), we define the distance between $Y_R(t)$ and $Y_E(t)$ (Recall (2.44) for the definition of $Y_E(t)$) by

$$d(t) \triangleq \text{dist}\{Y_R(t), \ Y_E(t)\} = \inf_{y_1 \in Y_R(t), y_2 \in Y_E(t)} \|y_1 - y_2\|_Y \quad \text{for all } t \in T_E.$$
(3.139)

Now we introduce an optimization problem:

$$d^* \triangleq \inf_{t \in T_E} d(t).$$
(3.140)

The connection between admissible controls and the reachable sets is presented in the following theorem.

Theorem 3.8 *The problem* $(TP)_{min}^{Q_S, Q_E}$ *has an admissible control if and only if the following conditions hold:*

(i) $d^* = 0$;
(ii) The optimization problem (3.140) admits a solution;
(iii) There exists $t_E \in T_E$, *which is a solution of the problem (3.140), so that*

$$\inf_{(y_1, y_2) \in Y_R(t_E) \times Y_E(t_E)} \|y_1 - y_2\|_Y$$
(3.141)

admits a solution.

Proof Necessity. Assume that u is an admissible control (associated with an admissible tetrad $(0, y_0, t_E, u)$) to $(TP)_{min}^{Q_S, Q_E}$. Then

$$y(t_E; 0, y_0, u) \in Y_R(t_E) \cap Y_E(t_E).$$
(3.142)

By (3.142), (3.139), and (3.140), we see that

$$0 = d^* = \text{dist}\{Y_R(t_E), \ Y_E(t_E)\}.$$

From the above and (3.142), we find that the conditions (i), (ii), and (iii) hold.
    Sufficiency. Assume that the conditions (i), (ii), and (iii) hold. Then there is $t_E \in T_E$, $y_1^* \in Y_R(t_E)$ and $y_2^* \in Y_E(t_E)$ so that

$$d(t_E) = 0 \quad \text{and} \quad \|y_1^* - y_2^*\|_Y = \text{dist}\{Y_R(t_E), Y_E(t_E)\}.$$
(3.143)

By (3.139) and (3.143), we find that $y_1^* = y_2^*$. Hence $y_1^* \in Y_E(t_E)$. Since $y_1^* \in Y_R(t_E)$, there exists $y_0 \in Y_S$ and $u \in L(0, t_E; \mathbb{U})$ so that $y_1^* = y(t_E; 0, y_0, u)$, which, combined with the fact that $y_1^* \in Y_E(t_E)$, implies that $(0, y_0, t_E, u)$ is an admissible tetrad of the problem $(TP)_{min}^{Q_S, Q_E}$. Hence, u is an admissible control to this problem. Thus, we end the proof of Theorem 3.8.   □

We now give some applications of Theorem 3.8 by the following three examples.

Example 3.2 Consider the time optimal control problem $(TP)^{Q_S,Q_E}_{min}$, where the controlled system is as:

$$\dot{y}(t) = y(t) + u(t), \quad t \in (0, +\infty),$$

and where $\mathbb{U} \triangleq [-1, 1]$, $Q_S \triangleq \{(0, 2)\}$ and $Q_E \triangleq (0, +\infty) \times \{0\}$. It is clear that $(\mathscr{A}_{TP})$ holds for this case. Moreover, by (2.43) and (2.44), we have that

$$T_E = (0, +\infty) \quad \text{and} \quad Y_E(t) = \{0\} \quad \text{for all } t \in (0, +\infty). \tag{3.144}$$

Meanwhile, it follows from (3.137) and (3.136) that

$$Y_R(t) = \left\{ \left(2 + \int_0^t e^{-s} u(s) ds \right) e^t \mid u \in L(0, t; \mathbb{U}) \right\} \quad \text{for all } t \in (0, +\infty).$$

This, together with (3.139) and (3.144), implies that

$$d(t) = \inf_{u \in L(0,t;\mathbb{U})} \left(2 + \int_0^t e^{-s} u(s) ds \right) e^t \quad \text{for all } t \in (0, +\infty),$$

from which, one can easily check that

$$d(t) = \left(2 + \int_0^t e^{-s} (-1) ds \right) e^t = 1 + e^t \quad \text{for all } t \in (0, +\infty).$$

This, combined with (3.140), indicates that $d^* = 2$. So the condition (i) in Theorem 3.8 does not hold. Thus, by Theorem 3.8, we find that the problem $(TP)^{Q_S,Q_E}_{min}$ has no admissible control.

Besides, for this example, we can directly check that $(TP)^{Q_S,Q_E}_{min}$ has no admissible control. Indeed, since

$$y(t; 0, 2, u) = \left(2 + \int_0^t e^{-s} u(s) ds \right) e^t \geq 2 > 0$$

for all $t \in (0, +\infty)$ and $u \in L(0, t; \mathbb{U})$,

$(TP)^{Q_S,Q_E}_{min}$ has no admissible control. Now we end Example 3.2.

Example 3.3 Consider the time optimal control problem $(TP)^{Q_S,Q_E}_{min}$, where the controlled system is as:

$$\dot{y} = u(t) y(t), \quad t \in (0, +\infty),$$

and where $\mathbb{U} \triangleq [-1, 1]$, $Q_S \triangleq \{(0, 2)\}$ and $Q_E \triangleq (0, +\infty) \times \{0\}$. It is clear that $(\mathscr{A}_{TP})$ (given at the beginning of Chapter 2) holds. Moreover, by (2.43) and (2.44), we have that

$$T_E = (0, +\infty) \quad \text{and} \quad Y_E(t) = \{0\} \quad \text{for all} \quad t \in (0, +\infty). \tag{3.145}$$

It follows from (3.137) and (3.136) that

$$Y_R(t) = \left\{ 2e^{\int_0^t u(s)ds} \mid u \in L(0, t; \mathbb{U}) \right\} \quad \text{for all} \quad t \in (0, +\infty).$$

This, together with (3.139) and (3.145), implies that

$$d(t) = \inf_{u \in L(0,t;\mathbb{U})} 2e^{\int_0^t u(s)ds} = 2e^{-t} \quad \text{for all} \quad t \in (0, +\infty), \tag{3.146}$$

which, combined with (3.140), indicates that $d^* = 0$, i.e., the condition (i) in Theorem 3.8 holds.

However, (ii) in Theorem 3.8 does not hold for this case. Indeed, by (3.146), one can easily check that for any $t \in (0, +\infty)$, $d(t) > d^*$. So the optimization problem (3.140) has no solution. Thus, by Theorem 3.8, this problem $(TP)_{min}^{Q_S, Q_E}$ has no admissible control.

Besides, we can directly check that this problem $(TP)_{min}^{Q_S, Q_E}$ has no admissible control. This can be seen from what follows:

$$y(t; 0, 2, u) = 2e^{\int_0^t u(s)ds} > 0 \quad \text{for all} \quad t \in (0, +\infty) \quad \text{and} \quad u \in L(0, t; \mathbb{U}).$$

We now end Example 3.3.

Example 3.4 Let $\Omega \subseteq \mathbb{R}^d$, $d \geq 1$ be a bounded domain with a C^2 boundary $\partial\Omega$, and $\omega \subseteq \Omega$ is a nonempty open subset with its characteristic function χ_ω. Consider the following controlled heat equation:

$$\begin{cases} \partial_t y - \Delta y = \chi_\omega u, & \text{in } \Omega \times (0, +\infty), \\ y = 0, & \text{on } \partial\Omega \times (0, +\infty). \end{cases} \tag{3.147}$$

We can put the Equation (3.147) into our framework (2.1) in the same manner as that used to deal with the Equation (3.107) (with $a \equiv 0$). In this way, the Equation (3.147) can be rewritten as

$$\dot{y} - \Delta y = \chi_\omega u, \quad t \in (0, +\infty). \tag{3.148}$$

Consider the time optimal control problem $(TP)_{min}^{Q_S, Q_E}$, where the controlled system is (3.148), and where

$$Q_S \triangleq \{(0, 0)\}; \ Q_E \triangleq (0, +\infty) \times \{z \in L^2(\Omega) \backslash H_0^1(\Omega) \mid \|z\|_{L^2(\Omega)} \leq r\}, \text{ with } r > 0;$$

$$\mathbb{U} \triangleq \{u \in L^2(\Omega) \mid \|u\|_{L^2(\Omega)} \leq \rho\}, \quad \text{with} \quad \rho > 0.$$

It is clear that $(\mathscr{A}_{TP})$ (given at the beginning of Chapter 2) holds. Moreover, by (2.43) and (2.44), we have that, for all $t \in (0, +\infty)$,

$$T_E = (0, +\infty) \quad \text{and} \quad Y_E(t) = \{z \in L^2(\Omega) \setminus H_0^1(\Omega) \mid \|z\|_{L^2(\Omega)} \leq r\}. \qquad (3.149)$$

Meanwhile, it follows from (3.137) and (3.136) that

$$Y_R(t) = \left\{ \int_0^t e^{(t-s)\Delta} \chi_\omega u(s) ds \mid u \in L(0, t; \mathbb{U}) \right\} \quad \text{for all } t \in (0, +\infty).$$

This, together with (3.139) and (3.149), implies that

$$\begin{aligned} d(t) &= \inf_{u \in L(0,t;\mathbb{U}), y_2 \in Y_E(t)} \left\| \int_0^t e^{(t-s)\Delta} \chi_\omega u(s) ds - y_2 \right\|_{L^2(\Omega)} \\ &\leq \inf_{y_2 \in Y_E(t)} \|y_2\|_{L^2(\Omega)} = 0 \quad \text{for all } t \in (0, +\infty). \end{aligned} \qquad (3.150)$$

By (3.150) and (3.140), one can easily check that $d^* = 0$ and that every $t \in (0, +\infty)$ is a solution to the problem (3.140). So the conditions (i) and (ii) in Theorem 3.8 hold.

However, the condition (iii) in Theorem 3.8 does not hold for this case. Indeed, by contradiction, suppose that (iii) was true. Then there would exist $t_E \in (0, +\infty)$, $y_1 \in Y_R(t_E)$ and $y_2 \in Y_E(t_E)$ so that $\|y_1 - y_2\|_{L^2(\Omega)} = d(t_E)$. This, together with (3.150), implies that $y_1 = y_2$. However,

$$y_1 \in Y_R(t_E) \subseteq H_0^1(\Omega) \quad \text{and} \quad y_2 \notin H_0^1(\Omega).$$

Thus, we obtain a contradiction. Hence, by Theorem 3.8, this problem $(TP)_{min}^{Q_S, Q_E}$ has no admissible control.

Besides, for this example, we can directly check that $(TP)_{min}^{Q_S, Q_E}$ has no admissible control. Indeed, since

$$y(t; 0, 0, u) = \int_0^t e^{(t-s)\Delta} \chi_\omega u(s) ds \in H_0^1(\Omega)$$

$$\text{for any } t \in (0, +\infty) \text{ and } u \in L(0, t; \mathbb{U}),$$

we see that

$$(t, y(t; 0, 0, u)) \notin Q_E \quad \text{for all } t \in (0, +\infty) \text{ and } u \in L(0, t; \mathbb{U}).$$

This indicates that the problem $(TP)_{min}^{Q_S, Q_E}$ has no admissible control. We end Example 3.4.

Remark 3.2 We would like to mention that conditions (i), (ii), and (iii) in Theorem 3.8 are not verifiable in many cases. In the next subsection, we will present some properties on reachable sets and the function $d(\cdot)$ for some special cases.

3.3.2   Reachable Sets of Linear ODEs

Reachable sets are very important in studies of the existence of admissible control. Unfortunately, our knowledge on reachable sets of the Equation (2.1) is quite limited. The main purpose of this subsection is to present some properties on reachable sets of the Equation (3.3), under the following assumptions:

$(L1)$ $\mathbb{U}$ is a nonempty compact subset of $\mathbb{R}^m$;
$(L2)$ Y_S is a nonempty, convex, and closed subset of $\mathbb{R}^n$.

Besides, we will give some properties on the function $d(\cdot)$ (defined by (3.139)).

We start with studying reachable sets of the Equation (3.3). Write $\Phi(\cdot, \cdot)$ for the evolution operator generated by $A(\cdot)$. Then the reachable set at the time t of the Equation (3.3) is as:

$$
Y_R(t) = \left\{ \Phi(t, 0)y_0 + \int_0^t \Phi(t, s)B(s)u(s)ds \ \Big| \ y_0 \in Y_S, \ u \in L(0, t; \mathbb{U}) \right\}.
$$
$$(3.151)$$

Write $co(\mathbb{U})$ for the convex hull of $\mathbb{U}$. Let

$$
Q_R(co(\mathbb{U})) \triangleq \left\{ (t, y(t; 0, y_0, u)) \in (0, +\infty) \times Y \ \Big| \ y_0 \in Y_S \text{ and } u \in L(0, t; co(\mathbb{U})) \right\}
$$

and

$$
Y_R(t; co(\mathbb{U})) \triangleq \left\{ y \in Y \ \Big| \ (t, y) \in Q_R(co(\mathbb{U})) \right\} \quad \text{for all } t \in (0, +\infty).
$$

Some properties on reachable sets of (3.3) are given in the next theorem.

Theorem 3.9 *The following statements are true:*

(i) Suppose that $(L1)$ holds. Then

$$
Y_R(t) = Y_R(t, co(\mathbb{U})) \quad \text{for all } t \in (0, +\infty). \tag{3.152}
$$

(ii) Suppose that $(L1)$ and $(L2)$ are true. Then for each $t \in (0, +\infty)$, $Y_R(t)$ is a convex and closed subset in $\mathbb{R}^n$.

Proof We prove the conclusions one by one.

(i) Suppose that $(L1)$ holds. Arbitrarily fix $t \in (0, +\infty)$ and $y_0 \in Y_S$. We define two sets

$$
Y_R(t, y_0) \triangleq \left\{ y(t; 0, y_0, u) \in Y \ \Big| \ u \in L(0, t; co(\mathbb{U})) \right\}
$$

and

$$
\widehat{Y}_R(t, y_0) \triangleq \left\{ y(t; 0, y_0, u) \in Y \ \Big| \ u \in L(0, t; \mathbb{U}) \right\}.
$$

We organize the proof by the following two steps:
Step 1. We show that

$$Y_R(t, y_0) = \widehat{Y}_R(t, y_0).$$

It is obvious that $Y_R(t, y_0) \supseteq \widehat{Y}_R(t, y_0)$. We now show that

$$Y_R(t, y_0) \subseteq \widehat{Y}_R(t, y_0). \tag{3.153}$$

Let $y \in Y_R(t, y_0)$. Then there is $u \in L(0, t; \text{co}(\mathbb{U}))$ so that

$$y = \Phi(t, 0)y_0 + \int_0^t \Phi(t, s)B(s)u(s)ds. \tag{3.154}$$

Since $u(s) \in \text{co}(\mathbb{U})$ for a.e. $s \in (0, t)$, according to the Carathéodory Theorem and the Measurable Selection Theorem (see Theorems 1.7 and 1.17), there are measurable functions $u_j(\cdot)$ and $\lambda_j(\cdot)$, $j = 0, 1, \ldots, m$, so that

$$\sum_{j=0}^m \lambda_j(s) = 1 \quad \text{and} \quad u(s) = \sum_{j=0}^m \lambda_j(s)u_j(s) \quad \text{for a.e. } s \in (0, t) \tag{3.155}$$

and so that

$$u_j(s) \in \mathbb{U} \quad \text{and} \quad \lambda_j(s) \geq 0 \quad \text{for a.e. } s \in (0, t) \quad \text{and for all} \quad j = 0, 1, \ldots, m. \tag{3.156}$$

It follows from (3.154) and (3.155) that

$$y = \Phi(t, 0)y_0 + \sum_{j=0}^m \int_0^t \lambda_j(s)\Phi(t, s)B(s)u_j(s)ds. \tag{3.157}$$

Furthermore, by the Lyapunov Theorem (see Theorem 1.15), there are measurable sets $E_j \subseteq (0, t)$, $j = 0, 1, \ldots, m$, so that

$$\sum_{j=0}^m \int_0^t \lambda_j(s)\Phi(t, s)B(s)u_j(s)ds = \sum_{j=0}^m \int_{E_j} \Phi(t, s)B(s)u_j(s)ds, \tag{3.158}$$

and so that

$$\bigcup_{j=0}^m E_j = (0, t) \quad \text{and} \quad E_i \bigcap E_j = \emptyset \quad \text{when} \quad i \neq j. \tag{3.159}$$

Write

$$\tilde{u}(s) \triangleq u_j(s) \quad \text{for a.e. } s \in E_j, \quad j = 0, 1, \ldots, m.$$

Then, by (3.156), (3.157), (3.158), and (3.159), we see that

$$y = \Phi(t, 0)y_0 + \int_0^t \Phi(t, s)B(s)\tilde{u}(s)\mathrm{d}s \in \widehat{Y}_R(t, y_0),$$

which leads to (3.153). Hence, we have proved that $Y_R(t, y_0) = \widehat{Y}_R(t, y_0)$.

Step 2. We show (3.152).

Note that

$$Y_R(t) = \bigcup_{y_0 \in Y_S} \widehat{Y}_R(t, y_0) \quad \text{and}$$

$$Y_R(t, \mathrm{co}(\mathbb{U})) = \bigcup_{y_0 \in Y_S} Y_R(t, y_0) \quad \text{for all } t \in (0, +\infty).$$

From these and the result in Step 1, (3.152) follows at once.

(ii) Suppose that $(L1)$ and $(L2)$ hold. Arbitrarily fix $t \in (0, +\infty)$. We first show the convexity of $Y_R(t)$. To this end, we let $y_1, y_2 \in Y_R(t)$ and $\lambda \in (0, 1)$. Then by (3.152), there exist two vectors $y_{0,1}, y_{0,2} \in Y_S$ and two functions $u_1, u_2 \in L(0, t; \mathrm{co}(\mathbb{U}))$, so that

$$y_1 = y(t; 0, y_{0,1}, u_1) \quad \text{and} \quad y_2 = y(t; 0, y_{0,2}, u_2). \tag{3.160}$$

Set

$$y_{0,\lambda} \triangleq \lambda y_{0,1} + (1 - \lambda)y_{0,2} \quad \text{and} \quad u_\lambda \triangleq \lambda u_1 + (1 - \lambda)u_2 \in L(0, t; \mathrm{co}(\mathbb{U})).$$

By $(L2)$, we have that $y_{0,\lambda} \in Y_S$. Then by (3.152), we deduce that

$$y_\lambda \triangleq \lambda y_1 + (1 - \lambda)y_2 = y(t; 0, y_{0,\lambda}, u_\lambda) \in Y_R(t; \mathrm{co}(\mathbb{U})) = Y_R(t),$$

which implies that $Y_R(t)$ is convex.

We next show that $Y_R(t)$ is closed. For this purpose, we let $\{y_k\}_{k\geq 1} \subseteq Y_R(t)$ satisfy that $y_k \to \widehat{y}$. By (3.152), there are two sequences $\{y_{0,k}\}_{k\geq 1} \subseteq Y_S$ and $\{u_k\}_{k\geq 1} \subseteq L(0, t; \mathrm{co}(\mathbb{U}))$ so that

$$y_k = \Phi(t, 0)y_{0,k} + \int_0^t \Phi(t, s)B(s)u_k(s)\mathrm{d}s \quad \text{for all } k \geq 1. \tag{3.161}$$

Since $\{u_k\}_{k\geq 1} \subseteq L(0, t; \mathrm{co}(\mathbb{U}))$, it follows by $(L1)$ that $\{u_k\}_{k\geq 1}$ is bounded in $L^2(0, t; \mathbb{R}^m)$. Meanwhile, since $y_k \to \widehat{y}$, it follows from $(L1)$, the boundedness of $\{u_k\}_{k\geq 1}$ and (3.161) that $\{y_{0,k}\}$ is bounded in $\mathbb{R}^n$. Then by $(L1)$ and $(L2)$, there exists a subsequence of $\{k\}_{k\geq 1}$, denoted in the same manner, so that for some $\widehat{y}_0 \in Y_S$ and $\widehat{u} \in L^2(0, t; \mathbb{R}^m)$,

$$y_{0,k} \to \widehat{y}_0 \quad \text{and} \quad u_k \to \widehat{u} \quad \text{weakly in } L^2(0, t; \mathbb{R}^m). \tag{3.162}$$

From the second conclusion in (3.162), we can use the Mazur Theorem (see Theorem 1.8) to find, for each $k \geq 1$, a finite sequence of nonnegative numbers $\{\alpha_{kj}\}_{j=1}^{J_k}$, with $\sum_{j=1}^{J_k} \alpha_{kj} = 1$, so that

$$\sum_{j=1}^{J_k} \alpha_{kj} u_{k+j} \to \widehat{u} \text{ strongly in } L^2(0, t; \mathbb{R}^m). \tag{3.163}$$

Since $\{u_{k+j}\}_{j=1}^{J_k} \subseteq L(0, t; \text{co}(\mathbb{U}))$, it follows by (3.163) that

$$\widehat{u} \in L(0, t; \text{co}\mathbb{U}). \tag{3.164}$$

Since $y_k \to \widehat{y}$ and because of (3.162), (3.152), and (3.164), we can pass to the limit for $k \to +\infty$ in (3.161) to get that

$$\widehat{y} = \Phi(t, 0)\widehat{y}_0 + \int_0^t \Phi(t, s)B(s)\widehat{u}(s)\mathrm{d}s = y(t; 0, \widehat{y}_0, \widehat{u}) \in Y_R(t, \text{co}(\mathbb{U})) = Y_R(t),$$

which implies that $Y_R(t)$ is closed.

This completes the proof of Theorem 3.9.   □

Remark 3.3 In infinitely dimensional cases, $Y_R(t)$ may be not convex (see Example 5.2 on page 306 of [15]).

At the end of this subsection, we will study the function $d(\cdot)$ (see (3.139)) in the framework that the controlled system is (3.3), the assumptions ($L1$) and ($L2$) (given at the beginning of Section 3.3.2) hold, and $Q_E = (0, +\infty) \times Y_E$, with $Y_E \subseteq \mathbb{R}^n$ a nonempty, convex, and compact subset. We begin with introducing the next J. von Neumann Theorem.

Theorem 3.10 ([1]) *Suppose that the following two conditions hold:*

(i) E and F are nonempty, convex, and compact subsets of $\mathbb{R}^n$;
(ii) The function $g(\cdot, \cdot) : \mathbb{R}^n \times \mathbb{R}^n \mapsto \mathbb{R}$ satisfies that for each $y \in F$, the function $g(\cdot, y) : \mathbb{R}^n \mapsto \mathbb{R}$ is convex and lower semi-continuous; and that for each $x \in E$, the function $g(x, \cdot) : \mathbb{R}^n \mapsto \mathbb{R}$ is concave and upper semi-continuous.

Then there exists a saddle point $(x^*, y^*) \in E \times F$ so that

$$\min_{x \in E} \max_{y \in F} g(x, y) = g(x^*, y^*) = \max_{y \in F} \min_{x \in E} g(x, y). \tag{3.165}$$

Proposition 3.2 *Suppose that the following assumptions hold: (i) $\mathbb{U}$ satisfies ($L1$). (ii) $Q_E = (0, +\infty) \times Y_E$ and $Y_E \subseteq \mathbb{R}^n$ is nonempty, convex, and compact. (iii) Y_S satisfies ($L2$) and is bounded. Then for each $t > 0$,*

$$d(t) = \max_{\psi \in B_1(0)} \Big[\min_{y_0 \in Y_S} \langle \Phi(t, 0)^* \psi, y_0 \rangle_{\mathbb{R}^n} - \max_{y_1 \in Y_E} \langle \psi, y_1 \rangle_{\mathbb{R}^n}$$

$$+ \int_0^t \min_{u \in \mathbb{U}} \langle B(s)^* \Phi(t, s)^* \psi, u \rangle_{\mathbb{R}^m} ds \Big].$$

Here, $B_1(0)$ is the closed unit ball in $\mathbb{R}^n$.

Proof Arbitrarily fix $t > 0$. Since Y_S is a bounded subset of $\mathbb{R}^n$, by $(L1)$ and $(L2)$, we can apply Theorem 3.9 to see that $Y_R(t)$ is convex and compact. This, together with the convexity and compactness of Y_E, implies that the following set:

$$Y_{RE}(t) \triangleq Y_R(t) - Y_E(t)$$

is also convex and compact. Hence, it follows from (3.139) that

$$d(t) = \inf_{z \in Y_{RE}(t)} \|z\|_{\mathbb{R}^n} = \min_{z \in Y_{RE}(t)} \|z\|_{\mathbb{R}^n} = \min_{z \in Y_{RE}(t)} \max_{\psi \in B_1(0)} \langle \psi, z \rangle_{\mathbb{R}^n}.$$

Then, we apply Theorem 3.10, with $E = Y_{RE}(t)$, $F = B_1(0)$ and $g(x, y) = \langle x, y \rangle_{\mathbb{R}^n}$, $x, y \in \mathbb{R}^n$, to get that

$$d(t) = \max_{\psi \in B_1(0)} \min_{z \in Y_{RE}(t)} \langle \psi, z \rangle_{\mathbb{R}^n}$$

$$= \max_{\substack{\psi \in B_1(0) \\ u \in L(0,t;\mathbb{U})}} \min_{y_0 \in Y_S, y_1 \in Y_E} \left\langle \psi, \Phi(t, 0)y_0 + \int_0^t \Phi(t, s)B(s)u(s)ds - y_1 \right\rangle_{\mathbb{R}^n}.$$

$$(3.166)$$

Meanwhile, since $B(\cdot) \in L^\infty_{loc}(0, +\infty, \mathbb{R}^{n \times m})$, for any $\varepsilon > 0$, there exists $B_\varepsilon(\cdot) \in C([0, t]; \mathbb{R}^{n \times m})$ so that

$$\int_0^t \|B_\varepsilon(s) - B(s)\|_{\mathbb{R}^{n \times m}} ds \leq \varepsilon.$$

This, together with the Measurable Selection Theorem (see Theorem 1.17), implies that

$$\min_{u \in L(0,t;\mathbb{U})} \int_0^t \langle B(s)^* \Phi(t, s)^* \psi, u(s) \rangle_{\mathbb{R}^m} ds$$

$$= \int_0^t \min_{u \in \mathbb{U}} \langle B(s)^* \Phi(t, s)^* \psi, u \rangle_{\mathbb{R}^m} ds.$$

$$(3.167)$$

From (3.166) and (3.167), the desired result follows at once. This ends the proof of Proposition 3.2. □

Corollary 3.5 *Suppose that*

$$\mathbb{U} = B_r(0), \quad \text{with } r \in (0, +\infty); \quad Y_S = \{y_0\} \subseteq \mathbb{R}^n \setminus \{0\} \quad \text{and} \quad Y_E = \{0\}.$$

Then $(TP)_{min}^{Q_S,Q_E}$ has an admissible control if and only if there is $t^* > 0$ so that

$$\langle \Phi(t^*, 0)^* \psi, y_0 \rangle_{\mathbb{R}^n} \leq r \int_0^{t^*} \|B(s)^* \Phi(t^*, s)^* \psi\|_{\mathbb{R}^m} ds \text{ for all } \psi \in B_1(0).$$
(3.168)

Proof Since $Y_R(t)$ is compact and $Y_E(t) = \{0\}$ for any $t \in (0, +\infty)$, by Theorem 3.8, we can easily check that $(TP)_{min}^{Q_S,Q_E}$ has an admissible control if and only if there is $t^* > 0$ so that $d(t^*) = 0$. This, together with Proposition 3.2, implies that $(TP)_{min}^{Q_S,Q_E}$ has an admissible control if and only if

$$\langle \Phi(t^*, 0)^* \psi, y_0 \rangle_{\mathbb{R}^n} \leq r \int_0^{t^*} \|B(s)^* \Phi(t^*, s)^* \psi\|_{\mathbb{R}^m} ds \text{ for all } \psi \in B_1(0).$$

This ends the proof of Corollary 3.5. □

Remark 3.4 When $A(\cdot) \equiv A$ and $B(\cdot) \equiv B$, we can prove that (3.168), with $t^* > 0$, is equivalent to

$$r > N(y_0).$$
(3.169)

Here, $N(y_0)$ is given by (3.48). Hence, in this special case, Corollary 3.5 coincides with Corollary 3.3.

We first show that (3.168), with $t^* > 0$, implies (3.169). Observe that in this case, (3.168), with $t^* > 0$, is equivalent to what follows:

$$\langle e^{t^* A^*} \psi, y_0 \rangle_{\mathbb{R}^n} \leq r \int_0^{t^*} \|B^* e^{(t^*-s)A^*} \psi\|_{\mathbb{R}^m} ds \text{ for all } \psi \in \mathbb{R}^n.$$
(3.170)

Since (3.170) is equivalent to the inequality:

$$\langle \psi, y_0 \rangle_{\mathbb{R}^n} \leq r \int_0^{t^*} \|B^* e^{-sA^*} \psi\|_{\mathbb{R}^m} ds \text{ for all } \psi \in \mathbb{R}^n,$$
(3.171)

we see that when $t^* > 0$, (3.168) is equivalent to (3.171).

Now, we suppose that (3.168) holds for some $t^* > 0$. Then (3.171) is true for this t^*. Define

$$\mathscr{A}_1 \triangleq \{B^* e^{-sA^*} \psi, \ s \in (0, t^*) \mid \psi \in \mathbb{R}^n\} \subseteq L^1(0, t^*; \mathbb{R}^m)$$

and

$$\mathscr{A}_2 \triangleq \{B^* e^{-sA^*} \psi, \ s \in (t^*, 2t^*) \mid \psi \in \mathbb{R}^n\} \subseteq L^1(t^*, 2t^*; \mathbb{R}^m).$$

It is obvious that $\mathscr{A}_1$ and $\mathscr{A}_2$ are finite dimensional. Since $s \rightarrow B^* e^{-sA^*}$, $s \in (0, +\infty)$, is an analytic function, we can easily check that the following operator is well defined and linear:

$$\mathscr{F} : \mathscr{A}_2 \mapsto \mathscr{A}_1$$

$$B^* e^{-sA^*} \psi, \ s \in (t^*, 2t^*) \rightarrow B^* e^{-sA^*} \psi, \ s \in (0, t^*).$$

Hence, there exists a positive constant c_0 so that

$$\int_{t^*}^{2t^*} \| B^* e^{-sA^*} \psi \|_{\mathbb{R}^m} \, ds \geq c_0 \int_0^{t^*} \| B^* e^{-sA^*} \psi \|_{\mathbb{R}^m} \, ds \ \text{ for all } \ \psi \in \mathbb{R}^n.$$

This, along with (3.171), yields that

$$\begin{aligned}
& r \int_0^{+\infty} \| B^* e^{-sA^*} \psi \|_{\mathbb{R}^m} \, ds \\
& \geq r \int_0^{t^*} \| B^* e^{-sA^*} \psi \|_{\mathbb{R}^m} \, ds + r \int_{t^*}^{2t^*} \| B^* e^{-sA^*} \psi \|_{\mathbb{R}^m} \, ds \\
& \geq (1 + c_0) \langle \psi, y_0 \rangle_{\mathbb{R}^n} \ \text{ for all } \ \psi \in \mathbb{R}^n.
\end{aligned} \tag{3.172}$$

By (3.48) and (3.172), we get that $r \geq (1 + c_0) N(y_0)$, from which, (3.169) follows at once.

We next show that (3.169) implies (3.168) for some $t^* > 0$. Suppose that (3.169) holds. Then, by (3.47), we have that

$$r > \langle y_0, \psi \rangle_{\mathbb{R}^n} \Big/ \int_0^{+\infty} \| B^* e^{-sA^*} \psi \|_{\mathbb{R}^m} \, ds \ \text{ for all } \ \psi \in \mathbb{R}^n. \tag{3.173}$$

Then, it follows from (3.31), (3.47), and (3.173) that

$$\langle y_0, \psi \rangle_{\mathbb{R}^n} \leq 0 \ \text{ for all } \ \psi \in N_{A,B},$$

which indicates that

$$y_0 \in N_{A,B}^\perp. \tag{3.174}$$

By making use of (3.31) and (3.173) again, we get that

$$r \int_0^{+\infty} \| B^* e^{-sA^*} \psi \|_{\mathbb{R}^m} \, ds > \langle y_0, \psi \rangle_{\mathbb{R}^n} \ \text{ for all } \ \psi \in N_{A,B}^\perp \cap \partial B_1(0). \tag{3.175}$$

Now for an arbitrarily fixed $\widehat{\psi} \in N_{A,B}^\perp \cap \partial B_1(0)$, we see from (3.175) that there is $t_{\widehat{\psi}} \in (0, +\infty)$ so that

$$r \int_0^{t_{\widehat{\psi}}} \| B^* e^{-sA^*} \widehat{\psi} \|_{\mathbb{R}^m} \, ds > \langle y_0, \widehat{\psi} \rangle_{\mathbb{R}^n}. \tag{3.176}$$

Since the following two functions are continuous on $\mathbb{R}^n$:

$$\psi \to \int_0^{t_{\widehat{\psi}}} \|B^* e^{-sA^*} \psi\|_{\mathbb{R}^m} \, ds, \quad \psi \in \mathbb{R}^n; \quad z \to \langle y_0, z \rangle_{\mathbb{R}^n}, \; z \in \mathbb{R}^n,$$

there exists a positive constant $\delta(\widehat{\psi})$ so that

$$r \int_0^{t_{\widehat{\psi}}} \|B^* e^{-sA^*} \varphi\|_{\mathbb{R}^m} \, ds > \langle y_0, \varphi \rangle_{\mathbb{R}^n} \quad \text{for all } \varphi \in U_{\delta(\widehat{\psi})}, \tag{3.177}$$

where

$$U_{\delta(\widehat{\psi})} \triangleq \{ \varphi \in N_{A,B}^{\perp} \cap \partial B_1(0) \mid \|\varphi - \widehat{\psi}\|_{\mathbb{R}^n} < \delta(\widehat{\psi}) \}.$$

Meanwhile, by the finite covering theorem, there are $\psi_1, \psi_2, \dots, \psi_{\ell_0} \in N_{A,B}^{\perp} \cap \partial B_1(0)$ so that

$$N_{A,B}^{\perp} \cap \partial B_1(0) \subseteq \bigcup_{i=1}^{\ell_0} U_{\delta(\psi_i)}. \tag{3.178}$$

Let

$$t^* \triangleq \max\{ t_{\psi_i} \mid 1 \le i \le \ell_0 \}. \tag{3.179}$$

Then it follows by (3.178) that for each $\psi \in N_{A,B}^{\perp} \cap \partial B_1(0)$, there exists $1 \le i_0 \le \ell_0$ so that $\psi \in U_{\delta(\psi_{i_0})}$. Hence, by (3.177), we have that

$$
\begin{aligned}
\langle y_0, \psi \rangle_{\mathbb{R}^n} &< r \int_0^{t_{\psi_{i_0}}} \|B^* e^{-sA^*} \psi\|_{\mathbb{R}^m} \, ds \\
&\le r \int_0^{t^*} \|B^* e^{-sA^*} \psi\|_{\mathbb{R}^m} \, ds \quad \text{for all } \psi \in N_{A,B}^{\perp} \cap \partial B_1(0).
\end{aligned} \tag{3.180}
$$

Moreover, for each $\psi \in \mathbb{R}^n$, there are two vectors $\psi_1 \in N_{A,B}$ and $\psi_2 \in N_{A,B}^{\perp}$ so that

$$\psi = \psi_1 + \psi_2.$$

This, together with (3.174), (3.180), and (3.31), implies that

$$\langle y_0, \psi \rangle_{\mathbb{R}^n} = \langle y_0, \psi_2 \rangle_{\mathbb{R}^n} \le r \int_0^{t^*} \|B^* e^{-sA^*} \psi_2\|_{\mathbb{R}^m} \, ds = r \int_0^{t^*} \|B^* e^{-sA^*} \psi\|_{\mathbb{R}^m} \, ds.$$

From the latter, we see that (3.171), where t^* is given by (3.179), is true. We end Remark 3.4.

3.4   Existence of Optimal Controls

In this section, we will present ways deriving the existence of optimal controls of $(TP)_{min}^{Q_S,Q_E}$ from the existence of admissible controls for $(TP)_{min}^{Q_S,Q_E}$. From Theorem 1.13 and its proof, we can conclude the following procedure to show the existence of solutions for the optimization problem (1.21): The first step is to find a minimizing sequence. (In this step, we need to show the existence of minimizing sequences.) The second step is to find a convergent subsequence (from the minimizing sequence) in some suitable topology. The last step is to show that the limit of the above subsequence is a solution of the optimization problem.

We can borrow the above-mentioned procedure to show the existence of optimal controls for some time optimal control problems, provided that these problems have admissible controls. We will explain this in detail in the forthcoming subsection.

3.4.1   Existence of Optimal Controls for Regular Cases

In this subsection, we study the existence of optimal controls for the problem $(TP)_{min}^{Q_S,Q_E}$. This problem is as follows:

$$t_E^* \triangleq \inf_{(0,y_0,t_E,u)\in\mathscr{A}_{ad}} t_E, \tag{3.181}$$

where the admissible tetrad set is defined by

$$\mathscr{A}_{ad} \triangleq \big\{ (0, y_0, t_E, u) \in Q_S \times (0, +\infty) \times L(0, t_E; \mathbb{U}) :$$

$$(t_E, y(t_E; 0, y_0, u)) \in Q_E \big\}, \tag{3.182}$$

with $Q_S = \{0\} \times Y_S$. We will present a procedure to show the existence of optimal controls for the problem (3.181), under the assumption that $\mathscr{A}_{ad} \neq \emptyset$. The first step is to study the admissible tetrad set. Since $\mathscr{A}_{ad} \neq \emptyset$, there are only two possibilities on the set $\mathscr{A}_{ad}$: Either it has finitely many elements or it has infinite elements. In the case that $\mathscr{A}_{ad}$ has finitely many elements, it is obvious that the problem (3.181) has an optimal control. And the procedure is finished. In the case when $\mathscr{A}_{ad}$ has infinite elements, there is a sequence of admissible tetrads $\{(0, y_{0,\ell}, t_{E,\ell}, u_\ell)\}_{\ell\geq 1}$ so that $\{t_{E,\ell}\}_{\ell\geq 1}$ is monotone decreasing and converges to t^*. We call such a sequence of admissible tetrads a minimizing sequence of the problem (3.181). The second step is to find a subsequence of the minimizing sequence so that it converges in some topology. More precisely, we need to show that there exists a subsequence of $\{\ell\}_{\ell\geq 1}$, still denoted in the same way, so that for some $(0, \widehat{y_0}) \in Q_S$ and $\widehat{u} \in L(0, t_E^*; \mathbb{U})$,

$$y(\cdot; 0, y_{0,\ell}, u_\ell) \to y(\cdot; 0, \widehat{y_0}, \widehat{u}) \quad \text{in some suitable topology.}$$

The final step is to show that $(t_E^*, y(t_E^*; 0, \widehat{y_0}, \widehat{u})) \in Q_E$.

The following lemma is quoted from [15] (see Lemma 3.2 on page 104 of [15]). It plays an important role in the study of the existence of optimal controls for the problem (3.181).

Lemma 3.9 Let $\{e^{\widehat{A}t}\}_{t\geq 0}$ be a compact C_0 semigroup on a Banach space Z. Let $p > 1$ and $t_0 \in (0, +\infty)$. Then the operator $\mathscr{L} : L^p(0, t_0; Z) \mapsto C([0, t_0]; Z)$ defined by

$$\mathscr{L}(u)(t) \triangleq \int_0^t e^{\widehat{A}(t-s)} u(s)\, ds \quad \text{for all } t \in [0, t_0] \tag{3.183}$$

is compact.

The following assumptions will be effective in the rest of this subsection.

($\mathscr{H}1$) $A : \mathscr{D}(A) \subseteq Y \mapsto Y$ generates a compact semigroup $\{e^{At}\}_{t\geq 0}$ on Y.

($\mathscr{H}2$) $f : (0, +\infty) \times Y \times U \mapsto Y$ is measurable in (t, y, u); uniformly continuous in u locally at (t_0, y_0, u_0) for almost every $t_0 \in (0, +\infty)$, for each $y_0 \in Y$ and each $u_0 \in U$; and continuous in y uniformly with respect to $u \in U$ for almost every $t \in (0, +\infty)$.

($\mathscr{H}3$) The set $f(t, y, \mathbb{U}) \triangleq \{f(t, y, u)|u \in \mathbb{U}\}$ is convex and closed.

($\mathscr{H}4$) $\|f(t, y, u)\|_Y \leq l(t)\|y\|_Y + \varphi(t)$ for all $t \in (0, +\infty)$ and $u \in \mathbb{U}$, where $l(\cdot) \in L^\infty(0, +\infty)$ and $\varphi(\cdot) \in L^p(0, +\infty)$ are nonnegative and $p > 1$.

($\mathscr{H}5$) Y_S is a nonempty convex, bounded, and closed subset of Y.

($\mathscr{H}6$) Q_E is a nonempty closed subset of $(0, +\infty) \times Y$.

The main result of this subsection is as follows.

Theorem 3.11 Suppose that $(\mathscr{H}1) - (\mathscr{H}6)$ hold. Assume that $\mathscr{A}_{ad}$ (given by (3.182)) is not empty. Then the problem (3.181) has an optimal control.

Proof Since $\mathscr{A}_{ad} \neq \emptyset$, we have that $t_E^* < +\infty$. Thus, there is a sequence $\{(0, y_{0,k}, t_{E,k}, u_k)\}_{k\geq 1}$ in $\mathscr{A}_{ad}$ so that

$$t_{E,k} \searrow t_E^* \tag{3.184}$$

and

$$y_{0,k} \in Y_S, \quad u_k \in L(0, t_{E,k}; \mathbb{U}), \quad \left(t_{E,k}, y(t_{E,k}; 0, y_{0,k}, u_k)\right) \in Q_E \quad \text{for all } k \geq 1. \tag{3.185}$$

By the first conclusion in (3.185) and by ($\mathscr{H}5$), we can find a subsequence of $\{k\}_{k\geq 1}$, denoted in the same manner, so that

$$y_{0,k} \to \widehat{y}_0 \quad \text{weakly in } Y \text{ for some } \widehat{y}_0 \in Y. \tag{3.186}$$

Claim One: We have that

$$\widehat{y}_0 \in Y_S. \tag{3.187}$$

Indeed, since Y_S is convex and closed in Y, it is weakly closed in Y, which, along with (3.186), leads to (3.187). This ends the proof of Claim One.

We now let, for each $k \in \mathbb{N}^+$,

$$y_k(t) \triangleq y\left(t; 0, y_{0,k}, u_k\right) \quad \text{for each } t \in [0, t_{E,k}] \tag{3.188}$$

and

$$f_k(t) \triangleq f\left(t, y_k(t), u_k(t)\right) \quad \text{for a.e. } t \in (0, t_{E,k}). \tag{3.189}$$

Then we have that for each $k \in \mathbb{N}^+$,

$$y_k(t) = e^{At} y_{0,k} + \int_0^t e^{A(t-\tau)} f_k(\tau) \, d\tau \quad \text{for each } t \in [0, t_{E,k}]. \tag{3.190}$$

Claim Two: There is $\widehat{f} \in L^p(0, t_E^*; Y)$ and a subsequence of $\{k\}_{k \geq 1}$, denoted in the same way, so that

$$f_k \to \widehat{f} \quad \text{weakly in } L^p(0, t_E^*; Y), \tag{3.191}$$

and so that

$$y_k(t_E^*) \to \widehat{y}(t_E^*) \quad \text{strongly in } Y, \tag{3.192}$$

where $\widehat{y}(\cdot)$ is the function defined by

$$\widehat{y}(t) \triangleq e^{At} \widehat{y}_0 + \int_0^t e^{A(t-\tau)} \widehat{f}(\tau) \, d\tau \quad \text{for all } t \in [0, t_E^*]. \tag{3.193}$$

To this end, we first use (3.190), (3.189), ($\mathscr{H}1$), and ($\mathscr{H}4$) to get that for all $k \in \mathbb{N}^+$ and $t \in [0, t_{E,k}]$,

$$\begin{aligned} \|y_k(t)\|_Y &\leq \left\| e^{At} y_{0,k} \right\|_Y + \int_0^t \left\| e^{A(t-\tau)} f_k(\tau) \right\|_Y \, d\tau \\ &\leq C e^{Ct} \|y_{0,k}\|_Y + \int_0^t C e^{C(t-\tau)} \left[l(\tau) \|y_k(\tau)\|_Y + \varphi(\tau) \right] d\tau. \end{aligned} \tag{3.194}$$

(Here and throughout the proof of this theorem, C denotes a generic positive constant independent of k.) Since $y_{0,k} \in Y_S$ for all $k \geq 1$ (see (3.185)) and Y_S is bounded (see ($\mathscr{H}5$)), we can apply Gronwall's inequality to (3.194) to get that

$$\|y_k(t)\|_Y \leq C \quad \text{for all } t \in [0, t_{E,k}] \text{ and all } k \geq 1. \tag{3.195}$$

By (3.189), (3.195), and ($\mathscr{H}$4), we find that $\{f_k\}_{k\geq 1}$ is bounded in $L^p(0, t_E^*; Y)$. Thus, there is $\widehat{f} \in L^p(0, t_E^*; Y)$ and a subsequence of $\{k\}_{k\geq 1}$, denoted in the same way, satisfying (3.191).

Next, by (3.191), ($\mathscr{H}$1), and Lemma 3.9, there is a subsequence of $\{k\}_{k\geq 1}$, denoted in the same manner, so that

$$\int_0^t e^{A(t-\tau)} f_k(\tau)\, d\tau \to \int_0^t e^{A(t-\tau)} \widehat{f}(\tau)\, d\tau \quad \text{strongly in } C([0, t_E^*]; Y).$$
(3.196)

Meanwhile, by ($\mathscr{H}$1) and (3.186), we can find a subsequence of $\{k\}_{k\geq 1}$, denoted in the same way, so that

$$e^{At_E^*} y_{0,k} \to e^{At_E^*} \widehat{y}_0 \quad \text{strongly in } Y.$$
(3.197)

Then, by (3.190), (3.196), (3.193), and (3.197), we can easily obtain (3.192). This ends the proof of Claim Two.

Claim Three: we have that

$$(t_E^*, \widehat{y}(t_E^*)) \in Q_E, \quad \text{where } \widehat{y}(\cdot) \text{ is given by} \quad (3.193).$$
(3.198)

In fact, by (3.188), ($\mathscr{H}$1), ($\mathscr{H}$4), and (3.195), we have that

$$\begin{aligned}
&\|y_k(t_{E,k}) - \widehat{y}(t_E^*)\|_Y \\
&\leq \left\| e^{A(t_{E,k}-t_E^*)} y_k(t_E^*) - \widehat{y}(t_E^*) \right\|_Y + \int_{t_E^*}^{t_{E,k}} \left\| e^{A(t_{E,k}-\tau)} f(\tau, y_k(\tau), u_k(\tau)) \right\|_Y d\tau \\
&\leq \left\| e^{A(t_{E,k}-t_E^*)} (y_k(t_E^*) - \widehat{y}(t_E^*)) \right\|_Y + \left\| e^{A(t_{E,k}-t_E^*)} \widehat{y}(t_E^*) - \widehat{y}(t_E^*) \right\|_Y \\
&\quad + \int_{t_E^*}^{t_{E,k}} Ce^{C(t_{E,k}-\tau)} [l(\tau) + \varphi(\tau)]\, d\tau \\
&\leq Ce^{C(t_{E,k}-t_E^*)} \|y_k(t_E^*) - \widehat{y}(t_E^*)\|_Y + \left\| e^{A(t_{E,k}-t_E^*)} \widehat{y}(t_E^*) - \widehat{y}(t_E^*) \right\|_Y \\
&\quad + C\int_{t_E^*}^{t_{E,k}} [l(\tau) + \varphi(\tau)]\, d\tau.
\end{aligned}$$

This, together with (3.192) and (3.184), yields that

$$y_k(t_{E,k}) \to \widehat{y}(t_E^*) \quad \text{strongly in } Y.$$
(3.199)

Meanwhile, by (3.185) and (3.188), we have that

$$(t_{E,k}, y_k(t_{E,k})) \in Q_E.$$

From this, ($\mathscr{H}$6), (3.184), and (3.199), one can easily obtain (3.198). This ends the proof of Claim Three.

Claim Four: There is $\widehat{u} \in L(0, t_E^*; \mathbb{U})$ so that

$$y(t; 0, \widehat{y}_0, \widehat{u}) = \widehat{y}(t) \quad \text{for each } t \in [0, t_E^*]. \tag{3.200}$$

In fact, by (3.191), we can use the Mazur Theorem (see Theorem 1.8) to find, for each $k \geq 1$, a finite sequence of nonnegative numbers $\{\alpha_{kj}\}_{j=1}^{J_k}$, with $\sum_{j=1}^{J_k} \alpha_{kj} = 1$, so that

$$g_k \triangleq \sum_{j=1}^{J_k} \alpha_{kj} f_{k+j} \to \widehat{f} \quad \text{strongly in } L^p(0, t_E^*; Y). \tag{3.201}$$

Meanwhile, by the same way as that used to prove (3.192), we can verify that for any $t \in (0, t_E^*]$, there exists a subsequence of $\{k\}_{k \geq 1}$, still denoted in the same way, so that

$$y_k(t) \to \widehat{y}(t). \tag{3.202}$$

By ($\mathscr{H}2$) and (3.189), we see that there is a subset $E \subseteq (0, t_E^*)$, with $|E| = 0$ (i.e., the measure of E is zero), so that for each $t \in (0, t_E^*) \setminus E$, $f(t, \cdot, u)$ is continuous uniformly with respect to u, and so that

$$f_k(t) = f(t, y_k(t), u_k(t)) \quad \text{for all } t \in (0, t_E^*) \setminus E. \tag{3.203}$$

Arbitrarily fix $t \in (0, t_E^*) \setminus E$. Then by the aforementioned continuity of $f(t, \cdot, u)$ and by (3.202), we find that for an arbitrarily fixed $\varepsilon \in (0, 1)$, there exists a positive integer $k_0(t, \varepsilon)$ so that

$$\|f(t, y_k(t), u) - f(t, \widehat{y}(t), u)\|_Y \leq \varepsilon \quad \text{for all } k \geq k_0(t, \varepsilon) \text{ and } u \in U.$$

This, along with (3.203) and (3.185), yields that for all $k \geq k_0(t, \varepsilon)$,

$$f_{k+j}(t) = f(t, y_{k+j}(t), u_{k+j}(t)) \in f(t, \widehat{y}(t), u_{k+j}(t)) + \varepsilon B_1(0)$$
$$\subseteq f(t, \widehat{y}(t), \mathbb{U}) + \varepsilon B_1(0). \tag{3.204}$$

Here $B_1(0)$ denotes the closed ball in Y centered at 0 and of radius 1. Since the set $f(t, \widehat{y}(t), \mathbb{U})$ is convex and closed (see ($\mathscr{H}3$)), it follows from (3.204) that

$$g_k(t) \triangleq \sum_{j=1}^{J_k} \alpha_{kj} f_{k+j}(t) \in f(t, \widehat{y}(t), \mathbb{U}) + \varepsilon B_1(0) \quad \text{for all } k \geq k_0(t, \varepsilon). \tag{3.205}$$

Since $t \in (0, t_E^*) \setminus E$ and $\varepsilon \in (0, 1)$ was arbitrarily fixed, it follows by (3.201), (3.205), and the closedness of $f(t, \widehat{y}(t), \mathbb{U})$ that

$$\widehat{f}(t) \in f(t, \widehat{y}(t), \mathbb{U}) \quad \text{for a.e. } t \in (0, t_E^*).$$

From this and the condition $(\mathscr{H}2)$, we can apply the Filippov Lemma (see Corollary 1.1) to find $\widehat{u} \in L(0, t^*; \mathbb{U})$ so that

$$\widehat{f}(t) = f(t, \widehat{y}(t), \widehat{u}(t)) \quad \text{for a.e. } t \in (0, t_E^*).$$

This, together with (3.193), leads to (3.200). This ends the proof of Claim Four.

Finally, by (3.187), (3.198), and (3.200), we see that $(0, \widehat{y}_0, t_E^*, \widehat{u}) \in \mathscr{A}_{ad}$. This completes the proof of Theorem 3.11.   □

3.4.2 Existence of Optimal Controls for Blowup Case

The time optimal control problem studied in this subsection is a special case of the problem (2.18). Namely, it is a minimal blowup time control problem. We will derive the existence of optimal controls for this problem, under the assumption that the set of admissible controls is not empty. We begin with introducing this problem. Let $y_0 \in \mathbb{R}^n$ be arbitrarily fixed. Consider the following controlled system:

$$\begin{cases} \dfrac{dy}{dt} = f(t, y(t)) + B(t)u(t), & t \in (0, +\infty), \\ y(0) = y_0. \end{cases} \qquad (3.206)$$

Here, $u \in L(0, +\infty; \mathbb{U})$, $B(\cdot)$, $f(\cdot, \cdot)$, and $\mathbb{U}$ satisfy the following assumptions:

(S1) $B(\cdot) \in L^1_{loc}(0, +\infty; \mathbb{R}^{n \times m})$;
(S2) $f : [0, +\infty) \times \mathbb{R}^n \mapsto \mathbb{R}^n$ satisfies that for each $y \in \mathbb{R}^n$, $f(\cdot, y)$ is measurable on $[0, +\infty)$, and that for any $r > 0$, there exists a positive constant M_r so that

$$\|f(t, y_1) - f(t, y_2)\|_{\mathbb{R}^n} \leq M_r \|y_1 - y_2\|_{\mathbb{R}^n},$$

$$\text{for all } y_1, y_2 \in B_r(0) \text{ and } t \geq 0.$$

Here $B_r(0)$ denotes a closed ball in $\mathbb{R}^n$, centered at 0 and of radius r;
(S3) $\mathbb{U}$ is a nonempty bounded, convex, and closed set in $\mathbb{R}^m$.

We write $y(\cdot; 0, y_0, u)$ for the solution of (3.206). The minimal blowup time control problem under study is as:

$$t_E^* \triangleq \inf_{(0, y_0, t_E, u) \in \mathscr{A}_{ad}} t_E, \qquad (3.207)$$

where the set of admissible tetrads $\mathscr{A}_{ad}$ is given by

$$\mathscr{A}_{ad} \triangleq \big\{ (0, y_0, t_E, u) \in \{(0, y_0)\} \times (0, +\infty) \times L(0, t_E; \mathbb{U}) \ \big|$$

$$\lim_{t \to t_E-} \|y(t; 0, y_0, u|_{(0,t)})\|_{\mathbb{R}^n} = +\infty \big\}.$$

One can easily check that the problem (3.207) can be put into the framework of the problem (2.18).

Lemma 3.10 *Suppose that (S1)–(S3) hold. Let* $\{u_\ell\}_{\ell \geq 1} \subseteq L(0, +\infty; \mathbb{U})$ *satisfy that*

$$u_\ell \to u \quad \text{weakly star in } L^\infty(0, +\infty; \mathbb{R}^m) \quad \text{for some } u \in L(0, +\infty; \mathbb{U}). \tag{3.208}$$

Further assume that $y(\cdot; 0, y_0, u)$ *exists on* $[0, t_0^*]$ *for some* $t_0^* \in (0, +\infty)$. *Then there is* $\delta \in (0, +\infty)$ *and* $\ell_0 \in \mathbb{N}^+$ *so that* $\{y(\cdot; 0, y_0, u_\ell)\}_{\ell \geq \ell_0}$ *is bounded in* $C([0, t_0^* + \delta]; \mathbb{R}^n)$.

Proof Let $y(\cdot; 0, y_0, u)$ exist on $[0, t_0^*]$ for some $t_0^* \in (0, +\infty)$. Then there is $\delta \in (0, +\infty)$ so that $y(\cdot; 0, y_0, u) \in C([0, t_0^* + \delta]; \mathbb{R}^n)$. Let

$$y(t) \triangleq y(t; 0, y_0, u), \quad t \in [0, t_0^* + \delta],$$

and let

$$r \triangleq \max_{0 \leq t \leq t_0^* + \delta} \|y(t)\|_{\mathbb{R}^n} + 1 \quad \text{and} \quad \varepsilon_0 \triangleq r e^{-M_{2r}(t_0^* + \delta)}.$$

By (S1), one can easily check that the following function $b(\cdot)$ is absolutely continuous on $[0, t_0^* + \delta]$:

$$t \to b(t) \triangleq \int_0^t \|B(s)\|_{\mathcal{L}(\mathbb{R}^m, \mathbb{R}^n)} ds, \quad t \geq 0.$$

Thus, there are t_i, $i = 0, \ldots, k_0$ (for some $k_0 \in \mathbb{N}^+$), with

$$0 \triangleq t_0 < t_1 < t_2 < \cdots < t_{k_0} = t_0^* + \delta,$$

so that

$$\max_{0 \leq j \leq k_0 - 1} \int_{t_j}^{t_{j+1}} \|B(s)\|_{\mathcal{L}(\mathbb{R}^m, \mathbb{R}^n)} ds \leq \frac{\varepsilon_0}{4(c_0 + 1)}, \tag{3.209}$$

with $c_0 \triangleq \sup\{\|u\|_{\mathbb{R}^m} \mid u \in \mathbb{U}\}$. Meanwhile, by (S1) and (3.208), there is a positive integer ℓ_0 so that when $\ell \geq \ell_0$,

$$\left\| \int_0^{t_j} B(s)(u_\ell(s) - u(s)) ds \right\|_{\mathbb{R}^n} \leq \frac{\varepsilon_0}{2} \quad \text{for all } 1 \leq j \leq k_0.$$

This, along with (3.209), yields that

$$\max_{t \in [0, t_0^* + \delta]} \left\| \int_0^t B(s)(u_\ell(s) - u(s)) ds \right\|_{\mathbb{R}^n} \leq \varepsilon_0 \quad \text{for all } \ell \geq \ell_0. \tag{3.210}$$

Now we claim that

$$\max_{t \in [0, t_0^* + \delta]} \|y(t; 0, y_0, u_\ell)\|_{\mathbb{R}^n} < 2r \quad \text{for all } \ell \geq \ell_0. \tag{3.211}$$

By contradiction, we suppose that (3.211) was not true. Then there would be a positive integer $\widehat{\ell} \geq \ell_0$ and $\widehat{t} \in (0, t_0^* + \delta]$ so that

$$\|y(t; 0, y_0, u_{\widehat{\ell}})\|_{\mathbb{R}^n} < 2r \quad \text{for all } t \in [0, \widehat{t}); \quad \text{and} \quad \|y(\widehat{t}; 0, y_0, u_{\widehat{\ell}})\|_{\mathbb{R}^n} = 2r. \tag{3.212}$$

Let

$$y_{\widehat{\ell}}(t) \triangleq y(t; 0, y_0, u_{\widehat{\ell}}), \quad t \in [0, \widehat{t}].$$

Then by (3.206), (3.212), (S2), and (3.210), we get that

$$\begin{aligned}
&\|y_{\widehat{\ell}}(t) - y(t)\|_{\mathbb{R}^n} \\
&= \left\| \int_0^t [f(s, y_{\widehat{\ell}}(s)) - f(s, y(s)) + B(s)(u_{\widehat{\ell}}(s) - u(s))] ds \right\|_{\mathbb{R}^n} \\
&\leq M_{2r} \int_0^t \|y_{\widehat{\ell}}(s) - y^*(s)\|_{\mathbb{R}^n} ds + \varepsilon_0 \quad \text{for all } t \in [0, \widehat{t}],
\end{aligned}$$

which, combined with Gronwall's inequality, indicates that

$$\|y_{\widehat{\ell}}(\widehat{t}) - y(\widehat{t})\|_{\mathbb{R}^n} \leq \varepsilon_0 e^{M_{2r} \widehat{t}} \leq r. \tag{3.213}$$

Since $\|y(\widehat{t})\|_{\mathbb{R}^n} \leq r - 1$, it follows by (3.213) that

$$\|y_{\widehat{\ell}}(\widehat{t})\|_{\mathbb{R}^n} \leq \|y_{\widehat{\ell}}(\widehat{t}) - y(\widehat{t})\|_{\mathbb{R}^n} + \|y(\widehat{t})\|_{\mathbb{R}^n} \leq 2r - 1.$$

This contradicts (3.212). Hence, (3.211) holds. Thus we end the proof of Lemma 3.10. $\qquad\square$

Theorem 3.12 *Suppose that (S1)–(S3) hold. Assume that $\mathscr{A}_{ad} \neq \emptyset$. Then the problem (3.207) has an optimal control.*

Proof Since $\mathscr{A}_{ad} \neq \emptyset$, we have that $t_E^* < +\infty$. Thus, there exist two sequences $\{t_{E,\ell}\}_{\ell \geq 1} \subseteq (0, +\infty)$ and $\{u_\ell\}_{\ell \geq 1} \subseteq L(0, t_{E,\ell}; \mathbb{U})$ so that

$$\lim_{t \to t_{E,\ell}-} \|y(t; 0, y_0, u_\ell|_{(0,t)})\|_{\mathbb{R}^n} = +\infty \tag{3.214}$$

and

$$t_{E,\ell} \searrow t_E^*. \tag{3.215}$$

Arbitrarily fix $u_0 \in \mathbb{U}$. Define

$$\widehat{u}_\ell(t) \triangleq \begin{cases} u_\ell(t), & t \in (0, t_{E,\ell}), \\ u_0, & t \in [t_{E,\ell}, +\infty). \end{cases}$$

Then we have that

$$y(t; 0, y_0, \widehat{u}_\ell) = y(t; 0, y_0, u_\ell) \quad \text{for all } t \in [0, t_{E,\ell}). \tag{3.216}$$

Meanwhile, by (S3), there exists a subsequence of $\{\ell\}_{\ell \geq 1}$, denoted in the same manner, so that

$$\widehat{u}_\ell \to u^* \quad \text{weakly star in } L^\infty(0, +\infty; \mathbb{R}^m) \quad \text{for some } u^* \in L^\infty(0, +\infty; \mathbb{U}). \tag{3.217}$$

We now claim that

$$\lim_{t \to t_E^{*-}} \|y(t; 0, y_0, u^*|_{(0,t)})\|_{\mathbb{R}^n} = +\infty. \tag{3.218}$$

By contradiction, suppose that (3.218) was not true. Then $y(\cdot; 0, y_0, u^*)$ would exist on $[0, t_E^*]$. Thus, by Lemma 3.10 and (3.217), there are two positive constants δ and ℓ_0 so that $\{y(\cdot; 0, y_0, \widehat{u}_\ell)\}_{\ell \geq \ell_0}$ is bounded in $C([0, t_E^* + \delta]; \mathbb{R}^n)$. Hence, by (3.215) and (3.216), there are two positive constants $\ell_1 \geq \ell_0$ and C so that

$$\max_{t \in [0, t_{E,\ell})} \|y(t; 0, y_0, u_\ell)\|_{\mathbb{R}^n} \leq C \quad \text{for all } \ell \geq \ell_1,$$

which contradicts (3.214). Hence, (3.218) is true. From (3.218), we find that u^* is an optimal control for the problem (3.207). This completes the proof of Theorem 3.12. $\qquad\square$

Miscellaneous Notes

The existence of admissible controls and optimal controls is an important subject in the studies of time optimal control problems. It is independent of other issues, such as Pontryagin's maximum principle, the bang-bang property, and so on. In many cases, the existence of optimal controls can be derived from the existence of admissible controls, through using some standard methods which are introduced in Section 3.4.1. These methods are summarized from the previous works in this direction (see, for instance, [2, 3, 20, 22, 28], and [23]).

The existence of admissible controls can be treated as a kind of constraint controllability. Indeed, an admissible control is a control (in a constraint set) driving the corresponding solution (of a system) to a given target in some time, while the general controllability is to ask for a control, without any constraint, so that the corresponding solution (of a system) reaches a given target set at a fixed ending time. Hence, we study the existence of admissible controls of time optimal control problems from three different viewpoints: the controllability, the minimal norm control problems, and reachable sets.

Several notes on Section 3.1 are given in order:

- Many materials in Section 3.1 are based on some parts of [5, 12, 24], and [25] (which are about finite dimensional cases) and some parts of [23] and [30] (which are about infinite dimensional cases). Some materials in Section 3.1 (see, for instance, Theorem 3.4) are developed in this monograph, based on known results and methods in some existing literatures.
- Materials in Section 3.1.3 are based on [23]. There, the existence of admissible controls is global, i.e., the initial state can be arbitrarily given. (Notice that the controlled system is a semi-linear heat equation.) In general, it is not easy to get the global null or approximate controllability for nonlinear equations in a fixed time interval. For such works, we refer readers to [6, 7] and [4]. It should be more difficult to obtain the global controllability for nonlinear equations with constraint controls even in a mobile time interval, in other words, it should be harder to get the existence of admissible controls for nonlinear equations. To the best of our knowledge, such studies are very limited (see, for instance, [30]). Our global result (obtained in Section 3.1.3) is due to the particularity of the semi-linear term f in the Equation (3.59). An interesting question is as follows: Can we have the existence of admissible controls when $f(r) \lesssim r|\ln r|^{3/2}$? This growth rate appeared in both [4] and [7].

For Section 3.2, we would like to mention what follows:

- The usual minimal norm control problem is to ask for a control having the minimal norm among all such controls that drive the corresponding solutions of a controlled system from a given initial state to a given target in a fixed time interval. Such problems have been widely studied (see, for instance, [8–10], and [26]). The norm optimal control problem studied in Section 3.2.1 is a generalization of the above-mentioned minimal norm control problem. We devote this generalization to the current monograph.
- The materials in Section 3.2.2 are essentially taken from [32] (see also [26] and [31]).

The materials in Section 3.3 are based on some development of the related materials in Section 5 of Chapter 7 in [15]. About Section 3.4, we would like to mention the following facts:

- The materials in Section 3.4.2 are simplified from the materials in [18]. More general results on minimal blowup time control problems are presented in [19].
- It is worth mentioning the paper [16], where the author studied some minimal quenching time control problems of some ODEs. It is the first paper studying such interesting problems, which have some connections with, but differ from the minimal time control problems.

For more results about the existence of admissible controls and optimal controls, we would like to mention [13, 14, 21, 27, 29], and [17].

At the end of this miscellaneous note, we would like to present several open problems:

- Given infinite dimensional system (A, B) with the ball-type control constraint, can we find a criterion on the existence of admissible controls (such as some counterpart of Corollary 3.2)? For a finite or infinite dimensional system (A, B) with other kinds of control constraints, what is a criterion on the existence of admissible controls?
- For controlled systems with some state constraints (for instance, heat equations with nonnegative states), can we find some reasonable condition to ensure the existence of admissible controls?

References

1. J.P. Aubin, *Optima and Equilibria, An Introduction to Nonlinear Analysis*, Translated from the French by Stephen Wilson. Graduate Texts in Mathematics, vol. 140 (Springer, Berlin, 1993)
2. V. Barbu, *Analysis and Control of Nonlinear Infinite-Dimensional Systems*. Mathematics in Science and Engineering, vol. 190 (Academic, Boston, MA, 1993)
3. V. Barbu, The time optimal control of Navier-Stokes equations. Syst. Control Lett. **30**, 93–100 (1997)
4. V. Barbu, Exact controllability of the superlinear heat equation. Appl. Math. Optim. **42**, 73–89 (2000)
5. R. Conti, Teoria del controllo e del controllo ottimo, UTET, Torino (1974)
6. J.M. Coron, A.V. Fursikov, Global exact controllability of the 2D Navier-Stokes equations on a manifold without boundary. Russian J. Math. Phys. **4**, 429–446 (1996)
7. T. Duyckaerts, X. Zhang, E. Zuazua, On the optimality of the observability inequalities for parabolic and hyperbolic systems with potentials. Ann. Inst. H. Poincaré, Anal. Non Linéaire **25**, 1–41 (2008)
8. C. Fabre, J.P. Puel, E. Zuazua, Approximate controllability of the semilinear heat equation. Proc. R. Soc. Edinb. A **125**, 31–61 (1995)
9. H.O. Fattorini, *Infinite Dimensional Linear Control Systems, the Time Optimal and Norm Optimal Problems*. North-Holland Mathematics Studies, vol. 201 (Elsevier Science B.V., Amsterdam, 2005)
10. F. Gozzi, P. Loreti, Regularity of the minimum time function and minimum energy problems: the linear case. SIAM J. Control Optim. **37**, 1195–1221 (1999)
11. J.B. Hiriart-Urruty, C. Lemaréchal, *Fundamentals of Convex Analysis* (Springer, Berlin, 2001)
12. R.E. Kalman, Mathematical description of linear dynamical system. J. SIAM Control A **1**, 152–192 (1963)
13. K. Kunisch, L. Wang, The bang-bang property of time optimal controls for the Burgers equation. Discrete Contin. Dyn. Syst. **34**, 3611–3637 (2014)
14. K. Kunisch, L. Wang, Bang-bang property of time optimal controls of semilinear parabolic equation. Discrete Contin. Dyn. Syst. **36**, 279–302 (2016)
15. X. Li, J. Yong, *Optimal Control Theory for Infinite-Dimensional Systems*, Systems & Control: Foundations & Applications (Birkhäuser Boston, Boston, MA, 1995)
16. P. Lin, Quenching time optimal control for some ordinary differential equations. J. Appl. Math. Art. ID 127809, 13 pp. (2014)
17. P. Lin, S. Luan, Time optimal control for some ordinary differential equations with multiple solutions. J. Optim. Theory Appl. **173**, 78–90 (2017)

18. P. Lin, G. Wang, Blowup time optimal control for ordinary differential equations. SIAM J. Control Optim. **49**, 73–105 (2011)
19. H. Lou, J. Wen, Y. Xu, Time optimal control problems for some non-smooth systems. Math. Control Relat. Fields **4**, 289–314 (2014)
20. Q. Lü, G. Wang, On the existence of time optimal controls with constraints of the rectangular type for heat equations. SIAM J. Control Optim. **49**, 1124–1149 (2011)
21. S. Micu, L.E. Temereancǎ, A time-optimal boundary controllability problem for the heat equation in a ball. Proc. R. Soc. Edinb. A **144**, 1171–1189 (2014)
22. K.D. Phung, G. Wang, X. Zhang, On the existence of time optimal controls for linear evolution equations. Discrete Contin. Dyn. Syst. Ser. B **8**, 925–941 (2007)
23. K.D. Phung, L. Wang, C. Zhang, Bang-bang property for time optimal control of semilinear heat equation. Ann. Inst. H. Poincaré Anal. Non Linéaire **31**, 477–499 (2014)
24. W.E. Schmitendorf, B.R. Barmish, Null controllability of linear system with constrained controls. SIAM J. Control Optim. **18**, 327–345 (1980)
25. E.D. Sontag, *Mathematical Control Theory: Deterministic Finite-Dimensional Systems*. Texts in Applied Mathematics, 2nd edn., vol. 6 (Springer, New York, 1998)
26. M. Tucsnak, G. Wang, C. Wu, Perturbations of time optimal control problems for a class of abstract parabolic systems. SIAM J. Control Optim. **54**, 2965–2991 (2016)
27. L.J. A-Vázquez, F.J. Fernández, A. Martínez, Analysis of a time optimal control problem related to the management of a bioreactor. ESAIM Control Optim. Calc. Var. **17**, 722–748 (2011)
28. G. Wang, The existence of time optimal control of semilinear parabolic equations. Syst. Control Lett. **53**, 171–175 (2004)
29. L. Wang, G. Wang, The optimal time control of a phase-field system. SIAM J. Control Optim. **42**, 1483–1508 (2003)
30. L. Wang, Q. Yan, Bang-bang property of time optimal null controls for some semilinear heat equation. SIAM J. Control Optim. **54**, 2949–2964 (2016)
31. G. Wang, Y. Zhang, Decompositions and bang-bang problems. Math. Control Relat. Fields **7**, 73–170 (2017)
32. G. Wang, Y. Xu, Y. Zhang, Attainable subspaces and the bang-bang property of time optimal controls for heat equations. SIAM J. Control Optim. **53**, 592–621 (2015)

Chapter 4
Maximum Principle of Optimal Controls

In this chapter, we will present some first-order necessary conditions on time optimal controls for some evolution systems. Such necessary conditions are referred to as the Pontryagin Maximum Principle (or Pontryagin's maximum principle), which provides some information on optimal controls. Differing from analysis methods (see [3, 12, 14] and [4]), there are other geometric ways to approach the Pontryagin Maximum Principle (see, for instance, [1] and [16]). Usually, these geometric methods are based on the use of separation theorems and representation theorems. In the first three sections of this chapter, we introduce geometric methods through studying the problem $(TP)_{min}^{Q_S, Q_E}$ in several different cases. Three different kinds of Pontryagin's maximum principles for $(TP)_{min}^{Q_S, Q_E}$ are given in order. They are respectively called as: the classical Pontryagin Maximum Principle, the local Pontryagin Maximum Principle, and the weak Pontryagin Maximum Principle. These Pontryagin's maximum principles are obtained by separating different objects: separating the target from the reachable set in the state space at the optimal time; separating a reachable set from a controllable set in the state space before the optimal time; separating the target from the reachable set in the reachable space at the optimal time. We then discuss the classical Pontryagin Maximum Principle and the local Pontryagin Maximum Principle for $(TP)_{max}^{Q_S, Q_E}$ in the final section, where the methods are similar to those used for $(TP)_{min}^{Q_S, Q_E}$.

4.1 Classical Maximum Principle of Minimal Time Controls

This section studies the problem $(TP)_{min}^{Q_S, Q_E}$, under the following framework $(\mathscr{A}_{TP})$:

(i) The state space Y and the control space U are real separable Hilbert spaces.

© Springer International Publishing AG, part of Springer Nature 2018
G. Wang et al., *Time Optimal Control of Evolution Equations*, Progress
in Nonlinear Differential Equations and Their Applications 92,
https://doi.org/10.1007/978-3-319-95363-2_4

(ii) The controlled system is as:

$$\dot{y}(t) = Ay(t) + D(t)y(t) + B(t)u(t), \quad t \in (0, +\infty), \tag{4.1}$$

where $A : \mathscr{D}(A) \subseteq Y \mapsto Y$ generates a C_0 semigroup $\{e^{At}\}_{t \geq 0}$ on Y; $D(\cdot) \in L^\infty(0, +\infty; \mathscr{L}(Y))$ and $B(\cdot) \in L^\infty(0, +\infty; \mathscr{L}(U, Y))$. Let $\{\Phi(t, s) : t \geq s \geq 0\}$ be the evolution system generated by $A + D(\cdot)$ over Y. Given $\tau \geq 0$, $y_0 \in Y$ and $u \in L^\infty(\tau, +\infty; U)$, denote by $y(\cdot; \tau, y_0, u)$ the unique mild solution of (4.1) over $[\tau, +\infty)$, with the initial condition $y(\tau) = y_0$, i.e., for each $T > \tau$, $y(\cdot; \tau, y_0, u)|_{[\tau, T]}$ is the mild solution of (4.1) over $[\tau, T]$, with the initial condition $y(\tau) = y_0$. (The existence and uniqueness of such solutions are ensured by Proposition 1.3.)

(iii) The control constraint set $\mathbb{U}$ is nonempty, bounded, convex, and closed in U.

(iv) Let $Q_S = \{0\} \times Y_S$ and $Q_E = (0, +\infty) \times Y_E$, where Y_S and Y_E are two nonempty, bounded, convex, and closed in Y so that $Y_S \cap Y_E = \emptyset$.

We always assume that the problem $(TP)_{min}^{Q_S,Q_E}$ *has an optimal control.* The aim of this section is to derive the classical Pontryagin Maximum Principle of $(TP)_{min}^{Q_S,Q_E}$ by separating the target from the reachable set (at the optimal time) in the state space.

Remark 4.1 Several notes are given in order.

(i) The Pontryagin Maximum Principle of $(TP)_{min}^{Q_S,Q_E}$ is a kind of necessary condition for an optimal control. It is indeed an Euler equation associated with a minimizer of a variational problem. Thus, in studies of this subject, it is not necessary to assume the existence of optimal control in general. In this chapter, for the sake of convenience, we make this assumption.

(ii) Recall what follows: When $(0, y_0^*, t_E^*, u^*)$ is an optimal tetrad for $(TP)_{min}^{Q_S,Q_E}$, u^* is called an optimal control and t_E^* is called the optimal time (or the minimal time). We write $y^*(\cdot)$ for $y(\cdot; 0, y_0^*, u^*)$, and call it as the corresponding optimal trajectory (or the optimal state).

(iii) Throughout this chapter, we will use t_E^* to denote the optimal time of $(TP)_{min}^{Q_S,Q_E}$.

4.1.1   Geometric Intuition

Let us recall the reachable set at the time t (with $0 < t < +\infty$) for the system (4.1) (see Section 3.3.1):

$$Y_R(t) \triangleq \left\{ y(t; 0, y_0, u) \mid y_0 \in Y_S, \ u \in L(0, t; \mathbb{U}) \right\}. \tag{4.2}$$

We define a distance function $d(\cdot)$ between $Y_R(\cdot)$ and the target Y_E as follows:

$$d(t) \triangleq \text{dist}\{Y_R(t), Y_E\} = \inf_{y_1 \in Y_R(t), y_2 \in Y_E} \|y_1 - y_2\|_Y, \quad t \in [0, t_E^*]. \tag{4.3}$$

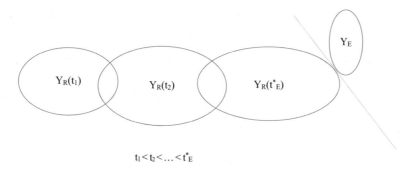

$$t_1 < t_2 < \ldots < t_E^*$$

Fig. 4.1 Evolution of reachable sets

Here, we denote $Y_R(0) \triangleq Y_S$. From geometric intuition, we observe what follows: At the initial time $t = 0$, the set Y_S is away from the target Y_E (since we assumed that $Y_E \cap Y_S = \emptyset$). Thus, we have that $d(0) > 0$. As time t goes on, the reachable set $Y_R(t)$ is getting closer and closer to the target Y_E. The first time t at which $Y_R(t)$ hits Y_E is exactly the optimal time t_E^*. (Here, we used the assumption that $(TP)_{min}^{Q_S, Q_E}$ has an optimal control.) Hence, t_E^* is the first zero point of the function $d(\cdot)$. Since $Y_R(t_E^*)$ is convex (see Lemma 4.3), there is a hyperplane so that $Y_R(t_E^*)$ and Y_E stay on two sides of this hyperplane respectively. Mathematically, we say that this hyperplane separates $Y_R(t_E^*)$ and Y_E. In this book, we prefer to say that a normal vector of this hyperplane separates $Y_R(t_E^*)$ and Y_E. In finitely dimensional cases, such hyperplane exists, provided that $Y_R(t_E^*)$ is convex. For infinitely dimensional cases, the convexity of $Y_R(t_E^*)$ cannot ensure the existence of such hyperplanes. The above-mentioned separation will play an important role in deriving the Pontryagin Maximum Principle for $(TP)_{min}^{Q_S, Q_E}$ (Figure 4.1).

4.1.2   Classical Maximum Principle

First of all, we give the definition of the classical Pontryagin Maximum Principle for the problem $(TP)_{min}^{Q_S, Q_E}$.

Definition 4.1 Several definitions are given in order.

(i) An optimal control u^* (associated with an optimal tetrad $(0, y_0^*, t_E^*, u^*)$) to $(TP)_{min}^{Q_S, Q_E}$ is said to satisfy the classical Pontryagin Maximum Principle, if there exists $\varphi(\cdot) \in C([0, t_E^*]; Y)$, with $\varphi(\cdot) \neq 0$ (i.e., φ is not a zero function), so that

$$\dot{\varphi}(t) = -A^*\varphi(t) - D(t)^*\varphi(t) \quad \text{for a.e. } t \in (0, t_E^*); \tag{4.4}$$

$$H(t, y^*(t), u^*(t), \varphi(t)) = \max_{u \in \mathbb{U}} H(t, y^*(t), u, \varphi(t)) \quad \text{for a.e. } t \in (0, t_E^*),$$

$$\text{(4.5)}$$

where $y^* \in C([0, t_E^*]; Y)$ is the corresponding optimal trajectory, and where

$$H(t, y, u, \varphi) \triangleq \langle \varphi, D(t)y + B(t)u \rangle_Y \tag{4.6}$$

for a.e. $t \in (0, t_E^*)$ and all $(y, u, \varphi) \in Y \times U \times Y$;

$$\langle \varphi(0), \ y_0 - y_0^* \rangle_Y \leq 0 \quad \text{for each} \ \ y_0 \in Y_S; \tag{4.7}$$

and

$$\langle \varphi(t_E^*), \ z - y^*(t_E^*) \rangle_Y \geq 0 \quad \text{for each} \ \ z \in Y_E. \tag{4.8}$$

Here, (4.4) is called the dual (or adjoint, or co-state) equation; (4.5) is called the maximum condition; the function $H(\cdot)$ defined by (4.6) is called the Hamiltonian (associated with $(TP)_{min}^{Q_S, Q_E}$); (4.7) and (4.8) are called the transversality conditions.

(ii) The problem $(TP)_{min}^{Q_S, Q_E}$ is said to satisfy the classical Pontryagin Maximum Principle, if any optimal control satisfies the classical Pontryagin Maximum Principle.

Recall Definition 1.7 for the separability. We have the following result:

Theorem 4.1 *The problem $(TP)_{min}^{Q_S, Q_E}$ satisfies the classical Pontryagin Maximum Principle if and only if $Y_R(t_E^*)$ and Y_E are separable in Y.*

Remark 4.2 Several notes are given in order.

(i) From Theorem 4.1, we see that the key to get the classical Pontryagin Maximum Principle is to find conditions so that $Y_R(t_E^*)$ and Y_E are separable in Y.

(ii) The proof of Theorem 4.1 is based on the classical separation (of reachable set at the optimal time and the target) and the representation formula in the next lemma, which is indeed Proposition 5.7 in Chapter 2 of [12]. We omit its proof.

Lemma 4.1 *Let $T > 0$ and $\psi_T \in Y$. Let ψ be the solution of the equation: $\dot{\psi}(t) = -A^* \psi(t) - D(t)^* \psi(t), t \in (0, T)$, with $\psi(T) = \psi_T$. Then for each $y_0 \in Y$ and each $u \in L(0, T; \mathbb{U})$, the following representation formula holds:*

$$\langle y(T; 0, y_0, u), \psi_T \rangle_Y = \langle y_0, \psi(0) \rangle_Y + \int_0^T \langle B(s)^* \psi(s), u(s) \rangle_U \, ds. \tag{4.9}$$

We now are in a position to prove Theorem 4.1.

Proof (Proof of Theorem 4.1) We first show the sufficiency. Assume that $Y_R(t_E^*)$ and Y_E are separable in Y. Let $(0, y_0^*, t_E^*, u^*)$ be an optimal tetrad to $(TP)_{min}^{Q_S, Q_E}$. Let

y^* be the corresponding optimal trajectory. It follows by Definition 1.7 that there is $\varphi_0 \in Y$ with $\|\varphi_0\|_Y = 1$ and $C \in \mathbb{R}$ so that

$$\langle z_1, \varphi_0 \rangle_Y \geq C \geq \langle z_2, \varphi_0 \rangle_Y \quad \text{for all} \ \ z_1 \in Y_R(t_E^*) \ \text{and} \ z_2 \in Y_E. \tag{4.10}$$

Since

$$y^*(t_E^*) \in Y_R(t_E^*) \cap Y_E,$$

it follows from (4.10) that

$$\min_{z_1 \in Y_R(t_E^*)} \langle z_1, \varphi_0 \rangle_Y = \langle y^*(t_E^*), \varphi_0 \rangle_Y = \max_{z_2 \in Y_E} \langle z_2, \varphi_0 \rangle_Y. \tag{4.11}$$

Let φ be the solution to the Equation (4.4) with $\varphi(t_E^*) = -\varphi_0$. Then by Lemma 4.1, we see that for each $y_0 \in Y_S$ and $u \in L(0, t_E^*; \mathbb{U})$,

$$\langle y(t_E^*; 0, y_0, u), -\varphi_0 \rangle_Y = \langle y_0, \varphi(0) \rangle_Y + \int_0^{t_E^*} \langle B(t)^* \varphi(t), u(t) \rangle_U \, dt. \tag{4.12}$$

By the first equality of (4.11) and (4.12), we get that

$$\langle y_0^*, \varphi(0) \rangle_Y = \max_{y_0 \in Y_S} \langle y_0, \varphi(0) \rangle_Y \tag{4.13}$$

and

$$\max_{u(\cdot) \in L(0, t_E^*; \mathbb{U})} \int_0^{t_E^*} \langle B(t)^* \varphi(t), u(t) \rangle_U \, dt = \int_0^{t_E^*} \langle B(t)^* \varphi(t), u^*(t) \rangle_U \, dt. \tag{4.14}$$

We now claim that

$$\langle B(t)^* \varphi(t), u^*(t) \rangle_U = \max_{u \in \mathbb{U}} \langle B(t)^* \varphi(t), u \rangle_U \quad \text{a.e.} \ t \in (0, t_E^*). \tag{4.15}$$

Indeed, since U is separable, by Proposition 1.1, there exists a countable subset $\mathbb{U}_0 = \{u_\ell\}_{\ell \geq 1}$ so that $\mathbb{U}_0$ is dense in $\mathbb{U}$. From (4.14) it follows that for each $u \in L(0, t_E^*; \mathbb{U})$,

$$\int_0^{t_E^*} [\langle B(t)^* \varphi(t), u^*(t) \rangle_U - \langle B(t)^* \varphi(t), u(t) \rangle_U] \, dt \geq 0. \tag{4.16}$$

For each $u_\ell \in \mathbb{U}_0$, we define the following function:

$$g_\ell(t) \triangleq \langle B(t)^* \varphi(t), u^*(t) \rangle_U - \langle B(t)^* \varphi(t), u_\ell \rangle_U, \quad t \in (0, t_E^*).$$

Then $g_\ell(\cdot) \in L^1(0, t_E^*)$. Thus, there exists a measurable set $E_\ell \subseteq (0, t_E^*)$ with $|E_\ell| = t_E^*$, so that any point in E_ℓ is a Lebesgue point of $g_\ell(\cdot)$. Namely,

$$\lim_{\delta \to 0+} \frac{1}{\delta} \int_{t-\delta}^{t+\delta} |g_\ell(s) - g_\ell(t)| ds = 0 \quad \text{for each } t \in E_\ell.$$

Now, for any $t \in E_\ell$ and $\delta > 0$, we define

$$u_\delta(s) \triangleq \begin{cases} u^*(s), & \text{if } s \in (0, t_E^*) \setminus B_\delta(t), \\ u_\ell, & \text{if } s \in (0, t_E^*) \cap B_\delta(t). \end{cases}$$

Then, by (4.16), we get that

$$\int_{(0, t_E^*) \cap B_\delta(t)} [\langle B(s)^* \varphi(s), u^*(s) \rangle_U - \langle B(s)^* \varphi(s), u_\ell \rangle_U] ds \geq 0.$$

Dividing the above by $\delta > 0$ and then sending $\delta \to 0$, we obtain that $g_\ell(t) \geq 0$. From this, we see that

$$\langle B(t)^* \varphi(t), u^*(t) \rangle_U - \langle B(t)^* \varphi(t), u_\ell \rangle_U \geq 0$$
$$\text{for all } t \in E \triangleq \bigcap_{\ell \geq 1} E_\ell \text{ and } u_\ell \in \mathbb{U}_0. \tag{4.17}$$

Since $\mathbb{U}_0$ is countable and dense in $\mathbb{U}$, by (4.17), we have that $|E| = t_E^*$ and that

$$\langle B(t)^* \varphi(t), u^*(t) \rangle_U - \langle B(t)^* \varphi(t), u \rangle_U \geq 0 \quad \text{for all } t \in E \text{ and } u \in \mathbb{U}.$$

From these, we obtain (4.15).

Finally, (4.5), (4.7), and (4.8) follow from (4.15), (4.13), and the second equality of (4.11), respectively. Then by Definition 4.1, u^* satisfies the classical Pontryagin Maximum Principle, consequently, so does $(TP)_{min}^{Q_S, Q_E}$. Hence, we have proved the sufficiency.

We next show the necessity. Suppose that $(TP)_{min}^{Q_S, Q_E}$ holds the classical Pontryagin Maximum Principle. Let $(0, y_0^*, t_E^*, u^*)$ be an optimal tetrad to $(TP)_{min}^{Q_S, Q_E}$. Then by Definition 4.1, there is $\varphi(\cdot) \in C([0, t_E^*]; Y)$ with $\varphi(\cdot) \neq 0$ so that (4.4)–(4.8) hold. Then by (4.7) and (4.5), we obtain that

$$\langle \Phi(t_E^*, 0)^* \varphi(t_E^*), y_0^* \rangle_Y = \max_{y_0 \in Y_S} \langle \Phi(t_E^*, 0)^* \varphi(t_E^*), y_0 \rangle_Y$$

and

$$\int_0^{t_E^*} \langle \Phi(t_E^*, s)^* \varphi(t_E^*), B(s) u^*(s) \rangle_Y ds = \int_0^{t_E^*} \max_{u \in \mathbb{U}} \langle \Phi(t_E^*, s)^* \varphi(t_E^*), B(s) u \rangle_Y ds.$$

From the above two equalities it follows that

$$
\left\langle \varphi(t_E^*), \Phi(t_E^*, 0)y_0^* + \int_0^{t_E^*} \Phi(t_E^*, s)B(s)u^*(s)ds \right\rangle_Y
$$

$$
\geq \left\langle \varphi(t_E^*), \Phi(t_E^*, 0)y_0 + \int_0^{t_E^*} \Phi(t_E^*, s)B(s)u(s)ds \right\rangle_Y
$$

for all $y_0 \in Y_S$ and $u \in L(0, t_E^*; \mathbb{U})$. This, along with the definition of $Y_R(t_E^*)$ (see (4.2)), implies that

$$
\left\langle \varphi(t_E^*), y^*(t_E^*) \right\rangle_Y \geq \sup_{z \in Y_R(t_E^*)} \left\langle \varphi(t_E^*), z \right\rangle_Y. \tag{4.18}
$$

Since $y^*(t_E^*) \in Y_R(t_E^*)$, it follows by (4.18) that

$$
\left\langle \varphi(t_E^*), y^*(t_E^*) \right\rangle_Y = \max_{z \in Y_R(t_E^*)} \left\langle \varphi(t_E^*), z \right\rangle_Y. \tag{4.19}
$$

Since $y^*(t_E^*) \in Y_E$, it follows from (4.8) and (4.19) that

$$
\min_{z \in Y_E} \left\langle \varphi(t_E^*), z \right\rangle_Y = \left\langle \varphi(t_E^*), y^*(t_E^*) \right\rangle_Y = \max_{z \in Y_R(t_E^*)} \left\langle \varphi(t_E^*), z \right\rangle_Y.
$$

This shows that $Y_R(t_E^*)$ and Y_E are separable in Y. Thus, we have proved the necessity.

Hence, we complete the proof of Theorem 4.1.   □

Remark 4.3 Several notes are given in order.

(i) The definition of the classical Pontryagin Maximum Principle (see Definition 4.1) can be extended easily to the case where the controlled system is the nonlinear system (2.1). In that case, we only need to define the Hamiltonian H in the following way:

$$
H(t, y, u, \varphi) = \langle \varphi, f(t, y, u) \rangle_Y \quad \text{for each} \ (t, y, u, \varphi) \in (0, t_E^*) \times Y \times U \times Y.
$$

(ii) It follows from (4.6) that the maximum condition (4.5) is indeed as:

$$
\langle u^*(t), B(t)^*\varphi(t) \rangle_U = \max_{u \in \mathbb{U}} \langle u, B(t)^*\varphi(t) \rangle_U \quad \text{for a.e.} \ t \in (0, t_E^*). \tag{4.20}
$$

The corresponding separating vector (from $Y_R(t_E^*)$ and Y_E) is $-\varphi(t_E^*) \triangleq \varphi_0$. It is also called a separating vector w.r.t. u^*. In general, $\varphi(\cdot) \neq 0$ cannot ensure that $B(\cdot)^*\varphi(\cdot) \neq 0$. When $B(t)^*\varphi(t) = 0$ for a.e. $t \in (0, t_E^*)$, (4.20) provides nothing about the optimal control u^*. In this case, φ_0 is called a non-qualified separating vector w.r.t. u^*. When $B(t)^*\varphi(t) \neq 0$ for a.e. $t \in (0, t_E^*)$, (4.20) may provide information for u^* over the whole interval $(0, t_E^*)$. In this case, φ_0 is called a qualified separating vector w.r.t. u^*. When $B(t)^*\varphi(t) \neq 0$ for a.e. $t \in G$, with $G \subseteq (0, t_E^*)$ a measurable set with $0 < |G| < t_E^*$, and $B(t)^*\varphi(t) = 0$ for a.e. $t \in (0, t_E^*) \setminus G$, (4.20) may provide information for u^* over G. In this case, φ_0 is called a semi-qualified separating vector w.r.t. u^*.

(iii) It may happen that for an optimal control, there are two separating vectors w.r.t. the optimal control. One is qualified, while another is non-qualified. These can be seen from the next Example 4.1.

(iv) When an optimal control holds the maximum condition, any separating vector is qualified, provided that the following condition ($\mathscr{H}_q$) is true: If a solution $\varphi(\cdot) \in C([0, t_E^*]; Y)$ to (4.4) satisfies that $B(\cdot)^*\varphi(\cdot) = 0$ over a subset G of positive measure, then $\varphi(t) = 0$ for a.e. $t \in (0, t_E^*)$.

(v) When $D(\cdot) = 0$, the operator A generates an analytic semigroup, $B(\cdot) \equiv B$ and the system (4.1) is L^∞-null controllable for each interval (i.e., for any $T_2 > T_1 \geq 0$ and $y_0 \in Y$, there exists $u \in L^\infty(T_1, T_2; U)$ so that $y(T_2; T_1, y_0, u) = 0$), the condition ($\mathscr{H}_q$) (see (iv) of Remark 4.3) holds. To prove it, two facts are given in order: First, for any solution φ to the adjoint equation with $\varphi(t_E^*) \in Y$, the function $t \to B^*\varphi(t)$ is real analytic over $[0, t_E^*]$. This follows from the analyticity of the semigroup. Second, by the null controllability, we observe from Theorem 1.20 that for any $t \in [0, t_E^*)$, there is a constant $C(t) > 0$ so that

$$\|\varphi(t)\|_Y \leq C(t) \int_t^{t_E^*} \|B^*\varphi(s)\|_U \, ds$$

for any solution φ to the adjoint equation with $\varphi(t_E^*) \in Y$. These two facts clearly imply the condition ($\mathscr{H}_q$) given in (iv) of this remark.

Example 4.1 Let

$$B = \begin{pmatrix} 1 \\ 0 \end{pmatrix}, \quad y_0 = \begin{pmatrix} 1 \\ 0 \end{pmatrix}.$$

Consider the problem $(TP)_{min}^{Q_S, Q_E}$, where the controlled system is as:

$$\dot{y}(t) = Bu(t), \quad t \in (0, +\infty), \quad \text{with } y(t) \in \mathbb{R}^2, u(t) \in \mathbb{R},$$

and where

$$Q_S = \{0\} \times \{y_0\}, \quad Q_E = (0, +\infty) \times \{0\}, \quad U = \{u \in \mathbb{R} : |u| \leq 1\}.$$

One can easily check that $(TP)_{min}^{Q_S, Q_E}$ can be put into the framework $(\mathscr{A}_{TP})$ (given at the beginning of Section 4.1). By direct calculations, we can find that

$$t_E^* = 1, \quad u^*(\cdot) \equiv -1, \quad y^*(t) = (1 - t, 0)^\top, \quad t \in [0, 1],$$

and that both $\varphi_0 = (0, 1)^\top$ and $\tilde{\varphi}_0 = (1, 0)^\top$ are separating vectors w.r.t. u^*, the first one is non-qualified and the second one is qualified. We end Example 4.1.

The next example shows that in certain cases, any separating vector w.r.t. any optimal control is qualified.

Example 4.2 Consider the problem $(TP)_{min}^{Q_S,Q_E}$, where the controlled system is (3.59) (with $f = 0$) and where

$$Q_S = \{0\} \times \{y_0\} \text{ (with } y_0 \in L^2(\Omega) \setminus B_r(0)); \quad Q_E = (0, +\infty) \times B_r(0); \quad \mathbb{U} = B_\rho(0).$$

One can easily check that $(TP)_{min}^{Q_S,Q_E}$ can be put into the framework $(\mathscr{A}_{TP})$ (given at the beginning of Section 4.1). Moreover, by the result in Remark 3.1 (after Theorem 3.5), $(TP)_{min}^{Q_S,Q_E}$ has an admissible control. Then using the similar arguments as those in the proof of Theorem 3.11, we can easily show that $(TP)_{min}^{Q_S,Q_E}$ has an optimal control.

Since the target is a closed ball, we can use the Hahn-Banach Theorem (see Theorem 1.11) to find that the target and the reachable set at t_E^* are separable in $L^2(\Omega)$. Then by Theorem 4.1, this problem holds the classical Pontryagin Maximum Principle. Besides, since this controlled equation is null controllable (see Theorems 1.22 and 1.21), and because $\{e^{At}\}_{t \geq 0}$ is an analytic semigroup, we find that any separating vector w.r.t. any optimal control is qualified. We now end Example 4.2.

The next example gives a non-qualified separating vector w.r.t. an optimal control.

Example 4.3 Let

$$A = \begin{pmatrix} -1 & 0 \\ 0 & -1 \end{pmatrix}; \quad B = \begin{pmatrix} 1 \\ 0 \end{pmatrix}; \quad E = \text{span} \begin{pmatrix} 0 \\ 1 \end{pmatrix}; \quad y_0 = \begin{pmatrix} 0 \\ 1 \end{pmatrix}.$$

Let $B_{1/10}(0)$ be the closed ball in $\mathbb{R}^2$, centered at the origin, and of radius $1/10$. Consider the problem $(TP)_{min}^{Q_S,Q_E}$, where the controlled system is as:

$$\dot{y}(t) = Ay(t) + Bu(t), \quad t \in (0, +\infty), \quad \text{with } y(t) \in \mathbb{R}^2, u(t) \in \mathbb{R},$$

and where

$$Q_S = \{0\} \times \{y_0\}, \quad Q_E = (0, +\infty) \times (E \cap B_{1/10}(0)), \quad \mathbb{U} = \{u \in \mathbb{R} : |u| \leq 1\}.$$

One can easily check that $(TP)_{min}^{Q_S,Q_E}$ can be put into the framework $(\mathscr{A}_{TP})$ (given at the beginning of Section 4.1).

After some direct computation, we find the following facts on this problem: First, $u^* \equiv 0$ is an optimal control and $\ln 10$ is the optimal time; Second, the vector

$(0, 1)^\top$ separates the target $E \cap B_{1/10}(0)$ from the reachable set at time $\ln 10$. Hence, by Theorem 4.1, this problem holds the classical Pontryagin Maximum Principle; Third, the solution φ to the equation:

$$\dot{\varphi}(t) = -A^* \varphi(t), \quad t \in [0, \ln 10]; \quad \varphi(\ln 10) = (0, -1)^\top,$$

satisfies that $B^* \varphi \equiv 0$ over $[0, \ln 10]$. Hence, the separating vector $(0, 1)^\top$ is non-qualified w.r.t. the optimal control u^*. We end Example 4.3.

The following example is taken from [9] (see Sections 2.6, 2.7 in Chapter 2 of [9]). We will provide some conclusions and omit their proofs.

Example 4.4 Consider the controlled equation:

$$\begin{cases} \partial_t y(x, t) = -\partial_x y(x, t) + u(x, t), & 0 \le x, t < +\infty, \\ y(x, 0) = y_0(x), \quad y(0, t) = 0, \end{cases} \tag{4.21}$$

where $u \in L^\infty(0, +\infty; L^2(0, +\infty))$ and $u(\cdot, t) \in B_1(0)$ for a.e. $t \in (0, +\infty)$. (Here, $B_1(0)$ is the closed unit ball in $L^2(0, +\infty)$.) This is a controlled transport equation which can be put into the framework (4.1) in the following manner: Let $Y = U = L^2(0, +\infty)$; let $B = I$ (the identity operator from $L^2(0, +\infty)$ to $L^2(0, +\infty)$); $D(t) \equiv 0$; let A be the operator on Y defined by

$$Ay(x) = -y'(x), \quad x \in (0, +\infty)$$

with the domain

$$\mathscr{D}(A) = \{\text{all absolutely continuous } y(\cdot), \text{ with } y'(\cdot) \in L^2(0, +\infty) \text{ and } y(0) = 0\}.$$

The operator A generates a C_0 semigroup $\{S(t)\}_{t \ge 0}$ which is expressed by

$$S(t)y(x) = \begin{cases} y(x - t) & as \ x \ge t, \\ 0 & as \ x < t. \end{cases}$$

Consider the problem $(TP)_{min}^{Q_S, Q_E}$, where the controlled system is (4.21) and where

$$Q_S = \{0\} \times \{0\}; \quad Q_E = (0, +\infty) \times \{\hat{y}\} \text{ with } \hat{y} \in L^2(0, +\infty) \text{ and } \hat{y} \ne 0; \quad \mathbb{U} = B_1(0).$$

One can easily check that $(TP)_{min}^{Q_S, Q_E}$ can be put into the framework $(\mathscr{A}_{TP})$ (given at the beginning of Section 4.1).

We are still in Example 4.4. The next proposition is exactly Theorem 2.7.1 in [9] (see Page 87 in [9]).

Proposition 4.1 *Given $T > 0$ and $0 < \delta < T$, there is $\hat{y} \in L^2(0, +\infty)$ so that the control $\hat{u} \in L^\infty(0, T; L^2(0, +\infty))$ driving 0 to $\hat{y}$ in the optimal time T satisfies that*

$$\widehat{u}(t) = \frac{S(T-t)^*\varphi_0}{\|S(T-t)^*\varphi_0\|_{L^2(0,+\infty)}}, \quad T-\delta < t \leq T, \tag{4.22}$$

with $\varphi_0 \in L^2(0, +\infty) \setminus \{0\}$ so that

$$\varphi(t) \triangleq \begin{cases} S(T-t)^*\varphi_0 \neq 0 & as \ T-\delta < t \leq T, \\ S(T-t)^*\varphi_0 = 0 & as \ 0 \leq t \leq T - \delta. \end{cases} \tag{4.23}$$

From Proposition 4.1, we see that for an arbitrarily fixed $T > 0$, we can choose a target $\widehat{y}$ so that the corresponding problem $(TP)_{min}^{Q_S,Q_E}$ has the optimal time T and an optimal control $\widehat{u}$ which satisfies (4.22) and (4.23). So $(0, 0, T, \widehat{u})$ is an optimal tetrad to this problem.

Meanwhile, by (4.22) and (4.23), we can easily see that

$$\langle \widehat{u}(t), \varphi(t) \rangle_{L^2(0,+\infty)} = \max_{u \in \mathbb{U}} \langle u, \varphi(t) \rangle_{L^2(0,+\infty)} \quad \text{for a.e. } t \in (0, T), \tag{4.24}$$

where $\varphi(t) \triangleq S(T-t)^*\varphi_0, t \in [0, T]$, is the solution of the adjoint equation with the terminal condition that $\varphi(T) = \varphi_0$. Since $\varphi(\cdot) \neq 0$ (see (4.23)), the equality (4.24) is exactly the classical maximum condition. Furthermore, from (4.23), we see that the separating vector φ_0 is not qualified but is semi-qualified w.r.t. $\widehat{u}$. We end Example 4.4.

At the end of this subsection, we will present an example to correct a possible error. This error is as: the derivative of any optimal trajectory $y^*(\cdot)$ at t_E^* should be a normal vector of some hyperplane separating $Y_R(t_E^*)$ and Y_E (see Figure 4.2).

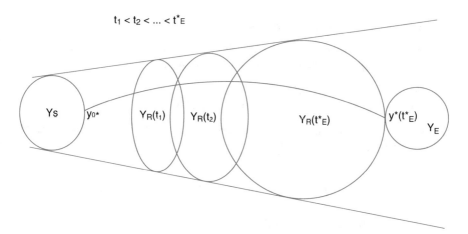

Fig. 4.2 Wrong geometric intuition

Example 4.5 Consider the problem $(TP)_{min}^{Q_S, Q_E}$, where the controlled system is as:

$$
\begin{cases}
\dfrac{d}{dt}\begin{pmatrix} y_1(t) \\ y_2(t) \end{pmatrix} = \begin{pmatrix} 1 \\ 0 \end{pmatrix} u(t), & t \in (0, +\infty), \\[3mm]
\begin{pmatrix} y_1(0) \\ y_2(0) \end{pmatrix} = \begin{pmatrix} 0 \\ 0 \end{pmatrix},
\end{cases}
\tag{4.25}
$$

$\mathbb{U} = [-1, 1]$, $Y_S = (0, 0)^\top$ and $Y_E = \{(y_1, y_2)^\top \mid (y_1 - 1)^2 + (y_2 - 1)^2 \le 1\}$.
One can easily check that $(TP)_{min}^{Q_S, Q_E}$ can be put into the framework $(\mathscr{A}_{TP})$ (given at the beginning of Section 4.1) in the current case. Furthermore, we can check that

$$
Y_R(t) = [-t, t] \times \{0\}, \quad t \in (0, +\infty); \quad t_E^* = 1;
$$

$$
u^*(t) = 1, \quad y^*(t) = (t, 0)^\top, \quad t \in (0, t_E^*).
$$

Any hyperplane, which separates $Y_R(t_E^*)$ and Y_E, can be written as

$$
\left\{ y \in \mathbb{R}^2 \mid \langle \psi, y \rangle_{\mathbb{R}^2} = 0 \right\}, \quad \text{with } \psi = c(0, 1)^\top \text{ and } c \in \mathbb{R} \setminus \{0\}.
$$

It is obvious that $\dot{y}^*(t_E^*) = (1, 0)^\top$, which is not parallel to the above ψ. Now we end Example 4.5.

4.1.3   Conditions on Separation in Classical Maximum Principle

The main result of this subsection is as follows:

Theorem 4.2 *For the problem* $(TP)_{min}^{Q_S, Q_E}$, *the set* $Y_R(t_E^*)$ *and the target* Y_E *are separable in* Y, *if one of the following conditions holds:*

(i) *The space* Y *is of finite dimension.*
(ii) *The space* Y *is of infinite dimension and* $Int(Y_R(t_E^*) - Y_E) \ne \emptyset$.

To prove Theorem 4.2, some preliminaries are needed. First, we introduce the following Hausdorff metric in $2^Y \setminus \{\emptyset\}$:

$$
\rho_H(Y_1, Y_2) \triangleq \frac{1}{2} \left\{ \sup_{y_1 \in Y_1} \text{dist}(y_1, Y_2) + \sup_{y_2 \in Y_2} \text{dist}(y_2, Y_1) \right\},
$$

$$
Y_1, Y_2 \subseteq Y, \ Y_1 \ne \emptyset, \ Y_2 \ne \emptyset.
$$

Lemma 4.2 *It holds that*

$$
\lim_{s \to t} \rho_H(Y_R(s), Y_R(t)) = 0 \ \text{for all } t \in (0, +\infty).
$$

Proof We only prove that

$$\lim_{s \to t+} \rho_H(Y_R(s), Y_R(t)) = 0.$$

The similar arguments can be applied to the case when $s \to t-$.

First, we claim that $Y_R(\cdot)$ is locally bounded, i.e., for any given $T \in (0, +\infty)$, there is $C(T) > 0$ so that

$$\|y(t; 0, y_0, u)\|_Y \le C(T) \quad \text{for all } t \in (0, T], \ y_0 \in Y_S \text{ and } u \in L(0, t; \mathbb{U}). \tag{4.26}$$

(Here and throughout the proof of this lemma, $C(T)$ denotes a generic positive constant dependent on T.) Indeed, by (iv) in $(\mathscr{A}_{TP})$ (given at the beginning of Section 4.1), there is a positive constant C so that $\sup_{y_0 \in Y_S} \|y_0\|_Y \le C$. (Here and throughout the proof of this lemma, C is a generic positive constant.) This, together with (ii) and (iii) in $(\mathscr{A}_{TP})$ (given at the beginning of Section 4.1), implies that for any $t \in (0, T]$, $y_0 \in Y_S$ and $u \in L(0, t; \mathbb{U})$,

$$\|y(t; 0, y_0, u)\|_Y = \left\| e^{tA} y_0 + \int_0^t e^{(t-\tau)A} \Big[D(\tau) y(\tau; 0, y_0, u) + B(\tau) u(\tau) \Big] d\tau \right\|_Y$$
$$\le C(T) + C(T) \int_0^t \|y(\tau; 0, y_0, u)\|_Y d\tau,$$

which, along with the Gronwall's inequality, yields (4.26).

Next, we arbitrarily fix $t \in (0, +\infty)$. Take $t < s < 2t$ and $y_1 \in Y_R(s)$. Then there is $\widehat{y}_0 \in Y_S$ and $u \in L(0, s; \mathbb{U})$ so that

$$y_1 = y(s; 0, \widehat{y}_0, u). \tag{4.27}$$

Denote $\widehat{y}(\cdot) \triangleq y(\cdot; 0, \widehat{y}_0, u)$. Then we have that

$$\widehat{y}(s) = e^{A(s-t)} \widehat{y}(t) + \int_t^s e^{A(s-\tau)} \Big[D(\tau) \widehat{y}(\tau) + B(\tau) u(\tau) \Big] d\tau. \tag{4.28}$$

Note that $\text{dist}(y_1, Y_R(t)) \le \|y_1 - \widehat{y}(t)\|_Y$. From this, (4.27), (4.28), and (ii) in $(\mathscr{A}_{TP})$ (given at the beginning of Section 4.1), it follows that

$$\text{dist}(y_1, Y_R(t))$$
$$\le \left\| e^{A(s-t)} \widehat{y}(t) - \widehat{y}(t) + \int_t^s e^{A(s-\tau)} \Big[D(\tau) \widehat{y}(\tau) + B(\tau) u(\tau) \Big] d\tau \right\|_Y \tag{4.29}$$
$$\le \left\| e^{A(s-t)} - I \right\|_{\mathscr{L}(Y)} \|\widehat{y}(t)\|_Y + C \int_t^s \|e^{A(s-\tau)}\|_{\mathscr{L}(Y)} (\|y(\tau; 0, \widehat{y}_0, u)\|_Y + 1) d\tau.$$

By (4.29) and (4.26), we obtain that

$$\text{dist}(y_1, Y_R(t)) \leq C(t)(\|e^{A(s-t)} - I\|_{\mathscr{L}(Y)} + s - t) \triangleq \gamma(s, t). \qquad (4.30)$$

Similarly, for any $y_2 \in Y_R(t)$, we can show that

$$\text{dist}(y_2, Y_R(s)) \leq \gamma(s, t),$$

which, combined with (4.30), indicates that

$$\rho_H(Y_R(s), Y_R(t)) \leq \gamma(s, t) \to 0 \quad \text{as} \quad s \to t+.$$

Hence, we end the proof of Lemma 4.2.   □

The next lemma presents some properties on the set $Y_R(t)$ with $t \in (0, +\infty)$.

Lemma 4.3 *For each* $t \in (0, +\infty)$, *the set* $Y_R(t)$ *is bounded, convex, and closed.*

Proof Arbitrarily fix $t \in (0, +\infty)$. The boundedness of $Y_R(t)$ follows from (4.26). The convexity of $Y_R(t)$ follows from (4.1), (4.2), (iii), and (iv) in $(\mathscr{A}_{TP})$ (given at the beginning of Section 4.1).

We now prove $Y_R(t)$ is closed. Let $\{y_k\}_{k\geq 1} \subseteq Y_R(t)$ satisfy that $y_k \to \widehat{y}$ strongly in Y. By (4.2), there are two sequences $\{y_{0,k}\}_{k\geq 1} \subseteq Y_S$ and $\{u_k\}_{k\geq 1} \subseteq L(0, t; \mathbb{U})$ so that

$$y_k = \Phi(t, 0)y_{0,k} + \int_0^t \Phi(t, s)B(s)u_k(s)\mathrm{d}s \quad \text{for all} \ \ k \geq 1. \qquad (4.31)$$

Here $\Phi(\cdot, \cdot)$ denotes the evolution operator generated by $A + D(\cdot)$. Since $\{y_{0,k}\}_{k\geq 1} \subseteq Y_S$ and $\{u_k\}_{k\geq 1} \subseteq L(0, t; \mathbb{U})$, by (i), (iii), and (iv) in $(\mathscr{A}_{TP})$ (given at the beginning of Section 4.1), there exists a subsequence of $\{k\}_{k\geq 1}$, denoted in the same manner, so that for some $\widehat{y}_0 \in Y_S$ and $\widehat{u} \in L^2(0, t; U)$,

$$y_{0,k} \to \widehat{y}_0 \ \text{weakly in} \ Y \ \text{and} \ u_k \to \widehat{u} \ \text{weakly in} \ L^2(0, t; U). \qquad (4.32)$$

From the second conclusion in (4.32), we can use the Mazur Theorem (see Theorem 1.8) to find, for each $k \geq 1$, a finite sequence of nonnegative numbers $\{\alpha_{kj}\}_{j=1}^{J_k}$ with $\sum_{j=1}^{J_k} \alpha_{kj} = 1$, so that

$$\sum_{j=1}^{J_k} \alpha_{kj} u_{k+j} \to \widehat{u} \ \text{strongly in} \ L^2(0, t; U). \qquad (4.33)$$

Since $\{u_{k+j}\}_{j=1}^{J_k} \subseteq L(0, t; \mathbb{U})$, it follows by (iii) in $(\mathscr{A}_{TP})$ and (4.33) that

$$\widehat{u} \in L(0, t; \mathbb{U}). \qquad (4.34)$$

Since $y_k \to \widehat{y}$ strongly in Y, and because of (4.31), (4.32), (4.34), and (ii) in $(\mathscr{A}_{TP})$ (given at the beginning of Section 4.1), we can pass to the limit for $k \to +\infty$ in (4.31) to get that

$$\widehat{y} = \Phi(t, 0)\widehat{y}_0 + \int_0^t \Phi(t, s)B(s)\widehat{u}(s)ds = y(t; 0, \widehat{y}_0, \widehat{u}) \in Y_R(t),$$

which implies that $Y_R(t)$ is closed.

Hence, we complete the proof of Lemma 4.3.   □

Proposition 4.2 *It holds that*

$$0 \in \partial\big(Y_R(t_E^*) - Y_E\big). \tag{4.35}$$

Proof Let

$$Z(t) \triangleq Y_R(t) - Y_E, \quad t \in [0, t_E^*]. \tag{4.36}$$

Two facts are given in order. Fact one: $Z(t)$ is convex and closed for each $t \in [0, t_E^*]$. Fact two: $Z(\cdot)$ is Hausdorff continuous. Fact one follows from Lemma 4.3 and (iv) in $(\mathscr{A}_{TP})$ (given at the beginning of Section 4.1). Fact two follows by Lemma 4.2 at once.

From the fact two, we find that for any $\delta \in (0, 1)$, there exists $t_\delta \in [0, t_E^*)$ so that

$$\rho_H(Z(t_\delta), Z(t_E^*)) \leq \delta/6. \tag{4.37}$$

Meanwhile, by the optimality of t_E^*, we see that $0 \notin Z(t_\delta)$. This, together with Theorem 1.11, and the convexity and the closedness of $Z(t_\delta)$ (see the fact one above), implies that there exists an $f \in Y$ with $\|f\|_Y = 1$ so that

$$\inf_{z \in Z(t_\delta)} \langle f, z \rangle_Y > 0. \tag{4.38}$$

Because

$$\|f\|_Y = \sup_{\|z\|_Y = 1} \langle f, z \rangle_Y \quad \text{and} \quad \|f\|_Y = 1,$$

there exists $\widehat{z} \in Y$ so that

$$\|\widehat{z}\|_Y = 1 \quad \text{and} \quad \langle f, \widehat{z} \rangle_Y \leq -1 + \delta/2. \tag{4.39}$$

By setting $z_\delta = \delta\widehat{z}$, (4.38), and (4.39), we obtain that

$$z_\delta \in B_\delta(0) \tag{4.40}$$

and

$$\langle f, z - z_\delta \rangle_Y \geq (1 - \delta/2)\delta \geq \delta/2 \ \text{for each} \ z \in Z(t_\delta).$$

From the latter inequality it follows that

$$\|z - z_\delta\|_Y \geq \delta/2 \ \text{for each} \ z \in Z(t_\delta),$$

which, combined with (4.37) and (4.36), indicates that

$$z_\delta \notin Z(t_E^*) = Y_R(t_E^*) - Y_E. \tag{4.41}$$

Since $0 \in Y_R(t_E^*) - Y_E$ and because δ was arbitrarily taken from $(0, 1)$, (4.35) follows from (4.40) and (4.41) at once.

Hence, we end the proof of Proposition 4.2.   □

We now turn to prove Theorem 4.2.

Proof (Proof of Theorem 4.2) Let $S \triangleq Y_R(t_E^*) - Y_E$ and $z_0 \triangleq 0$. By the definition of $Y_R(t_E^*)$ (see (4.2)) and $(\mathscr{A}_{TP})$ (given at the beginning of Section 4.1), one can easily check that S is a (nonempty) convex subset in Y. Meanwhile, it follows by Proposition 4.2 that $0 \in \partial S$. Thus, if (i) is true, then the desired separability follows from Theorem 1.9, while if (ii) is true, then the desired separability follows from Theorem 1.10. This ends the proof of Theorem 4.2.   □

We now explain the condition that $Int(Y_R(t_E^*) - Y_E) \neq \emptyset$. (See (ii) of Theorem 4.2.) This condition holds, if either $Y_R(t_E^*)$ or Y_E has a nonempty interior. (One can verify it directly.) For some problem $(TP)_{min}^{Q_S, Q_E}$, Y_E is already assumed to have a nonempty interior. In the case when $Int Y_E = \emptyset$, we need to consider the set $Y_R(t_E^*)$. The condition that $Int Y_R(t_E^*) \neq \emptyset$ is connected with some controllability for the controlled system. For instance, when Y_S is a singleton, this condition is equivalent to the exact controllability of the controlled system. The latter is too restrictive and is not satisfied by many controlled equations, such as the internally controlled heat equations. To relax the condition that $Int(Y_R(t_E^*) - Y_E) \neq \emptyset$, the concept of the finite codimensionality is introduced in [12]. For more details on this concept, we refer readers to Chapter 4 in [12].

We next give some examples, which may help us to understand Theorem 4.2 and Theorem 4.1 better.

Example 4.6 All minimal time control problems given in Examples 4.1, 4.2, 4.3, and 4.5 can be put into the framework $(\mathscr{A}_{TP})$ (given at the beginning of Section 4.1). And they have optimal controls. (These can be checked directly.) Thus we can apply Theorem 4.2 to find that the sets $Y_R(t_E^*)$ and Y_E in each of these problems are separable in Y. Then, according to Theorem 4.1, each of them satisfies the classical Pontryagin Maximum Principle. We end Example 4.6.

Example 4.7 Let $A(\cdot) = (a_{ij}(\cdot))_{n \times n}$ and $B(\cdot) = (b_{ij}(\cdot))_{n \times m}$ (where $a_{ij}(\cdot) \in L^\infty(0, +\infty)$ and $b_{ij}(\cdot) \in L^\infty(0, +\infty)$ for all $1 \le i \le n, 1 \le j \le m$. Let $y_0 \in \mathbb{R}^n$ and $y_1 \in \mathbb{R}^n$ satisfy that $y_0 \ne y_1$. Let $Y = \mathbb{R}^n$ and $U = \mathbb{R}^m$. Consider the problem $(TP)_{min}^{Q_S, Q_E}$, where the controlled system is as:

$$\dot{y}(t) = A(t)y(t) + B(t)u(t), \quad t \in (0, +\infty), \quad \text{with } y(t) \in \mathbb{R}^n, \ u(t) \in \mathbb{R}^m,$$

and where

$$Q_S = \{0\} \times \{y_0\}; \quad Q_E = (0, +\infty) \times \{y_1\}; \quad U = B_\rho(0) \text{ with } \rho > 0.$$

One can easily check that the corresponding minimal time control problem $(TP)_{min}^{Q_S, Q_E}$ can be put into the framework $(\mathscr{A}_{TP})$ (given at the beginning of Section 4.1).

We further assume that $(TP)_{min}^{Q_S, Q_E}$ has optimal controls. Then by Theorem 4.2, we obtain that the sets $Y_R(t_E^*)$ and Y_E in this problem are separable in Y. Hence, according to Theorem 4.1, $(TP)_{min}^{Q_S, Q_E}$ satisfies the classical Pontryagin Maximum Principle. We now end Example 4.7.

Example 4.8 Let the controlled system be the heat equation (3.107) with $a(\cdot, \cdot) \in C(\overline{\Omega} \times [0, +\infty)) \cap L^\infty(0, +\infty; L^\infty(\Omega))$. Let $Y = L^2(\Omega)$ and $U = L^2(\Omega)$. Let

$$Q_S = \{0\} \times \{y_0\}; \quad Q_E = (0, +\infty) \times B_r(0) \text{ and } \mathbb{U} = B_\rho(0),$$

where $y_0 \in L^2(\Omega) \setminus B_r(0)$, $r > 0$, and $\rho > 0$. One can easily check that the corresponding minimal time control problem $(TP)_{min}^{Q_S, Q_E}$ can be put into the framework $(\mathscr{A}_{TP})$ (given at the beginning of Section 4.1).

We further assume that $(TP)_{min}^{Q_S, Q_E}$ has optimal controls. Then by Theorem 4.2, we see that the sets $Y_R(t_E^*)$ and Y_E in this problem are separable in Y. Hence, according to Theorem 4.1, $(TP)_{min}^{Q_S, Q_E}$ satisfies the classical Pontryagin Maximum Principle. We end Example 4.8.

The final example gives a minimal time control problem satisfying what follows:

  (i) It is under the framework $(\mathscr{A}_{TP})$ (given at the beginning of Section 4.1);
 (ii) It has an optimal control;
(iii) It does not satisfy the classical Pontryagin Maximum Principle;
(iv) It satisfies the local Pontryagin Maximum Principle (which will be introduced in Definition 4.2 in the next section).

The proof of the above conclusion *(iv)* will be given in Example 4.10 of the next section. Because of the above *(iv)*, we introduce the local Pontryagin Maximum Principle in the next subsection.

Example 4.9 Consider the following controlled heat equation:

$$\begin{cases} \partial_t y - \Delta y = u & \text{in } \Omega \times (0, +\infty), \\ y = 0 & \text{on } \partial\Omega \times (0, +\infty), \end{cases} \tag{4.42}$$

where $\Omega = (0, \pi)$. As what we did in Section 3.2.2, the Equation (4.42) can be put into framework (4.1). Thus, (4.42) can be rewritten as

$$\dot{y} - \Delta y = u, \quad t \in (0, +\infty). \tag{4.43}$$

Let

$$e_k(x) \triangleq \sqrt{2/\pi} \sin kx, \quad x \in [0, \pi].$$

(It constitutes an orthonormal basis in $L^2(\Omega)$.) Let

$$z \triangleq \sum_{k=1}^{+\infty} b_k e_k, \quad \text{with } b_k \triangleq \int_0^1 \frac{1}{d(t)} e^{-2k^2(1-t)} dt \quad \text{for all } k \geq 1, \tag{4.44}$$

where

$$d(t) \triangleq \left(\sum_{k=1}^{+\infty} e^{-2k^2(1-t)} \right)^{\frac{1}{2}}, \quad t \in [0, 1). \tag{4.45}$$

Let

$$Y = U = L^2(\Omega); \quad Q_S \triangleq \{(0, 0)\}; \quad Q_E = (0, +\infty) \times \{z\}; \quad \mathbb{U} = \{u \in U \mid \|u\|_U \leq 1\}.$$

One can easily check that the corresponding minimal time control problem $(TP)_{min}^{Q_S, Q_E}$ can be put into the framework $(\mathscr{A}_{TP})$ (given at the beginning of Section 4.1). We will see later that this problem has an optimal control.

First we claim that there exists a control $\tilde{u} \in L(0, 1; \mathbb{U})$ so that

$$y(1; 0, 0, \tilde{u}) = z. \tag{4.46}$$

To this end, we define

$$\tilde{\varphi}(t) \triangleq \sum_{k=1}^{+\infty} e^{-k^2(1-t)} e_k, \quad t \in (-\infty, 1). \tag{4.47}$$

It is obvious that $\tilde{\varphi}(\cdot)$ satisfies

$$\begin{cases} \partial_t \tilde{\varphi} + \Delta\tilde{\varphi} = 0 & \text{in } \Omega \times (-\infty, 1), \\ \tilde{\varphi} = 0 & \text{on } \partial\Omega \times (-\infty, 1). \end{cases} \tag{4.48}$$

Since

$$\int_0^1 \|\widetilde{\varphi}(t)\|_{L^2(\Omega)} dt = \int_0^1 \left(\sum_{k=1}^{+\infty} e^{-2k^2(1-t)} \right)^{\frac{1}{2}} dt \leq \int_0^1 \left(\sum_{k=1}^{+\infty} e^{-2k(1-t)} \right)^{\frac{1}{2}} dt$$
$$= \int_0^1 \frac{e^{-(1-t)}}{\sqrt{1 - e^{-2(1-t)}}} dt < +\infty,$$

(4.49)

we see that $\widetilde{\varphi} \in L^1(0, 1; L^2(\Omega))$. Now, we define the following function:

$$\widetilde{u}(t) \triangleq \widetilde{\varphi}(t)/\|\widetilde{\varphi}(t)\|_{L^2(\Omega)} \quad \text{for a.e. } t \in (0, 1). \tag{4.50}$$

Then $\widetilde{u} \in L^\infty(0, 1; L^2(\Omega))$ and

$$y(1; 0, 0, \widetilde{u}) = \int_0^1 e^{(1-t)\Delta} \widetilde{\varphi}(t)/\|\widetilde{\varphi}(t)\|_{L^2(\Omega)} dt. \tag{4.51}$$

By (4.51), (4.44), and (4.45), we get that for each $k \geq 1$,

$$\langle y(1; 0, 0, \widetilde{u}) - z, e_k \rangle_{L^2(\Omega)} = \int_0^1 \langle e^{(1-t)\Delta} e_k, \widetilde{\varphi}(t)/d(t) \rangle_{L^2(\Omega)} dt - b_k$$
$$= \int_0^1 \langle e^{-k^2(1-t)} e_k, \widetilde{\varphi}(t)/d(t) \rangle_{L^2(\Omega)} dt - b_k.$$

This, together with (4.47) and (4.45), implies that

$$\langle y(1; 0, 0, \widetilde{u}) - z, e_k \rangle_{L^2(\Omega)} = 0 \quad \text{for each } k \geq 1,$$

which leads to (4.46).

Next, by (4.46) and using the similar arguments as those used in the proof of Theorem 3.11, we can find that $(TP)_{min}^{Q_S, Q_E}$ has an optimal control.

Finally, we prove the following CLAIM A: For any optimal tetrad $(0, 0, t_E^*, u^*)$, there is no $\varphi_{t_E^*}^* \in L^2(\Omega) \setminus \{0\}$ satisfying that for a.e. $t \in (0, t_E^*)$,

$$\langle \varphi^*(t), u^*(t) \rangle_{L^2(\Omega)} = \max_{u \in \mathbb{U}} \langle \varphi^*(t), u \rangle_{L^2(\Omega)} = \|\varphi^*(t)\|_{L^2(\Omega)}, \tag{4.52}$$

where $\varphi^*(\cdot) \triangleq \varphi(\cdot; t_E^*, \varphi_{t_E^*}^*)$. Here and throughout this example, we will use $\varphi(\cdot; T, \xi)$, with $T > 0$ and $\xi \in L^2(\Omega)$, to denote the solution of the following equation:

$$\begin{cases} \partial_t \varphi + \Delta \varphi = 0 & \text{in } \Omega \times (0, T), \\ \varphi = 0 & \text{on } \partial\Omega \times (0, T), \\ \varphi(T) = \xi & \text{in } \Omega. \end{cases}$$

By contradiction, suppose that CLAIM A did not hold. Then there would be an optimal tetrad $(0, 0, t_E^*, u^*)$ so that (4.52) is true for some $\varphi_{t_E^*}^* \in L^2(\Omega) \setminus \{0\}$. We will get a contradiction by the following five steps.

Step 1. We show that

$$|\langle z, \xi \rangle_{L^2(\Omega)}| \le \|\varphi(\cdot; t_E^*, \xi)\|_{L^1(0, t_E^*; L^2(\Omega))} \quad \text{for each } \xi \in L^2(\Omega). \tag{4.53}$$

Indeed, since

$$
\begin{aligned}
\langle z, \xi \rangle_{L^2(\Omega)} &= \langle y(t_E^*; 0, 0, u^*), \varphi(t_E^*; t_E^*, \xi) \rangle_{L^2(\Omega)} \\
&= \int_0^{t_E^*} \langle u^*(t), \varphi(t; t_E^*, \xi) \rangle_{L^2(\Omega)} dt \quad \text{for each } \xi \in L^2(\Omega),
\end{aligned}
\tag{4.54}
$$

and because $\|u^*(t)\|_{L^2(\Omega)} \le 1$ for a.e. $t \in (0, t_E^*)$, (4.53) follows immediately.

Step 2. We claim that

$$\sup_{\xi \in L^2(\Omega) \setminus \{0\}} \frac{\langle z, \xi \rangle_{L^2(\Omega)}}{\|\varphi(\cdot; t_E^*, \xi)\|_{L^1(0, t_E^*; L^2(\Omega))}} = 1. \tag{4.55}$$

Indeed, on one hand, it follows by (4.53) that

$$\sup_{\xi \in L^2(\Omega) \setminus \{0\}} \frac{\langle z, \xi \rangle_{L^2(\Omega)}}{\|\varphi(\cdot; t_E^*, \xi)\|_{L^1(0, t_E^*; L^2(\Omega))}} \le 1. \tag{4.56}$$

On the other hand, by (4.54) and (4.52), we obtain that

$$\langle z, \varphi_{t_E^*}^* \rangle_{L^2(\Omega)} = \int_0^{t_E^*} \langle u^*(t), \varphi^*(t) \rangle_{L^2(\Omega)} dt = \|\varphi^*\|_{L^1(0, t_E^*; L^2(\Omega))}, \tag{4.57}$$

which indicates that

$$\sup_{\xi \in L^2(\Omega) \setminus \{0\}} \frac{\langle z, \xi \rangle_{L^2(\Omega)}}{\|\varphi(\cdot; t_E^*, \xi)\|_{L^1(0, t_E^*; L^2(\Omega))}} \ge \frac{\langle z, \varphi_{t_E^*}^* \rangle_{L^2(\Omega)}}{\|\varphi^*\|_{L^1(0, t_E^*; L^2(\Omega))}} = 1.$$

This, along with (4.56), leads to (4.55).

Step 3. We prove that

$$t_E^* = 1. \tag{4.58}$$

By contradiction, we suppose that (4.58) was not true. Then, since $t_E^* \le 1$, we would have that $t_E^* < 1$. For each $\varepsilon > 0$, we define the following function:

$$\xi_\varepsilon \triangleq \sum_{k=1}^{+\infty} e^{-k^2 \varepsilon} e_k. \tag{4.59}$$

First, we show that

$$\lim_{\varepsilon \to 0+} \langle z, \xi_\varepsilon \rangle_{L^2(\Omega)} = \int_0^1 \langle \widetilde{u}(t), \widetilde{\varphi}(t) \rangle_{L^2(\Omega)} dt = \|\widetilde{\varphi}\|_{L^1(0,1;L^2(\Omega))}. \tag{4.60}$$

Indeed, by (4.46) and using the similar arguments as those used to show (4.54), we can verify that

$$\begin{aligned} \langle z, \xi_\varepsilon \rangle_{L^2(\Omega)} &= \langle y(1; 0, 0, \widetilde{u}), \varphi(1; 1, \xi_\varepsilon) \rangle_{L^2(\Omega)} \\ &= \int_0^1 \langle \widetilde{u}(t), \varphi(t; 1, \xi_\varepsilon) \rangle_{L^2(\Omega)} dt. \end{aligned} \tag{4.61}$$

Meanwhile, it follows from (4.47) and (4.59) that

$$\varphi(t; 1, \xi_\varepsilon) = \widetilde{\varphi}(t - \varepsilon) \quad \text{for each } t \in [0, 1], \tag{4.62}$$

which indicates that for a.e. $t \in (0, 1)$,

$$|\langle \widetilde{u}(t), \varphi(t; 1, \xi_\varepsilon) \rangle_{L^2(\Omega)}| \leq \|\widetilde{\varphi}(t - \varepsilon)\|_{L^2(\Omega)} \leq \|\widetilde{\varphi}(t)\|_{L^2(\Omega)}. \tag{4.63}$$

Since $\widetilde{\varphi} \in L^1(0, 1; L^2(\Omega)) \cap C([-1, 1); L^2(\Omega))$, by (4.61)–(4.63), we can apply the Lebesgue Dominated Convergence Theorem, as well as (4.50), to obtain (4.60).

Second, we show that

$$\|\varphi(\cdot; t_E^*, \xi_\varepsilon)\|_{L^1(0,t_E^*; L^2(\Omega))} = \int_{1-\varepsilon-t_E^*}^{1-\varepsilon} \|\widetilde{\varphi}(t)\|_{L^2(\Omega)} dt. \tag{4.64}$$

Indeed, by (4.47) and (4.59), we can check that

$$\varphi(t; t_E^*, \xi_\varepsilon) = \widetilde{\varphi}(1 - \varepsilon - t_E^* + t) \quad \text{for a.e. } t \in (0, t_E^*),$$

which leads to (4.64).

Next, since $\widetilde{\varphi} \in L^1(0, 1; L^2(\Omega))$, it follows by (4.64) that

$$\lim_{\varepsilon \to 0+} \|\varphi(\cdot; t_E^*, \xi_\varepsilon)\|_{L^1(0,t_E^*; L^2(\Omega))} = \|\widetilde{\varphi}\|_{L^1(1-t_E^*,1; L^2(\Omega))}. \tag{4.65}$$

Finally, by (4.60) and (4.65), we see that

$$\lim_{\varepsilon \to 0+} \frac{\langle z, \xi_\varepsilon \rangle_{L^2(\Omega)}}{\|\varphi(\cdot; t_E^*, \xi_\varepsilon)\|_{L^1(0,t_E^*; L^2(\Omega))}} = \frac{\|\widetilde{\varphi}\|_{L^1(0,1;L^2(\Omega))}}{\|\widetilde{\varphi}\|_{L^1(1-t_E^*,1; L^2(\Omega))}} > 1, \tag{4.66}$$

which contradicts (4.55). Hence, (4.58) is true.

Step 4. We show that

$$\widetilde{\varphi}(1) \in L^2(\Omega). \tag{4.67}$$

To this end, we first introduce the following problem:

$$(P_1) \qquad V_1 \triangleq \inf_{\xi \in L^2(\Omega)} \left\{ \frac{1}{2} \|\varphi(\cdot; 1, \xi)\|^2_{L^1(0,1; L^2(\Omega))} - \langle z, \xi \rangle_{L^2(\Omega)} \right\}. \qquad (4.68)$$

By (4.53) and (4.58), (P_1) is well defined. It follows from (4.68), (4.58), and (4.53) that

$$V_1 \geq -\frac{1}{2}. \qquad (4.69)$$

Meanwhile, by (4.68), (4.58), and (4.57), we obtain that

$$V_1 \leq \frac{1}{2} \|\varphi(\cdot; t_E^*, \varphi_{t_E^*}^* / \|\varphi^*\|_{L^1(0,t_E^*; L^2(\Omega))})\|^2_{L^1(0,t_E^*; L^2(\Omega))} \\ - \langle z, \varphi_{t_E^*}^* / \|\varphi^*\|_{L^1(0,t_E^*; L^2(\Omega))} \rangle_{L^2(\Omega)} = \frac{1}{2} - 1 = -\frac{1}{2}. \qquad (4.70)$$

From (4.69) and (4.70), we find that

$$V_1 = -\frac{1}{2}. \qquad (4.71)$$

Let

$$X \triangleq \{\varphi(\cdot; 1, \xi) \mid \xi \in L^2(\Omega)\} \quad \text{and} \quad Z \triangleq \overline{X}^{\|\cdot\|_{L^1(0,1; L^2(\Omega))}}.$$

Define a functional on Z by

$$F(\psi) \triangleq \langle \widetilde{u}, \psi \rangle_{L^\infty(0,1; L^2(\Omega)), L^1(0,1; L^2(\Omega))}, \qquad \psi \in Z. \qquad (4.72)$$

By (4.72) and (4.54), we can easily obtain that

$$F(\varphi(\cdot; 1, \xi)) = \langle z, \xi \rangle_{L^2(\Omega)} \quad \text{for each} \quad \xi \in L^2(\Omega). \qquad (4.73)$$

We next introduce the following problem:

$$(P_2) \qquad V_2 \triangleq \inf_{\psi \in Z} J(\psi), \qquad (4.74)$$

where $J(\cdot) : Z \mapsto \mathbb{R}$ is defined by

$$J(\psi) \triangleq \frac{1}{2} \|\psi\|^2_{L^1(0,1; L^2(\Omega))} - F(\psi), \qquad \psi \in Z. \qquad (4.75)$$

It follows from (4.68), (4.71), (4.73)–(4.75), and the density of X in Z that

$$V_2 = V_1 = -\frac{1}{2}. \qquad (4.76)$$

By (4.62) and (4.59), we have that

$$\widetilde{\varphi}(\cdot - \varepsilon) \in X. \tag{4.77}$$

Since $\widetilde{\varphi} \in L^1(0, 1; L^2(\Omega)) \cap C([-1, 1); L^2(\Omega))$, by the second inequality in (4.63) and (4.77), we can apply the Lebesgue Dominated Convergence Theorem to obtain that

$$\int_0^1 \|\widetilde{\varphi}(t - \varepsilon) - \widetilde{\varphi}(t)\|_{L^2(\Omega)} dt \to 0.$$

This implies that

$$\widetilde{\varphi} \in Z. \tag{4.78}$$

Meanwhile, by (4.45), (4.47), and (4.49), we have that

$$0 < \int_0^1 d(t) dt = \|\widetilde{\varphi}\|_{L^1(0,1;L^2(\Omega))} < +\infty. \tag{4.79}$$

This, together with (4.78), (4.75), and (4.72), implies that

$$J(\widetilde{\varphi}/\|d\|_{L^1(0,1)}) = \frac{1}{2} - \int_0^1 \langle \widetilde{u}, \widetilde{\varphi} \rangle_{L^2(\Omega)} dt / \|d\|_{L^1(0,1)},$$

which, combined with (4.50) and (4.79), indicates that

$$J(\widetilde{\varphi}/\|d\|_{L^1(0,1)}) = -\frac{1}{2}. \tag{4.80}$$

Besides, it follows from (4.75), (4.58), (4.73), and (4.70) that

$$J(\varphi(\cdot; t_E^*, \varphi_{t_E^*}^* / \|\varphi^*\|_{L^1(0,t_E^*; L^2(\Omega))})) = -\frac{1}{2}. \tag{4.81}$$

By (4.74), (4.76), (4.80), and (4.81), we see that the following two functions are solutions to the problem (P_2):

$$\widetilde{\varphi}(\cdot)/\|d\|_{L^1(0,1)} \quad \text{and} \quad \varphi(\cdot; t_E^*, \varphi_{t_E^*}^* / \|\varphi^*\|_{L^1(0,t_E^*; L^2(\Omega))}).$$

Since $J(\cdot)$ is strictly convex (see [18]), there is at most one solution to (P_2). Hence,

$$\widetilde{\varphi}(\cdot)/\|d\|_{L^1(0,1)} = \varphi(\cdot; t_E^*, \varphi_{t_E^*}^* / \|\varphi^*\|_{L^1(0,t_E^*; L^2(\Omega))}).$$

This, along with (4.58), leads to (4.67).

Step 5. We find a contradiction.

Obviously, (4.67) contradicts (4.47).

By Step 5, we see that CLAIM A is true. From it, we find that any optimal control to $(TP)_{min}^{Q_S,Q_E}$ does not satisfy the classical Pontryagin Maximum Principle. Furthermore, from (4.50), (4.46), and Step 3, we see that $(1, 0, 0, \tilde{u})$ is an optimal tetrad to $(TP)_{min}^{Q_S,Q_E}$. We now end Example 4.9.

4.2 Local Maximum Principle and Minimal Time Controls

In this section, we will study the problem $(TP)_{min}^{Q_S,Q_E}$ under the framework $(\mathscr{A}_{TP})$ (given at the beginning of Section 4.1). *As what we did in Section 4.1, we always assume that* $(TP)_{min}^{Q_S,Q_E}$ *has an optimal control.* The aim of this section is to obtain the local Pontryagin Maximum Principle of $(TP)_{min}^{Q_S,Q_E}$, through separating a reachable set from a controllable set in the state space before the optimal time.

Before giving the definition of the above-mentioned local Pontryagin Maximum Principle for the problem $(TP)_{min}^{Q_S,Q_E}$, we define, for any $0 \le t_1 < t_2 < +\infty$, the following two sets:

$$Y_C(t_1, t_2) \triangleq \left\{ y_0 \in Y \mid \exists u \in L(t_1, t_2; \mathbb{U}) \text{ so that } y(t_2; t_1, y_0, u) \in Y_E \right\}; \qquad (4.82)$$

$$\widehat{Y}_C(t_1, t_2) \triangleq \bigcup_{\tau \in (t_1, t_2)} Y_C(t_1, \tau). \qquad (4.83)$$

We call $Y_C(t_1, t_2)$ the controllable set in the interval (t_1, t_2) for the Equation (4.1).

Definition 4.2 Several definitions are given in order.

(i) An optimal control u^* (associated with an optimal tetrad $(0, y_0^*, t_E^*, u^*)$) to the problem $(TP)_{min}^{Q_S,Q_E}$ is said to satisfy the local Pontryagin Maximum Principle, if for any $T \in (0, t_E^*)$, there exists $\varphi_T(\cdot) \in C([0, T]; Y)$, with $\varphi_T(\cdot) \neq 0$ (i.e., φ_T is not a zero function), so that

$$\dot{\varphi}_T(t) = -A^*\varphi_T(t) - D(t)^*\varphi_T(t) \text{ for a.e. } t \in (0, T); \qquad (4.84)$$

$$H(t, y^*(t), u^*(t), \varphi_T(t)) = \max_{u \in \mathbb{U}} H(t, y^*(t), u, \varphi_T(t)) \text{ for a.e. } t \in (0, T), \qquad (4.85)$$

where $y^* \in C([0, t_E^*]; Y)$ is the corresponding optimal trajectory and

$$H(t, y, u, \varphi) \triangleq \langle \varphi, D(t)y + B(t)u \rangle_Y \qquad (4.86)$$

for a.e. $t \in (0, t_E^*)$ and all $(y, u, \varphi) \in Y \times U \times Y$. Besides,

$$\langle \varphi_T(0), y_0 - y_0^* \rangle_Y \le 0 \text{ for each } y_0 \in Y_S, \qquad (4.87)$$

and

$$\langle \varphi_T(T),\ z - y^*(T)\rangle_Y \geq 0 \quad \text{for each} \quad z \in Y_C(T, t_E^*). \tag{4.88}$$

Here (4.84) is called the dual (or adjoint, or co-state) equation; (4.85) is called the maximum condition; the function $H(\cdot)$ defined by (4.86) is called the Hamiltonian (associated with $(TP)_{min}^{Q_S, Q_E}$); (4.87) and (4.88) are called the transversality condition.

(ii) The problem $(TP)_{min}^{Q_S, Q_E}$ is said to satisfy the local Pontryagin Maximum Principle, if any optimal control satisfies the local Pontryagin Maximum Principle.

Remark 4.4 For the problem $(TP)_{min}^{Q_S, Q_E}$, the classical Pontryagin Maximum Principle cannot imply the local Pontryagin Maximum Principle. The reason is that a nontrivial solution to the adjoint equation on $[0, t_E^*]$ may be identically zero over $[0, T]$ for some $T < t_E^*$. This can be seen from Proposition 4.1 and (4.24). We will see from Corollary 4.2 that under some condition, the classical Pontryagin Maximum Principle implies the local Pontryagin Maximum Principle.

Example 4.10 The minimal time control problem $(TP)_{min}^{Q_S, Q_E}$ in Example 4.9 satisfies the local Pontryagin Maximum Principle. This can be verified by two steps as follows:

Step 1. This problem has the unique optimal tetrad $(0, 0, 1, \tilde{u})$.

Let $\tilde{u}$ and $\tilde{\varphi}$ be given by (4.50) and (4.47), respectively. In Example 4.9, we already see that $(0, 0, 1, \tilde{u})$ is an optimal tetrad of the minimal time control problem. To show the desired uniqueness, we arbitrarily fix an optimal control u^*. Then

$$y(1; 0, 0, u^*) = z \quad \text{and} \quad \|u^*\|_{L^\infty(0,1; L^2(\Omega))} \leq 1. \tag{4.89}$$

Because of (4.46), it follows by the equality in (4.89) that

$$\int_0^1 \langle \tilde{u}(t) - u^*(t), \tilde{\varphi}(t)\rangle_{L^2(\Omega)} dt = 0. \tag{4.90}$$

From (4.50), (4.90), and the inequality in (4.89), we obtain that

$$u^*(t) = \tilde{\varphi}(t)/\|\tilde{\varphi}(t)\|_{L^2(\Omega)} \quad \text{for a.e.} \quad t \in (0, 1),$$

which indicates that $u^*(t) = \tilde{u}(t)$ for a.e. $t \in (0, 1)$.

Step 2. $(TP)_{min}^{Q_S, Q_E}$ satisfies the local Pontryagin Maximum Principle.

By Definition 4.2 and Step 1, it suffices to show that $(0, 0, 1, \tilde{u})$ satisfies the local Pontryagin Maximum Principle. For this purpose, we arbitrarily fix $T \in (0, 1)$. Define

$$\varphi_T(t) \triangleq \tilde{\varphi}(t), \quad t \in [0, T].$$

It is obvious that $\varphi_T(\cdot) \in C([0, T]; L^2(\Omega))$ satisfies $\varphi_T(\cdot) \neq 0$ and (4.84)–(4.87) (see (4.47)–(4.50)). The reminder is to show (4.88). Arbitrarily take $\widehat{z} \in Y_C(T, 1)$. According to (4.82), there exists $\widehat{u} \in L^\infty(T, 1; L^2(\Omega))$ so that

$$y(1; T, \widehat{z}, \widehat{u}) = z \quad \text{and} \quad \|\widehat{u}\|_{L^\infty(T,1;L^2(\Omega))} \leq 1. \tag{4.91}$$

Since $y(1; T, y(T; 0, 0, \widetilde{u}), \widetilde{u}) = z$, it follows from the equality in (4.91) that

$$y(1; T, \widehat{z} - y(T; 0, 0, \widetilde{u}), \widehat{u} - \widetilde{u}) = 0,$$

which, combined with (4.50) and the inequality in (4.91), indicates that

$$-\langle \widetilde{\varphi}(T), \widehat{z} - y(T; 0, 0, \widetilde{u}) \rangle_{L^2(\Omega)} = \int_T^1 \langle \widehat{u}(t) - \widetilde{u}(t), \widetilde{\varphi}(t) \rangle_{L^2(\Omega)} dt \leq 0.$$

Then

$$\langle \varphi_T(T), \widehat{z} - y(T; 0, 0, \widetilde{u}) \rangle_{L^2(\Omega)} \geq 0 \quad \text{for each } \widehat{z} \in Y_C(T, 1).$$

This implies that (4.88) holds for this case.

From the above, we see that $(TP)_{min}^{Q_S, Q_E}$ satisfies the local Pontryagin Maximum Principle. We now end Example 4.10.

4.2.1   Some Properties on Controllable Sets and Reachable Sets

Recall (4.2), (4.82), and (4.83) for the definitions of sets $Y_R(t)$, $Y_C(t_1, t_2)$ and $\widehat{Y}_C(t_1, t_2)$. In this subsection, we will present some basic properties related to these subsets. First, by (4.82), (4.83), and the definition of the optimal time of $(TP)_{min}^{Q_S, Q_E}$, one can easily prove the next Proposition 4.3.

Proposition 4.3 *Time t^* is the optimal time t_E^* to $(TP)_{min}^{Q_S, Q_E}$ if and only if one of the following statements holds:*

(i) $Y_R(t^*) \cap Y_E \neq \emptyset$ and $Y_R(t_1) \cap Y_C(t_1, t_2) = \emptyset$ for all $0 \leq t_1 < t_2 < t^*$.
(ii) $Y_S \cap Y_C(0, t^*) \neq \emptyset$ and $Y_S \cap Y_C(0, t) = \emptyset$ for all $t \in (0, t^*)$.
(iii) $Y_S \cap Y_C(0, t^*) \neq \emptyset$ and $Y_S \cap \widehat{Y}_C(0, t^*) = \emptyset$.

Proposition 4.4 *Let $t \in [0, t_E^*)$. Then*

$$\begin{aligned}
&Y_R(t) \cap Y_C(t, t_E^*) \\
&= \left\{ y(t; 0, y_0^*, u^*) \mid (0, y_0^*, t_E^*, u^*) \text{ is an optimal tetrad for } (TP)_{min}^{Q_S, Q_E} \right\}
\end{aligned} \tag{4.92}$$

and

$$Y_R(t) \cap \widehat{Y}_C(t, t_E^*) = \emptyset. \tag{4.93}$$

Proof We only prove this proposition for the case when $t \in (0, t_E^*)$. For the case that $t = 0$, one can use the similar way, with a little modification, to show it.

First, we prove (4.92). The proof of "$\supseteq$" is obvious. We now turn to the proof of "$\subseteq$." To this end, we arbitrarily fix $z \in Y_R(t) \cap Y_C(t, t_E^*)$. Since $z \in Y_R(t)$, there is $y_0 \in Y_S$ and $u_1 \in L(0, t; \mathbb{U})$ so that

$$z = y(t; 0, y_0, u_1). \tag{4.94}$$

Meanwhile, since $z \in Y_C(t, t_E^*)$, there is $u_2 \in L(t, t_E^*; \mathbb{U})$ so that

$$y(t_E^*; t, z, u_2) \in Y_E. \tag{4.95}$$

Define the following control:

$$u(s) \triangleq \begin{cases} u_1(s), & s \in (0, t), \\ u_2(s), & s \in (t, t_E^*). \end{cases} \tag{4.96}$$

Then by (4.94)–(4.96), we can directly check that $(0, y_0, t_E^*, u)$ is an optimal tetrad for $(TP)_{min}^{Q_S, Q_E}$ and $z = y(t; 0, y_0, u)$. Hence, (4.92) is proved.

We now show (4.93). By contradiction, suppose that it was not true. Then, there would exist $\widehat{z} \in Y$ so that

$$\widehat{z} \in Y_R(t) \cap \widehat{Y}_C(t, t_E^*).$$

Since $\widehat{z} \in Y_R(t)$, there is $\widehat{y}_0 \in Y_S$ and $\widehat{u}_1 \in L(0, t; \mathbb{U})$ so that

$$\widehat{z} = y(t; 0, \widehat{y}_0, \widehat{u}_1). \tag{4.97}$$

Meanwhile, since $\widehat{z} \in \widehat{Y}_C(t, t_E^*)$, there is $t_0 \in (t, t_E^*)$ and $\widehat{u}_2 \in L(t, t_0; \mathbb{U})$ so that

$$y(t_0; t, \widehat{z}, \widehat{u}_2) \in Y_E. \tag{4.98}$$

Define the following control:

$$\widehat{u}(s) \triangleq \begin{cases} \widehat{u}_1(s), & s \in (0, t), \\ \widehat{u}_2(s), & s \in (t, t_0). \end{cases} \tag{4.99}$$

By (4.97)–(4.99), we can directly check that $(0, \widehat{y}_0, t_0, \widehat{u})$ is an admissible tetrad for $(TP)_{min}^{Q_S, Q_E}$, which contradicts that $t_0 < t_E^*$. Thus, (4.93) follows.

Hence, we end the proof of Proposition 4.4. $\qquad \square$

We next introduce the concept of viability domain and the related results.

Definition 4.3 Let S be a nonempty subset of Y. We call S a viability domain associated with (4.1) if for any $y_0 \in S$ and $\tau \in (0, +\infty)$, there is $u \in L(0, +\infty; \mathbb{U})$ so that

$$y(t; \tau, y_0, u) \in S \quad \text{for each } t \geq \tau.$$

Example 4.11 When $0 \in \mathbb{U}$, one can easily check that $\{0\}$ is a viability domain associated with (4.1).

Proposition 4.5 *The following statements are true:*

(i) *Let $0 \leq t < \tau < +\infty$. Then $Y_C(t, \tau)$ is a convex and closed subset in Y, and*

$$\widehat{Y}_C(t, t_1) \subseteq \widehat{Y}_C(t, t_2) \quad \text{for all } t < t_1 < t_2 < +\infty. \tag{4.100}$$

(ii) *Suppose that Y_E is a viability domain associated with (4.1). Then for any $t \geq 0$,*

$$Y_C(t, t_1) \subseteq Y_C(t, t_2) \quad \text{for all } t < t_1 < t_2 < +\infty. \tag{4.101}$$

Moreover, for any $\tau \in (t, +\infty)$ (with $t \geq 0$), $\widehat{Y}_C(t, \tau)$ is convex in Y and

$$\overline{\widehat{Y}_C(t, \tau)} \subseteq Y_C(t, \tau). \tag{4.102}$$

Proof Without loss of generality, we assume that the sets involved in the proof of this proposition are not empty.

(i) First, we prove the convexity of $Y_C(t, \tau)$. To this end, let $z_1, z_2 \in Y_C(t, \tau)$. Then there are $u_1, u_2 \in L(t, \tau; \mathbb{U})$ so that

$$y(\tau; t, z_1, u_1) \in Y_E \quad \text{and} \quad y(\tau; t, z_2, u_2) \in Y_E. \tag{4.103}$$

For each $\lambda \in (0, 1)$, we write

$$z_\lambda \triangleq (1 - \lambda)z_1 + \lambda z_2 \quad \text{and} \quad u_\lambda \triangleq (1 - \lambda)u_1 + \lambda u_2. \tag{4.104}$$

From the convexity of $\mathbb{U}$, we see that

$$u_\lambda \in L(t, T; \mathbb{U}). \tag{4.105}$$

By (4.1), the convexity of Y_E, (4.104), and (4.103), we find that

$$y(\tau; t, z_\lambda, u_\lambda) = (1 - \lambda)y(\tau; t, z_1, u_1) + \lambda y(\tau; t, z_2, u_2) \in Y_E.$$

This, together with (4.105), implies that $z_\lambda \in Y_C(t, \tau)$. Hence, the set $Y_C(t, \tau)$ is convex.

Second, we show that $Y_C(t, \tau)$ is closed. For this purpose, we assume that $\{z_\ell\}_{\ell \geq 1} \subseteq Y_C(t, \tau)$ satisfies that

$$z_\ell \to z \quad \text{strongly in } Y. \tag{4.106}$$

Since $\{z_\ell\}_{\ell \geq 1} \subseteq Y_C(t, \tau)$, there exists a sequence $\{u_\ell\}_{\ell \geq 1} \subseteq L(t, \tau; \mathbb{U})$ so that

$$\Phi(\tau, t)z_\ell + \int_t^\tau \Phi(\tau, s)B(s)u_\ell(s)ds \in Y_E. \tag{4.107}$$

Here $\Phi(\cdot, \cdot)$ denotes the evolution operator generated by $A + D(\cdot)$. Using the similar arguments to those used in the proofs of the second result in (4.32) and (4.34), we can prove that there exists a subsequence of $\{\ell\}_{\ell \geq 1}$, still denoted by itself, and a function $\hat{u} \in L(t, \tau; \mathbb{U})$, so that

$$u_\ell \to \hat{u} \quad \text{weakly in } L^2(t, \tau; U). \tag{4.108}$$

Since Y_E is convex and closed, by (4.106) and (4.108), we can pass to the limit for $\ell \to +\infty$ in (4.107) to get that

$$y(\tau; t, z, \hat{u}) = \Phi(\tau, t)z + \int_t^\tau \Phi(\tau, s)B(s)\hat{u}(s)ds \in Y_E.$$

This indicates that the set $Y_C(t, \tau)$ is closed.

Finally, (4.100) follows from (4.83) directly.

(ii) First, we prove (4.101). For this purpose, we arbitrarily fix $z_1 \in Y_C(t, t_1)$. Then there is $u_1 \in L(t, t_1; \mathbb{U})$ so that

$$z_2 \triangleq y(t_1; t, z_1, u_1) \in Y_E. \tag{4.109}$$

Since Y_E is a viability domain associated with (4.1), it follows by Definition 4.3 and (4.109) that there exists $u_2 \in L(t_1, t_2; \mathbb{U})$ so that

$$z_3 \triangleq y(t_2; t_1, z_2, u_2) \in Y_E. \tag{4.110}$$

Denote

$$u(s) \triangleq \begin{cases} u_1(s), & s \in (t, t_1), \\ u_2(s), & s \in (t_1, t_2). \end{cases} \tag{4.111}$$

It follows from (4.109)–(4.111) and (4.1) that

$$u \in L(t, t_2; \mathbb{U}) \quad \text{and} \quad y(t_2; t, z_1, u) = z_3 \in Y_E.$$

These imply that $z_1 \in Y_C(t, t_2)$. Hence, (4.101) is true.

Second, since $Y_C(t, s)$ is convex for any $s > t$, it follows by (4.83) and (4.101) that $\widehat{Y}_C(t, \tau)$ is also convex for any $\tau > t$.

Finally, by (4.83) and (4.101), we find that

$$\widehat{Y}_C(t, \tau) \subseteq Y_C(t, \tau) \quad \text{for any } \tau > t. \tag{4.112}$$

Since $Y_C(t, \tau)$ is closed, (4.102) follows from (4.112) immediately.

Hence, we complete the proof of Proposition 4.5.   □

4.2.2 Separability and Local Maximum Principle

In this subsection, we show some connection between the local Pontryagin Maximum Principle and the separation of $Y_R(t)$ (see (4.2)) and $Y_C(t, t_E^*)$ (see (4.82)), with $t \in (0, t_E^*)$, in the state space Y (see Theorem 4.3). We also present some relationship between the classical Pontryagin Maximum Principle and the local Pontryagin Maximum Principle (see Corollary 4.2).

The main result of this subsection is as follows:

Theorem 4.3 *The problem* $(TP)_{min}^{Q_S, Q_E}$ *satisfies the local Pontryagin Maximum Principle if and only if* $Y_R(t)$ *and* $Y_C(t, t_E^*)$ *are separable in* Y *for any* $t \in (0, t_E^*)$.

Proof We first show the sufficiency. Assume that $Y_R(t)$ and $Y_C(t, t_E^*)$ are separable in Y for any $t \in (0, t_E^*)$. Let $(0, y_0^*, t_E^*, u^*)$ be an optimal tetrad to $(TP)_{min}^{Q_S, Q_E}$. Arbitrarily fix $T \in (0, t_E^*)$. Then by the similar arguments to those used in the proof of Theorem 4.1 (where Y_E and $Y_R(t_E^*)$ are replaced by $Y_C(T, t_E^*)$ and $Y_R(T)$, respectively), we can prove that there exists $\varphi_T(\cdot) \in C([0, T]; Y)$ with $\varphi_T(\cdot) \neq 0$ so that (4.84)–(4.88) hold. From this and Definition 4.2, we see that $(0, y_0^*, t_E^*, u^*)$ satisfies the local Pontryagin Maximum Principle, consequently, so does $(TP)_{min}^{Q_S, Q_E}$. Hence, we have proved the sufficiency.

We next show the necessity. Suppose that $(TP)_{min}^{Q_S, Q_E}$ satisfies the local Pontryagin Maximum Principle. Let $(0, y_0^*, t_E^*, u^*)$ be an optimal tetrad to $(TP)_{min}^{Q_S, Q_E}$. Then by Definition 4.2, we see that for any $T \in (0, t_E^*)$, there exists $\varphi_T(\cdot) \in C([0, T]; Y)$ with $\varphi_T(\cdot) \neq 0$ so that (4.84)–(4.88) hold. Then by the similar arguments as those used in the proof of Theorem 4.1, we can prove that for each $T \in (0, t_E^*)$, $Y_R(T)$ and $Y_C(T, t_E^*)$ are separable in Y. Hence, we have proved the necessity.

Thus, we complete the proof of Theorem 4.3.   □

Remark 4.5 From Theorem 4.3, we see that the key to derive the local Pontryagin Maximum Principle for $(TP)_{min}^{Q_S, Q_E}$ is to find conditions ensuring the separation of $Y_R(t)$ and $Y_C(t, t_E^*)$ in Y at each time $t \in (0, t_E^*)$.

We next discuss the relationship between the classical Pontryagin Maximum Principle and the local Pontryagin Maximum Principle. For this purpose, we need some preparations. First of all, we define

$$Y_C(t_E^*, t_E^*) \triangleq Y_E.$$

Lemma 4.4 *Let $T \in (0, t_E^*]$. Let $\phi \in Y$ be a nonzero vector separating $Y_R(T)$ from $Y_C(T, t_E^*)$ in Y. Let $\varphi(\cdot)$ be the solution to the equation:*

$$\begin{cases} \dot{\varphi}(t) = -\big[A^* + D(t)^*\big]\varphi(t), \ t \in (0, T), \\ \varphi(T) = \phi. \end{cases}$$

Then for any $t \in [0, T)$ (with $\varphi(t) \neq 0$), $\varphi(t)$ separates $Y_R(t)$ from $Y_C(t, t_E^*)$ in Y.

Proof Since ϕ separates $Y_R(T)$ from $Y_C(T, t_E^*)$ in Y, we have that

$$\langle z_1, \phi \rangle_Y \geq \langle z_2, \phi \rangle_Y \quad \text{for all} \ \ z_1 \in Y_R(T) \ \text{and} \ z_2 \in Y_C(T, t_E^*). \tag{4.113}$$

Arbitrarily fix $t \in [0, T)$ with $\varphi(t) \neq 0$. By (4.113), we see that

$$\left\langle \Phi(T, t)\Phi(t, 0)y_0 + \int_t^T \Phi(T, s)B(s)u_1(s)ds \right.$$
$$\left. + \int_0^t \Phi(T, s)B(s)u_2(s)ds - z_2, \phi \right\rangle_Y \geq 0$$

for all $y_0 \in Y_S$, $u_1 \in L(t, T; \mathbb{U})$, $u_2 \in L(0, t; \mathbb{U})$ and $z_2 \in Y_C(T, t_E^*)$. This yields that

$$\left\langle \Phi(t, 0)y_0 + \int_0^t \Phi(t, s)B(s)u_2(s)ds, \Phi(T, t)^*\phi \right\rangle_Y$$
$$+ \left\langle \int_t^T \Phi(T, s)B(s)u_1(s)ds - z_2, \phi \right\rangle_Y \geq 0 \tag{4.114}$$

for all $y_0 \in Y_S$, $u_1 \in L(t, T; \mathbb{U})$, $u_2 \in L(0, t; \mathbb{U})$ and $z_2 \in Y_C(T, t_E^*)$. Then, by (4.114), we find that

$$\inf_{\xi \in Y_R(t)} \langle \xi, \varphi(t) \rangle_Y$$
$$\geq \sup_{z_2 \in Y_C(T, t_E^*), u_1 \in L(t, T; \mathbb{U})} \left\langle z_2 - \int_t^T \Phi(T, s)B(s)u_1(s)ds, \phi \right\rangle_Y. \tag{4.115}$$

Next, for any $\eta \in Y_C(t, t_E^*)$, there is $\widehat{u} \in L(t, t_E^*; \mathbb{U})$ so that

$$\Phi(t_E^*, t)\eta + \int_t^{t_E^*} \Phi(t_E^*, s)B(s)\widehat{u}(s)ds \in Y_E. \tag{4.116}$$

Denote

$$\widehat{z}_2 \triangleq \Phi(T,t)\eta + \int_t^T \Phi(T,s)B(s)\widehat{u}_1(s)ds$$

$$\text{and } \widehat{u}_1(\cdot) = \widehat{u}(\cdot)\big|_{(t,T)} \in L(t,T;\mathbb{U}). \tag{4.117}$$

It follows from (4.116) that

$$\widehat{z}_2 \in Y_C(T,t_E^*). \tag{4.118}$$

By (4.118), the second conclusion in (4.117), and by (4.115) with $(z_2,u_1) = (\widehat{z}_2,\widehat{u}_1)$, we have that for any $\eta \in Y_C(t,t_E^*)$,

$$\inf_{\xi \in Y_R(t)} \langle \xi, \varphi(t) \rangle_Y \geq \langle \Phi(T,t)\eta, \phi \rangle_Y = \langle \eta, \varphi(t) \rangle_Y. \tag{4.119}$$

This yields that $\varphi(t)$ separates $Y_R(t)$ from $Y_C(t,t_E^*)$ in Y.

Hence, we end the proof of Lemma 4.4.   □

For each $t \in [0,t_E^*]$, we define the following subset:

$$\Psi_t \triangleq \{\psi \in Y \setminus \{0\} \mid \psi \text{ separates } Y_R(t) \text{ from } Y_C(t,t_E^*) \text{ in } Y\}. \tag{4.120}$$

When $Y_R(t)$ and $Y_C(t,t_E^*)$ are not separable in Y, we set $\Psi_t \triangleq \emptyset$. Based on Lemma 4.4, we have the following result.

Corollary 4.1 *Let $\Psi_t, t \in [0,t_E^*]$, be defined by (4.120). Then*

$$\left(\Phi(t,s)^*\Psi_t\right) \setminus \{0\} \subseteq \Psi_s, \quad \text{when } 0 \leq s \leq t \leq t_E^*.$$

Here, we agree that $\Phi(t,s)^*\Psi_t \triangleq \emptyset$, when $\Psi_t = \emptyset$.

Now we give the following assumption:

(H_B) $\varphi(\cdot) = 0$ in $(0,t_E^*)$, provided that $\varphi(t_0) = 0$ for some $t_0 \in [0,t_E^*]$ and $\varphi(\cdot)$ solves the following equation:

$$\begin{cases} \dot{\varphi}(t) = -\left[A^* + D(t)^*\right]\varphi(t), & t \in (0,t_E^*), \\ \varphi(t_E^*) \in Y. \end{cases} \tag{4.121}$$

Corollary 4.2 *Assume that the problem $(TP)_{min}^{Q_S,Q_E}$ satisfies the classical Pontryagin Maximum Principle. Suppose that (H_B) is true. Then $(TP)_{min}^{Q_S,Q_E}$ also satisfies the local Pontryagin Maximum Principle.*

Proof Since the problem $(TP)_{min}^{Q_S,Q_E}$ satisfies the classical Pontryagin Maximum Principle, by Theorem 4.1, Corollary 4.1, and (H_B), we observe that $Y_R(t)$ and $Y_C(t,t_E^*)$ are separable in Y for any $t \in (0,t_E^*)$. This, together with Theorem 4.3, implies the desired result. Thus, we end the proof.   □

Remark 4.6 When an optimal tetrad $(0, y_0^*, t_E^*, u^*)$ to $(TP)_{min}^{Q_S, Q_E}$ satisfies the classical Pontryagin Maximum Principle, the assumption (H_B) ensures the following geometric property for the corresponding $\varphi(\cdot)$: For each $t \in [0, t_E^*]$, $\varphi(t)$ is a vector separating $Y_R(t)$ from $Y_C(t, t_E^*)$ in Y. (This follows from Theorem 4.1, (H_B), and Corollary 4.1.)

Example 4.12 Consider all minimal time control problems in Examples 4.6, 4.7, and 4.8. One can directly check that they satisfy the classical Pontryagin Maximum Principle. Besides, the assumption (H_B) holds for all of the above problems. For the ODE cases in the above, (H_B) follows from the theory of existence and uniqueness for ODEs; For the heat equation cases, (H_B) follows from (i) of Remark 1.5 (after Theorem 1.22). Then by Corollary 4.2, all of the above problems satisfy the local Pontryagin Maximum Principle. We end Example 4.12.

4.2.3 Conditions on Separation in Local Maximum Principle

In this subsection, we will give two conditions to guarantee the separation of $Y_R(t)$ (see (4.2)) and $Y_C(t, t_E^*)$ (see (4.82)) for any $t \in (0, t_E^*)$. We begin with the next definition.

Definition 4.4 Equation (4.1) is said to be L^∞-null controllable at any interval if for any $0 \leq t_1 < t_2 < +\infty$ and $y_0 \in Y$, there is $u \in L^\infty(t_1, t_2; U)$ so that $y(t_2; t_1, y_0, u) = 0$.

The first main result of this subsection is the next theorem.

Theorem 4.4 *Assume that Equation (4.1) is time-invariant and L^∞-null controllable at any interval. Suppose that $Y_E = \{0\}$, and $0 \in Int\,U$. Then for any $t \in (0, t_E^*)$, $Y_R(t)$, and $Y_C(t, t_E^*)$ are separable in Y.*

Proof Arbitrarily fix $t \in (0, t_E^*)$. Since (4.1) is L^∞-null controllable at any interval, by Theorem 1.20, there exists a positive constant $C(t, t_E^*)$ (dependent on t and t_E^*) so that

$$\|\varphi(t)\|_Y \leq C(t, t_E^*) \int_t^{t_E^*} \|B^* \varphi(s)\|_U ds, \tag{4.122}$$

for any solution $\varphi(\cdot)$ to the following equation:

$$\begin{cases} \dot{\varphi}(s) = -\big[A^* + D^*\big]\varphi(s), & s \in (t, t_E^*), \\ \varphi(t_E^*) \in Y. \end{cases} \tag{4.123}$$

Meanwhile, since $0 \in Int\,U$, there is a positive constant δ so that

$$B_\delta(0) \subseteq U. \tag{4.124}$$

Denote

$$\widehat{\delta} \triangleq \delta / C(t, t_E^*). \tag{4.125}$$

For each $y_0 \in B_{\widehat{\delta}}(0)$, we define

$$N(y_0) \triangleq \min_{u \in L^\infty(t,t_E^*;U)} \left\{ \|u\|_{L^\infty(t,t_E^*;U)} \mid y(t_E^*; t, y_0, u) = 0 \right\}.$$

The rest of the proof will be organized by four steps.
Step 1. We claim that

$$N(y_0) \le C(t, t_E^*) \|y_0\|_Y \quad \text{for all} \quad y_0 \in B_{\widehat{\delta}}(0). \tag{4.126}$$

It suffices to show (4.126) for $y_0 \in B_{\widehat{\delta}}(0) \setminus \{0\}$. To this end, we set

$$X \triangleq \left\{ B^*\varphi(\cdot) \mid \varphi(\cdot) \text{ solves } (4.123) \right\} \subseteq L^1(t, t_E^*; U).$$

It is clear that X is a linear subspace of $L^1(t, t_E^*; U)$. We define a map $\mathscr{F} : X \mapsto \mathbb{R}$
by

$$\mathscr{F}(B^*\varphi(\cdot)) \triangleq -\langle \varphi(t), y_0/\|y_0\|_Y \rangle_Y. \tag{4.127}$$

We first claim that $\mathscr{F}$ is well defined. In fact, if $B^*\varphi(\cdot) = B^*\widetilde{\varphi}(\cdot)$ over (t, t_E^*), then
by (4.122), we obtain that $\varphi(t) = \widetilde{\varphi}(t)$ in Y. Hence, $\mathscr{F}$ is well defined. Besides, one
can easily check that $\mathscr{F}$ is linear. It follows from (4.127) and (4.122) that

$$|\mathscr{F}(B^*\varphi(\cdot))| \le C(t, t_E^*) \|B^*\varphi(\cdot)\|_{L^1(t,t_E^*;U)},$$

which yields that

$$\|\mathscr{F}\|_{\mathscr{L}(X,\mathbb{R})} \le C(t, t_E^*).$$

Thus, $\mathscr{F}$ is a bounded linear functional on X. By the Hahn-Banach Theorem (see
Theorem 1.5), there exists a bounded linear functional $G : L^1(t, t_E^*; U) \to \mathbb{R}$ so
that

$$G = \mathscr{F} \text{ on } X \tag{4.128}$$

and so that

$$\|G\|_{\mathscr{L}(L^1(t,t_E^*;U),\mathbb{R})} = \|\mathscr{F}\|_{\mathscr{L}(X,\mathbb{R})} \le C(t, t_E^*).$$

Then by making use of the Riesz Representation Theorem (see Theorem 1.4), there
is a function $\widehat{u} \in L^\infty(t, t_E^*; U)$ so that

$$G(f) = \langle \widehat{u}, f \rangle_{L^\infty(t,t_E^*;U),L^1(t,t_E^*;U)} \tag{4.129}$$

and so that

$$\|\widehat{u}\|_{L^\infty(t,t_E^*;U)} = \|G\|_{\mathscr{L}(L^1(t,t_E^*;U),\mathbb{R})} \leq C(t,t_E^*). \tag{4.130}$$

Now, by (4.129), (4.128), and (4.127), we see that

$$\langle \widehat{u}, B^*\varphi(\cdot) \rangle_{L^\infty(t,t_E^*;U),L^1(t,t_E^*;U)} = -\langle \varphi(t), y_0/\|y_0\|_Y \rangle_Y.$$

This implies that

$$y(t_E^*; t, y_0/\|y_0\|_Y, \widehat{u}) = 0,$$

which, combined with (4.130), indicates (4.126).

Step 2. We show that

$$0 \in Int(Y_C(t,t_E^*)). \tag{4.131}$$

By (4.126) and (4.125), we obtain that

$$N(y_0) \leq \delta \quad \text{for all } y_0 \in B_{\widehat{\delta}}(0).$$

This, together with (4.124), yields that

$$B_{\widehat{\delta}}(0) \subseteq Y_C(t,t_E^*),$$

which leads to (4.131).

Step 3. We show that

$$0 \in \partial(Y_R(t) - Y_C(t,t_E^*)). \tag{4.132}$$

For this purpose, we first prove that

$$Y_R(\tau) \cap Y_C(t,t_E^*) = \emptyset \quad \text{for all } \tau \in [0,t). \tag{4.133}$$

By contradiction, suppose that it was not true. Then there would exist $\widehat{\tau} \in [0,t)$ so that

$$Y_R(\widehat{\tau}) \cap Y_C(t,t_E^*) \neq \emptyset. \tag{4.134}$$

Since Equation (4.1) is time-invariant, we can directly check that

$$Y_C(t,t_E^*) = Y_C(\widehat{\tau}, t_E^* - t + \widehat{\tau}). \tag{4.135}$$

Now it follows from (4.134) and (4.135) that

$$Y_R(\widehat{\tau}) \cap Y_C(\widehat{\tau}, t_E^* - t + \widehat{\tau}) \neq \emptyset.$$

The latter implies that

$$t_E^* \leq t_E^* - t + \widehat{\tau} < t_E^*,$$

which leads to a contradiction. Hence, (4.133) is true.

Furthermore, it follows from (4.92) that

$$Y_R(t) \cap Y_C(t, t_E^*) \neq \emptyset. \tag{4.136}$$

Finally, by (4.136) and (i) in Proposition 4.5, we can use the similar arguments as those used in the proof of Proposition 4.2 (In (4.36), we denote $Z(s) \triangleq Y_R(s) - Y_C(t, t_E^*)$, $s \in [0, t]$.) to obtain (4.132).

Step 4. We show the conclusion in this theorem.

By (4.131), we have that

$$Int(Y_R(t) - Y_C(t, t_E^*)) \neq \emptyset. \tag{4.137}$$

Since $Y_R(t) - Y_C(t, t_E^*)$ is a nonempty convex subset of Y (see Lemma 4.3 and (i) in Proposition 4.5), the desired result follows by (4.137), (4.132), and Theorem 1.10.

Hence, we complete the proof of Theorem 4.4.   □

We next give some examples, which will be helpful for us to understand Theorem 4.4 and Theorem 4.3.

Example 4.13 Let $A \in \mathbb{R}^{n \times n}$ and $B \in \mathbb{R}^{n \times m}$ with $n, m \in \mathbb{N}^+$. Assume that (A, B) satisfies Kalmann controllability rank condition: $rank(B, AB, \ldots, A^{n-1}B) = n$. Let $Y \triangleq \mathbb{R}^n$, $U = \mathbb{R}^m$, and $y_0 \in \mathbb{R}^n$ with $y_0 \neq 0$. Consider the problem $(TP)_{min}^{Q_S, Q_E}$, where the controlled system is as:

$$\dot{y}(t) = Ay(t) + Bu(t), \quad t \in (0, +\infty),$$

and where

$$Q_S = \{0\} \times \{y_0\}; \quad Q_E = (0, +\infty) \times \{0\}; \quad \mathbb{U} = B_\rho(0) \text{ with } \rho > 0.$$

One can easily check that $(TP)_{min}^{Q_S, Q_E}$ can be put into the framework $(\mathscr{A}_{TP})$ (given at the beginning of Section 4.1).

We further assume that $(TP)_{min}^{Q_S, Q_E}$ has optimal controls. Then by Theorem 1.23 and Theorem 4.4, we see that for each $t \in (0, t_E^*)$, $Y_R(t)$ and $Y_C(t, t_E^*)$ are separable in Y. Hence, according to Theorem 4.3, $(TP)_{min}^{Q_S, Q_E}$ satisfies the local Pontryagin Maximum Principle. We end Example 4.13.

Example 4.14 Consider the problem $(TP)_{min}^{Q_S, Q_E}$, where the controlled system is the heat equation (3.59) (with $f = 0$); and where

$$Q_S = \{0\} \times \{y_0\}; \quad Q_E = (0, +\infty) \times \{0\} \text{ and } \mathbb{U} = B_\rho(0),$$

$$\text{with } y_0 \in L^2(\Omega) \setminus \{0\}, \ \rho > 0.$$

One can easily check that $(TP)_{min}^{Q_S, Q_E}$ can be put into the framework $(\mathscr{A}_{TP})$ (given at the beginning of Section 4.1) in the current case. Moreover, according to Remark 3.1 (after Theorem 3.5), $(TP)_{min}^{Q_S, Q_E}$ has admissible controls, which, along with the similar arguments as those in the proof of Theorem 3.11, leads to the existence of optimal controls. Then by Theorems 1.22 and 1.21, we can apply Theorem 4.4 to see that for each $t \in (0, t_E^*)$, $Y_R(t)$ and $Y_C(t, t_E^*)$ are separable in Y. Hence, according to Theorem 4.3, $(TP)_{min}^{Q_S, Q_E}$ satisfies the local Pontryagin Maximum Principle. We end Example 4.14.

The second main result of this subsection is as follows.

Theorem 4.5 *Suppose that Y_E is a viability domain associated with (4.1) (see Definition 4.3). Assume that*

$$Int\left(Y_C(t_1, t_2)\right) \neq \emptyset \text{ for all } 0 < t_1 < t_2 < t_E^*, \tag{4.138}$$

and

$$Y_C(t_1, t_2) = \overline{Y_C(t_1, t_2)} \text{ for all } 0 < t_1 < t_2 \leq t_E^*. \tag{4.139}$$

Then for any $t \in (0, t_E^*)$, (i) $Y_R(t)$ and $\widehat{Y}_C(t, t_E^*)$ are separable in Y; (ii) $Y_R(t)$ and $Y_C(t, t_E^*)$ are separable in Y.

Proof When (i) is proved, we have that for any $t \in (0, t_E^*)$, $Y_R(t)$ and $\overline{\widehat{Y}_C(t, t_E^*)}$ are separable in Y. This, along with (4.139), yields (ii).

The remainder is to show (i). Arbitrarily fix $t \in (0, t_E^*)$. First, since Y_E is a viability domain associated with (4.1), it follows by (4.83), (4.138), and (ii) in Proposition 4.5 that $\widehat{Y}_C(t, t_E^*)$ is a nonempty convex subset in Y. Second, it follows by Lemma 4.3 that $Y_R(t)$ is a nonempty convex subset in Y. Hence, we have that

$$\widehat{Y}_C(t, t_E^*) - Y_R(t) \text{ is a nonempty convex subset in } Y. \tag{4.140}$$

Third, it follows by (4.93), (4.83), and (4.138) that

$$0 \notin \widehat{Y}_C(t, t_E^*) - Y_R(t) \text{ and } Int\left(\widehat{Y}_C(t, t_E^*) - Y_R(t)\right) \neq \emptyset. \tag{4.141}$$

From (4.141), we find that

$$\text{either } 0 \in \partial\left(\widehat{Y}_C(t, t_E^*) - Y_R(t)\right) \tag{4.142}$$

$$\text{or } 0 \notin \overline{\widehat{Y}_C(t, t_E^*) - Y_R(t)}. \tag{4.143}$$

In the case when (4.142) is true, it follows by Theorem 1.10, (4.140), the second result in (4.141) and (4.142) that $Y_R(t)$ and $\widehat{Y}_C(t, t_E^*)$ are separable in Y. In the case that (4.143) is true, it follows by Theorem 1.11, (4.140), and (4.143), that $Y_R(t)$ and $\widehat{Y}_C(t, t_E^*)$ are separable in Y.

Hence, we end the proof of Theorem 4.5. $\square$

The next result is comparable with Corollary 4.1. To state it, we define, for each $t \in [0, t_E^*)$, the following set:

$$\widehat{\Psi}_t \triangleq \left\{ \psi \in Y \setminus \{0\} \mid \psi \text{ is a vector separating } Y_R(t) \text{ from } \widehat{Y}_C(t, t_E^*) \text{ in } Y \right\}. \tag{4.144}$$

Here, we agree that $\widehat{\Psi}_t \triangleq \emptyset$ and $\Phi(t, s)^* \widehat{\Psi}_t \triangleq \emptyset$, provided that $Y_R(t)$ and $\widehat{Y}_C(t, t_E^*)$ are not separable in Y.

Proposition 4.6 *Assume that Y_E is a viability domain associated with (4.1). Then*

$$\left(\Phi(t, s)^* \widehat{\Psi}_t \right) \setminus \{0\} \subseteq \widehat{\Psi}_s \text{ for all } 0 \leq s \leq t < t_E^*. \tag{4.145}$$

Proof Since Y_E is a viability domain associated with (4.1), by Definition 4.3, (4.83), and (4.82), we have that

$$\widehat{Y}_C(t, t_E^*) \neq \emptyset \text{ for all } t \in [0, t_E^*).$$

Arbitrarily fix s and t so that $0 \leq s < t < t_E^*$. Without loss of generality, we can assume that $\widehat{\Psi}_t \neq \emptyset$. Given $\psi \in \widehat{\Psi}_t$, it follows by (4.144) that

$$\left\langle \Phi(t, 0)y_0 + \int_0^t \Phi(t, r)B(r)u(r)dr, \psi \right\rangle_Y \geq \langle \eta, \psi \rangle_Y, \tag{4.146}$$

for all $y_0 \in Y_S$, $u \in L(0, t; \mathbb{U})$ and $\eta \in \widehat{Y}_C(t, t_E^*)$. Arbitrarily fix $\widehat{z} \in \widehat{Y}_C(s, t_E^*)$. By (4.83) and (4.82), there is $\tau \in (s, t_E^*)$ and $\widehat{v} \in L(s, \tau; \mathbb{U})$ so that

$$\Phi(\tau, s)\widehat{z} + \int_s^\tau \Phi(\tau, r)B(r)\widehat{v}(r)dr \in Y_E. \tag{4.147}$$

We next claim that there exists a function $\widehat{u} \in L(s, t; \mathbb{U})$ so that

$$\Phi(t, s)\widehat{z} + \int_s^t \Phi(t, r)B(r)\widehat{u}(r)dr \in \widehat{Y}_C(t, t_E^*). \tag{4.148}$$

It will be proved by the following three cases related to τ:

Case 1. $\tau > t$. In this case, we set

$$z \triangleq \Phi(t, s)\widehat{z} + \int_s^t \Phi(t, r)B(r)\widehat{v}(r)dr, \tag{4.149}$$

which, combined with (4.147), indicates that

$$\Phi(\tau, t)z + \int_t^\tau \Phi(\tau, r)B(r)\widehat{v}(r)dr \in Y_E.$$

This, together with (4.82) and (4.83), implies that

$$z \in Y_C(t, \tau) \subseteq \widehat{Y}_C(t, t_E^*). \tag{4.150}$$

By setting

$$\widehat{u}(\cdot) \triangleq \widehat{v}(\cdot)|_{(s,t)},$$

we obtain (4.148) from (4.150) and (4.149) for Case 1.

Case 2. $\tau = t$. In this case, we set

$$z \triangleq \Phi(\tau, s)\widehat{z} + \int_s^\tau \Phi(\tau, r)B(r)\widehat{v}(r)dr,$$

which, combined with (4.147), indicates that

$$z = \Phi(t, s)\widehat{z} + \int_s^t \Phi(t, r)B(r)\widehat{v}(r)dr \in Y_E. \tag{4.151}$$

Since Y_E is a viability domain associated with (4.1), by Definition 4.3 and (4.151), there exists a function $w \in L(t, (t + t_E^*)/2; \mathbb{U})$ so that

$$y((t + t_E^*)/2; t, z, w) \in Y_E.$$

This, along with (4.82) and (4.83), yields that

$$z \in Y_C(t, (t + t_E^*)/2) \subseteq \widehat{Y}_C(t, t_E^*). \tag{4.152}$$

By setting

$$\widehat{u}(\cdot) \triangleq \widehat{v}(\cdot),$$

we obtain (4.148) from (4.152) and (4.151) for Case 2.

Case 3. $\tau < t$. Since Y_E is a viability domain associated with (4.1), by Definition 4.3 and (4.147), there exists a function $w \in L(\tau, t; \mathbb{U})$ so that

$$y\left(t; \tau, \Phi(\tau, s)\widehat{z} + \int_s^\tau \Phi(\tau, r)B(r)\widehat{v}(r)dr, w\right) \in Y_E. \tag{4.153}$$

Set

$$\widehat{u}(r) \triangleq \begin{cases} \widehat{v}(r), & r \in (s, \tau), \\ w(r), & r \in (\tau, t). \end{cases}$$

Then $\widehat{u} \in L(s, t; \mathbb{U})$, and by (4.153), we get that

$$\Phi(t, s)\widehat{z} + \int_s^t \Phi(t, r)B(r)\widehat{u}(r)dr \in Y_E. \qquad (4.154)$$

By (4.154) and using the similar arguments as those used in Case 2, we obtain (4.148) for Case 3.

In summary, we conclude that (4.148) is true.

Finally, we show (4.145). It will be proved by the following two possibilities related to s: $s > 0$ or $s = 0$. When $s > 0$, for each $v \in L(0, s; \mathbb{U})$, we define

$$u(r) \triangleq \begin{cases} v(r), & r \in (0, s), \\ \widehat{u}(r), & r \in (s, t). \end{cases}$$

It is obvious that $u \in L(0, t; \mathbb{U})$. Then, by (4.146) and (4.148), we obtain that

$$\left\langle \Phi(s, 0)y_0 + \int_0^s \Phi(s, r)B(r)v(r)dr, \Phi(t, s)^*\psi \right\rangle_Y \geq \langle \widehat{z}, \Phi(t, s)^*\psi \rangle_Y$$

for all $y_0 \in Y_S$, $v \in L(0, s; \mathbb{U})$ and $\widehat{z} \in \widehat{Y}_C(s, t_E^*)$. Hence, (4.145) follows at once. When $s = 0$, replacing u in (4.146) by $\widehat{u}$, using (4.146) and (4.148), we find that

$$\langle y_0, \Phi(t, 0)^*\psi \rangle_Y \geq \langle \widehat{z}, \Phi(t, 0)^*\psi \rangle_Y \quad \text{for all } y_0 \in Y_S \text{ and } \widehat{z} \in \widehat{Y}_C(0, t_E^*).$$

Thus, (4.145) is true.

Hence, we complete the proof of Proposition 4.6. ☐

The next Proposition 4.7 will give a condition to ensure the condition (4.139) when

$$Y_E \triangleq \{0\} \text{ and } \mathbb{U} \triangleq B_r(0).$$

Here, $B_r(0)$ is the closed ball in U with center at the origin and of radius $r \in (0, +\infty)$. For this purpose, we define a minimum norm function $N(\cdot, \cdot; \cdot)$ from $\{(t, T) \in \mathbb{R}^2 : 0 \leq t < T < +\infty\} \times Y$ to $[0, +\infty]$ by

$$N(t, T; y_0) \triangleq \inf \{\|u\|_{L^\infty(t,T;U)} \mid y(T; t, y_0, u) = 0\},$$

$$\text{for all } 0 \leq t < T < +\infty, \ y_0 \in Y. \qquad (4.155)$$

When the set in the right-hand side of (4.155) is empty, we agree that $N(t, T; y_0) = +\infty$.

Proposition 4.7 *Let* $Y_E \triangleq \{0\}$ *and* $\mathbb{U} \triangleq B_r(0)$. *Suppose that for any* $t \in [0, +\infty)$ *and* $y_0 \in Y$, *the function*

$$N(t, \cdot; y_0) : (t, +\infty) \to [0, +\infty]$$

is left continuous. Then

$$Y_C(t, T) = \overline{\widehat{Y}_C(t, T)} \text{ for all } 0 \le t < T < +\infty.$$

Proof Arbitrarily fix t and T with $0 \le t < T < +\infty$. Since $Y_E = \{0\}$ and $0 \in \mathbb{U}$, it follows by Definition 4.3 (or Example 4.11) that Y_E is a viability domain associated with (4.1). Hence, by (ii) of Proposition 4.5, we see that in order to prove this proposition, it suffices to show that

$$Y_C(t, T) \subseteq \overline{\widehat{Y}_C(t, T)}. \tag{4.156}$$

To show (4.156), we first observe from (4.82) that $0 \in Y_C(t, T)$. So we have that $Y_C(t, T) \ne \emptyset$. Thus, we can arbitrarily fix $z \in Y_C(t, T)$. Then, by (4.82) (where $\mathbb{U} = B_r(0)$), there exists $u \in L(t, T; B_r(0))$ so that $y(T; t, z, u) = 0$. This, along with (4.155), indicates that

$$N(t, T; z) \le r. \tag{4.157}$$

Meanwhile, since $N(t, \cdot; z) : (t, +\infty) \to [0, +\infty]$ is left continuous, we find that for any $n \in \mathbb{N}^+$, there is $\delta_n \in (0, T - t)$ so that

$$N(t, \widehat{T}; z) \le N(t, T; z) + 1/n \text{ for all } \widehat{T} \in (T - \delta_n, T).$$

From the above and (4.157) it follows that

$$N(t, \widehat{T}; z) \le r + 1/n \text{ for all } \widehat{T} \in (T - \delta_n, T),$$

which, along with (4.155), implies that

$$N\left(t, \widehat{T}; \left(\frac{nr}{nr+1}\right)^2 z\right) < r \text{ for all } \widehat{T} \in (T - \delta_n, T). \tag{4.158}$$

By (4.158) and the definition of $\widehat{Y}_C(t, T)$ (see (4.83)), we have that

$$\left(\frac{nr}{nr+1}\right)^2 z \in \widehat{Y}_C(t, T).$$

This indicates that $z \in \overline{\widehat{Y}_C(t, T)}$. Hence, (4.156) is true. This completes the proof of Proposition 4.7. $\square$

The next two examples may help us to understand Proposition 4.7, Theorem 4.5, and Theorem 4.3.

Example 4.15 Let $Y = U = \mathbb{R}$ and let $\widehat{y}_0 \in \mathbb{R} \setminus \{0\}$. Consider the problem $(TP)_{min}^{Q_S, Q_E}$, where the controlled system is as:

$$\dot{y}(t) = -ty(t) + u(t), \quad t \in (0, +\infty),$$

and where

$$Q_S = \{0\} \times \{\widehat{y}_0\}; \quad Q_E = (0, +\infty) \times \{0\}; \quad \mathbb{U} = B_\rho(0) \text{ with } \rho > 0.$$

One can easily check that the above problem $(TP)_{min}^{Q_S,Q_E}$ can be put into the framework $(\mathscr{A}_{TP})$ (given at the beginning of Section 4.1). Moreover, by Theorem 3.3, one can easily check that $(TP)_{min}^{Q_S,Q_E}$ has admissible controls. Then, by the similar arguments as those in the proof of Theorem 3.11, one can show the existence of optimal controls.

After some calculations, we see that for any $t \in [0, +\infty)$ and $y_0 \in \mathbb{R}$,

$$N(t, T; y_0) = |y_0| \left(\int_t^T e^{\tau^2/2} d\tau \right)^{-1} \text{ for each } T > t,$$

which indicates that the function $N(t, \cdot; y_0) : (t, +\infty) \to [0, +\infty]$ is continuous. Then, by Proposition 4.7, we have that

$$Y_C(t, T) = \overline{\widehat{Y}_C(t, T)} \text{ for all } t < T < +\infty. \tag{4.159}$$

Meanwhile, one can easily check that

$$0 \in Int(Y_C(t, T)) \text{ for all } t < T < +\infty. \tag{4.160}$$

It follows from (4.159), (4.160), Example 4.11, and Theorem 4.5 that $Y_R(t)$ and $Y_C(t, t_E^*)$ are separable in Y for each $t \in (0, t_E^*)$. Hence, according to Theorem 4.3, $(TP)_{min}^{Q_S,Q_E}$ satisfies the local Pontryagin Maximum Principle. We end Example 4.15.

Example 4.16 Consider the problem $(TP)_{min}^{Q_S,Q_E}$, where the controlled system is the heat equation (3.107) (with $a(x, t) = a_1(x) + a_2(t)$, $a_1(\cdot) \in L^\infty(\Omega)$ and $a_2(\cdot) \in L^\infty(0, +\infty)$), and where

$$Q_S = \{0\} \times \{y_0\}, \quad Q_E = (0, +\infty) \times \{0\}, \quad \mathbb{U} = B_\rho(0), \text{ with } y_0 \in L^2(\Omega) \setminus \{0\}, \rho > 0.$$

One can easily check that the above $(TP)_{min}^{Q_S,Q_E}$ can be put into the framework $(\mathscr{A}_{TP})$ (given at the beginning of Section 4.1). We further assume that $(TP)_{min}^{Q_S,Q_E}$ has an optimal control. Notice that when $\rho > 0$ is large enough, this problem has an optimal control (see Theorem 3.7).

We now prove that $Y_R(t)$ and $Y_C(t, t_E^*)$ are separable in Y for each $t \in (0, t_E^*)$ in the current case. For this purpose, we will use a fact whose proof can be found from Example 5.4 and Proposition 5.3 in the next chapter of this book. This fact is as: *for any $t \in [0, +\infty)$ and $y_0 \in L^2(\Omega)$, the function $N(t, \cdot; y_0) : (t, +\infty) \to [0, +\infty]$ is continuous*. From this and Proposition 4.7, we have that

$$Y_C(t, T) = \overline{\widehat{Y}_C(t, T)} \quad \text{for all } t < T < +\infty. \tag{4.161}$$

Meanwhile, by Theorems 1.22 and 1.21, one can easily check that

$$0 \in Int(Y_C(t, T)) \quad \text{for all } t < T < +\infty. \tag{4.162}$$

Now, it follows from (4.161), (4.162), Example 4.11, and Theorem 4.5 that $Y_R(t)$ and $Y_C(t, t_E^*)$ are separable in Y for each $t \in (0, t_E^*)$. Hence, according to Theorem 4.3, $(TP)_{min}^{Q_S, Q_E}$ satisfies the local Pontryagin Maximum Principle. We end Example 4.16.

The final example of this subsection gives a minimal time control problem satisfying what follows:

(i) It is under the framework $(\mathscr{A}_{TP})$ (given at the beginning of Section 4.1);
(ii) It has an optimal control;
(iii) It satisfies neither the classical nor the local Pontryagin Maximum Principle;
(iv) It satisfies the weak Pontryagin Maximum Principle (which will be given in Definition 4.7).

The proof of the above conclusion (iv) will be given in Example 4.18 of the next section. Because of the above (iv), we introduce the weak Pontryagin Maximum Principle in the next section.

Example 4.17 Consider the following controlled equation:

$$\begin{cases} \partial_t y = x^2 u & \text{in } \Omega \times (0, +\infty), \\ y = 0 & \text{on } \partial\Omega \times (0, +\infty), \end{cases} \tag{4.163}$$

where $\Omega = (0, 1)$. We can put the Equation (4.163) into our framework (4.1) in the following manner: Let $Y = U = L^2(0, 1)$; let A be the operator on Y defined by $A = 0$ with $\mathscr{D}(A) = L^2(0, 1)$; let B be the operator on Y defined by $Bu = x^2 u$ for each $u \in L^2(0, 1)$.

Consider the problem $(TP)_{min}^{Q_S, Q_E}$, where the controlled equation is (4.163) and where

$$Q_S = \{0\} \times \{y_0\} \text{ with } y_0(x) = -\sqrt{3}x^3, \ x \in (0, 1); \quad Q_E = (0, +\infty) \times \{0\};$$
$$\mathbb{U} = B_1(0).$$

One can easily check that $(TP)_{min}^{Q_S, Q_E}$ can be put into the framework $(\mathscr{A}_{TP})$ (given at the beginning of Section 4.1) in the current case. Moreover, we will see that $(TP)_{min}^{Q_S, Q_E}$ has an optimal control later.

The aim now is to show that $(TP)_{min}^{Q_S, Q_E}$ satisfies neither the classical nor the local Pontryagin Maximum Principle. The proofs will be carried out by three steps.

Step 1. We show that $(0, y_0, 1, \widetilde{u})$ with

$$\widetilde{u}(x, t) \triangleq \sqrt{3}x, \quad (x, t) \in (0, 1) \times (0, 1), \tag{4.164}$$

is the unique optimal tetrad of $(TP)_{min}^{Q_S, Q_E}$.

Indeed, after some simple calculations, we can directly check that $(0, y_0, 1, \widetilde{u})$ is an admissible tetrad of $(TP)_{min}^{Q_S, Q_E}$. Then by the similar arguments as those in the proof of Theorem 3.11, one can easily show the existence of optimal controls.

We arbitrarily fix an optimal tetrad $(0, y_0, t_E^*, u^*)$. It follows that

$$y(t_E^*; 0, y_0, u^*) = 0 \quad \text{and} \quad \|u^*\|_{L^\infty(0, t_E^*; L^2(0,1))} \leq 1. \tag{4.165}$$

For each $k \in \mathbb{N}^+$, we define

$$z_k(x) \triangleq \sqrt{3}\chi_{(1/2k, 1)}(x)/x, \quad x \in (0, 1). \tag{4.166}$$

From (4.166) and the equality in (4.165), we get that for each $k \in \mathbb{N}^+$,

$$0 = \langle y(t_E^*; 0, y_0, u^*), z_k \rangle_{L^2(0,1)}$$

$$= [(8k^3)^{-1} - 1] + \int_0^{t_E^*} \int_{1/2k}^1 \sqrt{3}x \cdot u^*(x, t)\mathrm{d}x\mathrm{d}t,$$

which, combined with the inequality in (4.165), indicates that

$$\int_0^{t_E^*} \int_0^1 \sqrt{3}x \cdot u^*(x, t)\mathrm{d}x\mathrm{d}t = 1.$$

This, together with the inequality in (4.165), yields that

$$1 = \int_0^{t_E^*} \int_0^1 \sqrt{3}x \cdot u^*(x, t)\mathrm{d}x\mathrm{d}t \leq \int_0^{t_E^*} \|u^*(x, t)\|_{L^2(0,1)}\|\sqrt{3}x\|_{L^2(0,1)}\mathrm{d}t$$

$$\leq \int_0^{t_E^*} \|\sqrt{3}x\|_{L^2(0,1)}\mathrm{d}t = t_E^*. \tag{4.167}$$

Since $(0, y_0, t_E^*, u^*)$ is an optimal tetrad and $(0, y_0, 1, \widetilde{u})$ is an admissible tetrad, (4.167) indicates that

$$t_E^* = 1. \tag{4.168}$$

Furthermore, by (4.167) and (4.168), we obtain that

$$\int_0^1 \int_0^1 \sqrt{3}x \cdot u^*(x, t)\mathrm{d}x\mathrm{d}t = \int_0^1 \|u^*(x, t)\|_{L^2(0,1)}\|\sqrt{3}x\|_{L^2(0,1)}\mathrm{d}t$$

$$= \int_0^1 \|\sqrt{3}x\|_{L^2(0,1)}\mathrm{d}t.$$

Thus, we get that there exists a function $\lambda(\cdot) : (0, 1) \to [0, +\infty)$ so that for a.e. $t \in (0, 1)$,

$$u^*(x, t) = \lambda(t) \cdot \sqrt{3}x \quad \text{and} \quad \|u^*(x, t)\|_{L^2(0,1)} = 1.$$

From these, we conclude that for a.e. $(x, t) \in (0, 1) \times (0, 1)$,

$$u^*(x, t) = \sqrt{3}x. \tag{4.169}$$

Hence, by (4.164), (4.168), (4.169), and the optimality of $(0, y_0, t_E^*, u^*)$, we complete the proof of Step 1.

Step 2. We prove that $(TP)_{min}^{Q_S, Q_E}$ does not satisfy the local Pontryagin Maximum Principle.

By contradiction, we suppose that $(TP)_{min}^{Q_S, Q_E}$ did satisfy the local Pontryagin Maximum Principle. Arbitrarily fix $T \in (0, 1)$. According to Definition 4.2 and Step 1, there exists $\varphi_T \in L^2(0, 1) \setminus \{0\}$ so that

$$\int_0^1 x^2 \tilde{u}(x, t)\varphi_T(x)dx = \max_{\|v\|_{L^2(0,1)} \leq 1} \int_0^1 x^2 v(x)\varphi_T(x)dx \quad \text{for a.e. } t \in (0, T).$$

This, along with (4.164), yields that

$$\langle \sqrt{3}x, x^2\varphi_T \rangle_{L^2(0,1)} = \|x^2\varphi_T\|_{L^2(0,1)}. \tag{4.170}$$

Since

$$\langle \sqrt{3}x, x^2\varphi_T \rangle_{L^2(0,1)} \leq \|\sqrt{3}x\|_{L^2(0,1)}\|x^2\varphi_T\|_{L^2(0,1)} = \|x^2\varphi_T\|_{L^2(0,1)},$$

according to (4.170), there exists a constant $\lambda_T \in [0, +\infty)$ so that

$$x^2\varphi_T(x) = \lambda_T \cdot \sqrt{3}x \quad \text{for a.e. } x \in (0, 1).$$

This yields that

$$\varphi_T(x) = \sqrt{3}\lambda_T/x \quad \text{for a.e. } x \in (0, 1).$$

Since $\varphi_T \in L^2(0, 1)$, the above indicates that

$$\lambda_T = 0 \quad \text{and} \quad \varphi_T(\cdot) = 0,$$

which leads to a contradiction with $\varphi_T(\cdot) \neq 0$. Hence, we finish the proof of Step 2.

Step 3. We show that $(TP)_{min}^{Q_S, Q_E}$ does not satisfy the classical Pontryagin Maximum Principle.

Indeed, Step 3 follows from Step 2 and Corollary 4.2 at once. We now end Example 4.17.

4.3   Weak Maximum Principle and Minimal Time Controls

This section studies the problem $(TP)_{min}^{Q_S,Q_E}$ under the following framework $(\mathscr{H}_{TP})$:

(i) The state space Y and the control space U are real separable Hilbert spaces.
(ii) The controlled system is as:

$$\dot{y}(t) = Ay(t) + Bu(t), \quad t \in (0, +\infty), \tag{4.171}$$

where $A : \mathscr{D}(A) \subseteq Y \mapsto Y$ generates a C_0 semigroup $\{e^{At}\}_{t \geq 0}$ on Y; $B \in \mathscr{L}(U, Y)$. Given $\tau \geq 0$, $y_0 \in Y$ and $u \in L^\infty(\tau, +\infty; U)$, denote by $y(\cdot; \tau, y_0, u)$ the unique mild solution of (4.1) over $[\tau, +\infty)$, with the initial condition $y(\tau) = y_0$, i.e., for each $T > \tau$, $y(\cdot; \tau, y_0, u)|_{[0,T]}$ is the mild solution of (4.1) over $[\tau, T]$, with the initial condition $y(\tau) = y_0$. (The existence and uniqueness of such solutions are ensured by Proposition 1.3.)

(iii) The control constraint set $\mathbb{U} = B_M(0)$ (i.e., the closed ball in U with centered at 0 and of radius $M > 0$).

(iv) Let $Q_S = \{0\} \times \{\widehat{y}_0\}$ and $Q_E = (0, +\infty) \times \{0\}$, where $\widehat{y}_0 \in Y \setminus \{0\}$ is fixed.

Notice that $(\mathscr{H}_{TP})$ is covered by $(\mathscr{A}_{TP})$ (given at the beginning of Section 4.1). As what we did in the previous two sections, we always assume that the problem $(TP)_{min}^{Q_S,Q_E}$ has an optimal control.

   The aim of this section is to define a weak Pontryagin Maximum Principle for $(TP)_{min}^{Q_S,Q_E}$ and to derive the weak Pontryagin Maximum Principle by separating the target from the reachable set at the optimal time t_E^* in the reachable subspace $\mathscr{R}_{t_E^*}$ (see (4.172) below).

4.3.1   Separability and Weak Maximum Principle

First, for each $t \in (0, +\infty)$, we define

$$\mathscr{R}_t \triangleq \{y(t; 0, 0, u) \in Y \mid u \in L^\infty(0, t; U)\}, \tag{4.172}$$

which is endowed with the following norm

$$\|z\|_{\mathscr{R}_t} \triangleq \text{Min}\{\|u\|_{L^\infty(0,t;U)} \mid y(t; 0, 0, u) = z\}, \quad z \in \mathscr{R}_t. \tag{4.173}$$

We call (4.172) the reachable subspace of the system (4.171) at t. Let us recall the reachable set of the system (4.171) at time $t \in (0, +\infty)$ (see (4.2)):

$$Y_R(t) \triangleq \{y(t; 0, \widehat{y}_0, u) \mid u \in L(0, t; \mathbb{U})\}. \tag{4.174}$$

It is obvious that $Y_R(t_E^*) \subseteq \mathscr{R}_{t_E^*}$. We endow $Y_R(t_E^*)$ with the topology inherited from $\mathscr{R}_{t_E^*}$. In this subsection, we will establish the weak Pontryagin Maximum Principle of $(TP)_{min}^{Qs,QE}$ by separating the target $\{0\}$ from $Y_R(t_E^*)$ in $\mathscr{R}_{t_E^*}$. For this purpose, we need to introduce the following two definitions.

Definition 4.5 The sets $Y_R(t_E^*)$ and $\{0\}$ are separable in $\mathscr{R}_{t_E^*}$ if there exists $\psi \in (\mathscr{R}_{t_E^*})^* \setminus \{0\}$ so that

$$\sup_{z \in Y_R(t_E^*)} \langle \psi, z \rangle_{(\mathscr{R}_{t_E^*})^*, \mathscr{R}_{t_E^*}} \leq 0.$$

Definition 4.6 An element $\psi \in (\mathscr{R}_{t_E^*})^*$ is regular if there exists $f_\psi \in L^1(0, t_E^*; U)$ so that the following representation formula holds:

$$\langle \psi, y(t_E^*; 0, 0, u) \rangle_{(\mathscr{R}_{t_E^*})^*, \mathscr{R}_{t_E^*}} = \int_0^{t_E^*} \langle u(t), f_\psi(t) \rangle_U dt \tag{4.175}$$

for all $u \in L^\infty(0, t_E^*; U)$.

Remark 4.7 For each $T > 0$, the space $\mathscr{R}_T$ may be neither separable nor reflexive (even the system (4.171) has the L^∞-null controllability). Such examples can be found in [9] (see, for instance, [9, Theorem 2.8.8]). When $\mathscr{R}_T$ is not reflexive for some $T > 0$, the dual space of $\mathscr{R}_T$ is very large and some elements in $\mathscr{R}_T$ may be very "irregular." Some elements in this dual space may not have the representation formula (4.175).

It should be pointed out that f_ψ in Definition 4.6 has a good regularity, since it belongs to a special subspace (see Corollary 4.3). This fact can be deduced from the next lemma.

Lemma 4.5 *Let $T > 0$ and $f \in L^1(0, T; U)$. The following two statements are equivalent:*

(i) For any $u_1, u_2 \in L^\infty(0, T; U)$, with $y(T; 0, 0, u_1) = y(T; 0, 0, u_2)$, it holds that

$$\int_0^T \langle f(t), u_1(t) \rangle_U dt = \int_0^T \langle f(t), u_2(t) \rangle_U dt. \tag{4.176}$$

(ii) The function f belongs to the following space:

$$\mathscr{O}_T \triangleq \overline{X_T}^{\|\cdot\|_{L^1(0,T;U)}} \quad with \quad X_T \triangleq \{B^* e^{A^*(T-\cdot)} z|_{(0,T)} : z \in D(A^*)\}. \tag{4.177}$$

Proof (i)$\Rightarrow$(ii). By contradiction, suppose that f satisfied (i) but $f \notin \mathscr{O}_T$. Then, since $\mathscr{O}_T$ is a closed subspace of $L^1(0, T; U)$ (see (4.177)), it would follow by the Hahn-Banach Theorem and the Riesz Representation Theorem (see Theorems 1.11 and 1.4) that there is $\widehat{u} \in L^\infty(0, T; U)$ so that

$$\int_0^T \langle g(t), \widehat{u}(t)\rangle_U \, dt = 0 < \int_0^T \langle f(t), \widehat{u}(t)\rangle_U \, dt \quad \text{for all} \ g \in \mathscr{O}_T. \qquad (4.178)$$

By the first equality in (4.178) and the definition of $\mathscr{O}_T$, we have that for each $z \in D(A^*)$,

$$\langle z, y(T; 0, 0, \widehat{u})\rangle_Y = \int_0^T \langle B^* e^{A^*(T-t)} z, \widehat{u}(t)\rangle_U \, dt = 0,$$

which indicates that

$$y(T; 0, 0, \widehat{u}) = 0 = y(T; 0, 0, 0).$$

This, together with (i), implies that

$$\int_0^T \langle f(t), \widehat{u}(t)\rangle_U \, dt = \int_0^T \langle f(t), 0\rangle_U \, dt = 0,$$

which contradicts the second relation in (4.178). Hence, the conclusion (ii) is true.

(ii)$\Rightarrow$(i). Suppose that f satisfies (ii). Then for arbitrarily fixed two controls $u_1, u_2 \in L^\infty(0, T; U)$, with $y(T; 0, 0, u_1) = y(T; 0, 0, u_2)$, we have that

$$y(T; 0, 0, u_1 - u_2) = 0. \qquad (4.179)$$

By (4.179), we see that for each $z \in D(A^*)$,

$$\int_0^T \langle B^* e^{A^*(T-t)} z, (u_1 - u_2)(t)\rangle_U \, dt = \langle z, y(T; 0, 0, u_1 - u_2)\rangle_Y = 0.$$

Since $f \in \mathscr{O}_T$, the above, combined with a density argument, shows that

$$\int_0^T \langle f(t), (u_1 - u_2)(t)\rangle_U \, dt = 0,$$

which leads to (4.176).

Hence, we complete the proof of Lemma 4.5. $\square$

Corollary 4.3 *If* $(\psi, f_\psi) \in (\mathscr{R}_{t_E^*})^* \times L^1(0, t_E^*; U)$ *satisfies* (4.175), *then* $f_\psi \in \mathscr{O}_{t_E^*}$.

Proof By Lemma 4.5, it suffices to check that (i) in Lemma 4.5 (with T replaced by t_E^*) is true. But, the latter can be easily obtained from (4.175). This completes the proof of Corollary 4.3. $\square$

We now give the definition of the weak Pontryagin Maximum Principle for the problem $(TP)_{min}^{Q_S, Q_E}$.

Definition 4.7 Several definitions are given in order.

(i) An optimal control u^* (associated with an optimal tetrad $(0, \widehat{y}_0, t_E^*, u^*)$) to the problem $(TP)_{min}^{Q_S, Q_E}$ is said to satisfy the weak Pontryagin Maximum Principle, if there exists $f \in \mathscr{O}_{t_E^*}$ with $f(\cdot) \neq 0$ (i.e., f is not a zero function) so that

$$\langle u^*(t), f(t)\rangle_U = \max_{v \in \mathbb{U}} \langle v, f(t)\rangle_U \quad \text{for a.e. } t \in (0, t_E^*). \tag{4.180}$$

(ii) The problem $(TP)_{min}^{Q_S, Q_E}$ is said to satisfy the weak Pontryagin Maximum Principle, if any optimal control satisfies the weak Pontryagin Maximum Principle.

Example 4.18 Consider the problem $(TP)_{min}^{Q_S, Q_E}$ in Example 4.17. One can easily check that it can be put into the framework $(\mathscr{H}_{TP})$ (given at the beginning of Section 4.3). We have seen in Example 4.17 that this problem has an optimal control.

The aim now is to show that the above $(TP)_{min}^{Q_S, Q_E}$ satisfies the weak Pontryagin Maximum Principle. By Step 1 in Example 4.17 and Definition 4.7, we see that it suffices to show that the function

$$f(x, t) \triangleq \sqrt{3}x, \quad (x, t) \in (0, 1) \times (0, 1),$$

belongs to $\mathscr{O}_1$ (see (4.177)). To prove the later, we first observe that

$$x^2 \varphi_k \to \sqrt{3}x \quad \text{strongly in } L^1(0, 1; L^2(0, 1)), \tag{4.181}$$

where

$$\varphi_k(x, t) \triangleq \chi_{(1/2k, 1)}(x)\sqrt{3}/x, \quad (x, t) \in (0, 1) \times (0, 1). \tag{4.182}$$

Meanwhile, it follows from (4.182) and the definition of X_1 (see (4.177)) that $x^2 \varphi_k \in X_1$. This, together with (4.181) and (4.177), implies that $f \in \mathscr{O}_1$. Hence, $(TP)_{min}^{Q_S, Q_E}$ satisfies the weak Pontryagin Maximum Principle. We now end Example 4.18.

We next present the main result of this subsection.

Theorem 4.6 *Assume that the target $\{0\}$ and $Y_R(t_E^*)$ are separable in $\mathscr{R}_{t_E^*}$ by a regular vector $\psi \in (\mathscr{R}_{t_E^*})^* \setminus \{0\}$. Then $(TP)_{min}^{Q_S, Q_E}$ satisfies the weak Pontryagin Maximum Principle.*

Proof Since $\{0\}$ and $Y_R(t_E^*)$ are separable in $\mathscr{R}_{t_E^*}$ by a regular vector $\psi \in (\mathscr{R}_{t_E^*})^* \setminus \{0\}$, it follows respectively from Definitions 4.5 and 4.6 that

$$\sup_{z \in Y_R(t_E^*)} \langle \psi, z\rangle_{(\mathscr{R}_{t_E^*})^*, \mathscr{R}_{t_E^*}} \leq 0, \tag{4.183}$$

and that there exists $f_\psi \in L^1(0, t_E^*; U)$ so that

$$\langle \psi, y(t_E^*; 0, 0, u)\rangle_{(\mathscr{R}_{t_E^*})^*, \mathscr{R}_{t_E^*}} = \int_0^{t_E^*} \langle u(t), f_\psi(t)\rangle_U dt \ \text{ for all } \ u \in L^\infty(0, t_E^*; U).$$
(4.184)

Since

$$y(t_E^*; 0, \widehat{y}_0, u) = y(t_E^*; 0, 0, u) + e^{At_E^*}\widehat{y}_0 \ \text{ for all } \ u \in L^\infty(0, t_E^*; U),$$ (4.185)

it follows by (4.183), (4.174), and (4.184) that

$$\sup_{u \in L(0, t_E^*; U)} \int_0^{t_E^*} \langle u(t), f_\psi(t)\rangle_U dt + \langle \psi, e^{At_E^*}\widehat{y}_0\rangle_{(\mathscr{R}_{t_E^*})^*, \mathscr{R}_{t_E^*}} \le 0.$$ (4.186)

Let $(0, \widehat{y}_0, t_E^*, u^*)$ be an optimal tetrad to $(TP)_{min}^{Q_S, Q_E}$. Since $y(t_E^*; 0, \widehat{y}_0, u^*) = 0$, we see from (4.185) and (4.184) that

$$\langle \psi, e^{At_E^*}\widehat{y}_0\rangle_{(\mathscr{R}_{t_E^*})^*, \mathscr{R}_{t_E^*}} = -\langle \psi, y(t_E^*; 0, 0, u^*)\rangle_{(\mathscr{R}_{t_E^*})^*, \mathscr{R}_{t_E^*}}$$

$$= -\int_0^{t_E^*} \langle u^*(t), f_\psi(t)\rangle_U dt.$$

This, together with (4.186), implies that

$$\max_{u \in L(0, t_E^*; U)} \int_0^{t_E^*} \langle u(t), f_\psi(t)\rangle_U dt = \int_0^{t_E^*} \langle u^*(t), f_\psi(t)\rangle_U dt.$$ (4.187)

By (4.187), using the similar arguments as those used to show (4.15), we obtain that

$$\max_{\|v\|_U \le M} \langle v, f_\psi(t)\rangle_U = \langle u^*(t), f_\psi(t)\rangle_U \ \text{ a.e. } \ t \in (0, t_E^*).$$

Hence, we end the proof of Theorem 4.6.   □

Several notes on the weak Pontryagin Maximum Principle of $(TP)_{min}^{Q_S, Q_E}$ are given in the next remark.

Remark 4.8 Several notes are given in order.

(i) In the proof of Theorem 4.6, the regular vector ψ, which separates the target $\{0\}$ and $Y_R(t_E^*)$ in $\mathscr{R}_{t_E^*}$, plays an important role. Without the assumption of such "regularity," the weak Pontryagin Maximum Principle of $(TP)_{min}^{Q_S, Q_E}$ may not be true. (At least, we do not know how to prove it.) These lead us to introduce two representation theorems (see Theorems 4.7 and 4.8) and a special subspace of $\mathscr{R}_t^0$ (see (4.214)).

(ii) The function f in (4.180) may not grow like $B^*\varphi$, where φ is an adjoint solution with the terminal condition $\varphi(t_E^*) \in Y$. But in some cases (for example, the system (4.171) has the L^∞-null controllability over $(0, T)$ for each $T > 0$), the function f in (4.180) belongs to $\mathscr{O}_{t_E^*}$ which consists of all functions $\widehat{f}$ satisfying that

$$\begin{cases} \widehat{f}(t) = B^*\widehat{\varphi}(t) \quad \text{a.e.} \quad t \in (0, t_E^*), \\ \widehat{\varphi} \in C([0, t_E^*); Y), \quad B^*\widehat{\varphi} \in L^1(0, t_E^*; U), \\ \widehat{\varphi} \text{ satisfies the adjoint equation over } (0, t_E^*). \end{cases}$$

The internally controlled time-varying heat equation is such an example (see [18, Lemma 2.1]).

4.3.2 Two Representation Theorems Related to the Separability

The first representation theorem is as follows:

Theorem 4.7 *For each $T > 0$, there is a linear isometry Φ_T from $\mathscr{R}_T$ to $\mathscr{O}_T^*$ (the dual space of $\mathscr{O}_T$) so that for all $y_T \in \mathscr{R}_T$ (given by (4.172)) and $f \in \mathscr{O}_T$ (given by (4.177)),*

$$\langle y_T, f \rangle_{\mathscr{R}_T, \mathscr{O}_T} \triangleq \langle \Phi_T(y_T), f \rangle_{\mathscr{O}_T^*, \mathscr{O}_T} = \int_0^T \langle u(t), f(t) \rangle_U dt, \qquad (4.188)$$

where u is any control in $L^\infty(0, T; U)$ satisfying $y_T = y(T; 0, 0, u)$. Here, $\mathscr{O}_T$ is equipped with the norm $\| \cdot \|_{L^1(0,T;U)}$.

Proof Arbitrarily fix a $T > 0$. For each $y_T \in \mathscr{R}_T$, we define the following set:

$$\mathscr{U}_{y_T} \triangleq \{ v \in L^\infty(0, T; U) \mid y(T; 0, 0, v) = y_T \}. \qquad (4.189)$$

From (4.172) and (4.189) it follows that

$$\mathscr{U}_{y_T} \neq \emptyset \quad \text{for any} \quad y_T \in \mathscr{R}_T, \qquad (4.190)$$

and

$$y_T = y(T; 0, 0, v) \quad \text{for all} \quad y_T \in \mathscr{R}_T \quad \text{and} \quad v \in \mathscr{U}_{y_T}. \qquad (4.191)$$

By (4.190) and (4.191), we see that for each $y_T \in \mathscr{R}_T$, $z \in D(A^*)$ and $v \in \mathscr{U}_{y_T}$,

$$\begin{aligned} \langle y_T, z \rangle_Y &= \int_0^T \langle v(t), B^* e^{A^*(T-t)} z \rangle_U dt \\ &\leq \|v\|_{L^\infty(0,T;U)} \|B^* e^{A^*(T-\cdot)} z\|_{L^1(0,T;U)}. \end{aligned} \qquad (4.192)$$

Given $y_T \in \mathscr{R}_T$ and $v \in \mathscr{U}_{y_T}$, define a map $\mathscr{F}_{y_T} : X_T \mapsto \mathbb{R}$ in the following manner:

$$\mathscr{F}_{y_T}\left(B^* e^{A^*(T-\cdot)} z|_{(0,T)}\right) \triangleq \int_0^T \langle v(t),\, B^* e^{A^*(T-t)} z\rangle_U dt, \quad z \in D(A^*), \qquad (4.193)$$

where X_T is given by (4.177). By the equality in (4.192), we observe that the definition of $\mathscr{F}_{y_T}(\cdot)$ is independent of the choice of $v \in \mathscr{U}_{y_T}$. Thus it is well defined. Moreover, by (4.193), the inequality in (4.192) and (4.177), we find that $\mathscr{F}_{y_T}(\cdot)$ can be uniquely extended to be an element $\widetilde{\mathscr{F}}_{y_T} \in \mathscr{O}_T^*$ satisfying

$$\|\widetilde{\mathscr{F}}_{y_T}\|_{\mathscr{O}_T^*} \leq \|v\|_{L^\infty(0,T;U)} \quad \text{for all } v \in \mathscr{U}_{y_T}.$$

This, together with (4.173), implies that

$$\|\widetilde{\mathscr{F}}_{y_T}\|_{\mathscr{O}_T^*} \leq \inf\{\|u\|_{L^\infty(0,T;U)} \mid u \in \mathscr{U}_{y_T}\} = \|y_T\|_{\mathscr{R}_T} \quad \text{for all } y_T \in \mathscr{R}_T.$$
$$(4.194)$$

Meanwhile, we define a map $\Phi_T : \mathscr{R}_T \mapsto \mathscr{O}_T^*$ in the following manner:

$$\Phi_T(y_T) \triangleq \widetilde{\mathscr{F}}_{y_T}, \quad y_T \in \mathscr{R}_T. \qquad (4.195)$$

It is obvious that Φ_T is well defined and linear.

Next, we claim that

$$\Phi_T \text{ is an isometry from } \mathscr{R}_T \text{ to } \mathscr{O}_T^*. \qquad (4.196)$$

The proof of (4.196) will be carried out by three steps.

Step 1. We show that

$$\Phi_T : \mathscr{R}_T \mapsto \mathscr{O}_T^* \text{ is surjective.} \qquad (4.197)$$

Arbitrarily fix $g \in \mathscr{O}_T^*$. Since $\mathscr{O}_T \subseteq L^1(0,T;U)$ (see (4.177)), by the Hahn-Banach Theorem (see Theorem 1.5), there exists $\widetilde{g} \in (L^1(0,T;U))^*$ so that

$$\widetilde{g}(\psi) = g(\psi) \text{ for all } \psi \in \mathscr{O}_T; \text{ and } \|\widetilde{g}\|_{(L^1(0,T;U))^*} = \|g\|_{\mathscr{O}_T^*}. \qquad (4.198)$$

Then, by the Riesz Representation Theorem (see Theorem 1.4), there is $\widehat{v} \in L^\infty(0,T;U)$ so that

$$\widetilde{g}(w) = \langle \widehat{v},\, w\rangle_{L^\infty(0,T;U),L^1(0,T;U)} \text{ for all } w \in L^1(0,T;U),$$

and so that

$$\|\widetilde{g}\|_{(L^1(0,T;U))^*} = \|\widehat{v}\|_{L^\infty(0,T;U)}.$$

These, combined with (4.198), indicate that

$$\int_0^T \langle \widehat{v}(t), B^* e^{A^*(T-t)} z \rangle_U \, dt = g(B^* e^{A^*(T-\cdot)} z|_{(0,T)}) \quad \text{for all } z \in D(A^*),$$

(4.199)

and

$$\|\widehat{v}\|_{L^\infty(0,T;U)} = \|g\|_{\mathscr{O}_T^*}.$$

(4.200)

Write $\widehat{y}_T \triangleq y(T; 0, 0, \widehat{v})$. Then $\widehat{v} \in \mathscr{U}_{\widehat{y}_T}$ (see (4.189)). This, along with (4.199), (4.193), and (4.177), yields that

$$g = \widetilde{\mathscr{F}}_{\widehat{y}_T} \quad \text{in} \quad \mathscr{O}_T^*,$$

which, together with (4.195), leads to (4.197).

Step 2. We show that

$$\Phi_T : \mathscr{R}_T \mapsto \mathscr{O}_T^* \quad \text{is injective.}$$

(4.201)

Let $y_T \in \mathscr{R}_T$ satisfy that $\Phi_T(y_T) = 0$. By (4.195), we get that

$$\widetilde{\mathscr{F}}_{y_T} = 0 \quad \text{in} \quad \mathscr{O}_T^*,$$

which, combined with (4.193) and (4.192), indicates that

$$\langle y_T, z \rangle_Y = 0 \quad \text{for all } z \in D(A^*).$$

Since $D(A^*)$ is dense in Y, the above equality yields that $y_T = 0$. So (4.201) is true.

Step 3. We show that $\Phi_T : \mathscr{R}_T \mapsto \mathscr{O}_T^*$ preserves norms.

Let $g \in \mathscr{O}_T^*$. Then we have that $g = \widetilde{\mathscr{F}}_{\widehat{y}_T}$ in $\mathscr{O}_T^*$, where $\widehat{y}_T = \widehat{y}(T; 0, 0, \widehat{v})$ with $\widehat{v} \in L^\infty(0, T; U)$ satisfying (4.199) and (4.200). This, along with (4.173), yields that $\|\widehat{y}_T\|_{\mathscr{R}_T} \leq \|\widetilde{\mathscr{F}}_{\widehat{y}_T}\|_{\mathscr{O}_T^*}$. From this and (4.194), we see that Φ_T preserves norms.

Finally, the conclusion of the theorem follows from (4.196), (4.195), (4.193), and (4.177), immediately.

Hence, we complete the proof of Theorem 4.7. $\square$

Remark 4.9 Several notes are given in order.

 (i) For a time-varying heat equation, a similar result to Theorem 4.7 has been built up in [18, (i) of Theorem 1.4].
(ii) In Theorem 4.6, we assume that the target $\{0\}$ and $Y_R(t_E^*)$ are separable in $\mathscr{R}_{t_E^*}$ by a regular vector $\psi \in (\mathscr{R}_{t_E^*})^* \setminus \{0\}$. This regular vector corresponds to a function $f \in \mathscr{O}_{t_E^*}$ (see Definition 4.6 and Corollary 4.3). By Theorem 4.7, one can regard $\mathscr{R}_{t_E^*}$ as the dual space of $\mathscr{O}_{t_E^*}$. Then $(\mathscr{R}_{t_E^*})^* = (\mathscr{O}_{t_E^*})^{**} \supseteq \mathscr{O}_{t_E^*}$. In general, one cannot obtain that $(\mathscr{R}_{t_E^*})^* = \mathscr{O}_{t_E^*}$. How do we check the assumption in Theorem 4.6? This assumption says that

the target $\{0\}$ and $Y_R(t_E^*)$ are separable in $\mathscr{R}_{t_E^*}$

under the weak star topology $\sigma(\mathscr{R}_{t_E^*}, \mathscr{O}_{t_E^*})$.   (4.202)

It will be seen later that we cannot obtain (4.202) by using the result in Remark 1.1 (after Theorem 1.11). This will lead to the other representation theorem (see Theorem 4.8).

As we point out in (ii) of Remark 4.9, to get the weak Pontryagin Maximum Principle via Theorem 4.6, we must face the separation of the target $\{0\}$ and $Y_R(t_E^*)$ in $\mathscr{R}_{t_E^*}$ under the weak star topology $\sigma(\mathscr{R}_{t_E^*}, \mathscr{O}_{t_E^*})$. By (4.172), (4.173) and (4.174), one can directly check that

$$Y_R(t_E^*) = \{e^{At_E^*}\widehat{y_0}\} + \{z \in \mathscr{R}_{t_E^*} \mid \|z\|_{\mathscr{R}_{t_E^*}} \leq M\}.$$

If we want to use the result in Remark 1.1 (after Theorem 1.11) to obtain the aforementioned separation, we need to answer the following problem: *Is the unit closed ball in $\mathscr{R}_{t_E^*}$ has a nonempty interior in the weak star topology?* The next lemma is concerned with this problem.

Lemma 4.6 *Let $T > 0$ and $\mathscr{R}_T$ be given by (4.172). Let $B_{\mathscr{R}_T}$ be the unit closed ball in $\mathscr{R}_T$, i.e.,*

$$B_{\mathscr{R}_T} \triangleq \{y_T \in \mathscr{R}_T \mid \|y_T\|_{\mathscr{R}_T} \leq 1\}.$$

Assume that $B_{\mathscr{R}_T}$ has a nonempty interior in the weak star topology of $\mathscr{R}_T$ (i.e., $\sigma(\mathscr{R}_T, \mathscr{O}_T)$). Then $\mathscr{R}_T$ is of finite dimension.

Proof Assume that $B_{\mathscr{R}_T}$ has a nonempty interior with respect to the weak star topology $\sigma(\mathscr{R}_T, \mathscr{O}_T)$. Write $\overset{\circ*}{B}_{\mathscr{R}_T}$ for this interior. Take $y_T \in \overset{\circ*}{B}_{\mathscr{R}_T}$. By the definition of $\overset{\circ*}{B}_{\mathscr{R}_T}$, there exists $\{f_j\}_{j=1}^k \subseteq \mathscr{O}_T$, with a positive integer k, and $r_0 > 0$ so that

$$y_T \in \{y_T\} + \left\{z \in \mathscr{R}_T \mid \max_{1 \leq j \leq k} |f_j(z)| < r_0\right\} \subseteq B_{\mathscr{R}_T}. \quad (4.203)$$

From the latter it follows that

$$-y_T \in \{-y_T\} + \left\{z \in \mathscr{R}_T \mid \max_{1 \leq j \leq k} |f_j(z)| < r_0\right\} \subseteq B_{\mathscr{R}_T}. \quad (4.204)$$

By (4.203) and (4.204), we have that

$$0 \in \{y_T/2 + (-y_T)/2\} + \left\{z \in \mathscr{R}_T \mid \max_{1 \leq j \leq k} |f_j(z)| < r_0\right\} \subseteq B_{\mathscr{R}_T}.$$

Then

$$\left\{z \in \mathscr{R}_T \mid \max_{1 \le j \le k} |f_j(z)| < r_0\right\} \subseteq B_{\mathscr{R}_T}. \tag{4.205}$$

We now claim that

$$\|z\|_{\mathscr{R}_T} \le \frac{1}{r_0} \max_{1 \le j \le k} |f_j(z)| \quad \text{for all } z \in \mathscr{R}_T. \tag{4.206}$$

By contradiction, we suppose that (4.206) was not true. Then there would exist $z_0 \in \mathscr{R}_T$ so that

$$\|z_0\|_{\mathscr{R}_T} > \frac{1}{r_0} \max_{1 \le j \le k} |f_j(z_0)|. \tag{4.207}$$

It is obvious that $z_0 \neq 0$ in $\mathscr{R}_T$ and $2\frac{z_0}{\|z\|_{\mathscr{R}_T}} \notin B_{\mathscr{R}_T}$. Hence, by (4.205), we get that

$$2\frac{z_0}{\|z_0\|_{\mathscr{R}_T}} \notin \left\{z \in \mathscr{R}_T \mid \max_{1 \le j \le k} |f_j(z)| < r_0\right\},$$

which implies that

$$c_0 \triangleq \max_{1 \le j \le k} |f_j(z_0)| = \frac{1}{2}\|z_0\|_{\mathscr{R}_T} \max_{1 \le j \le k} |f_j(2z_0/\|z_0\|_{\mathscr{R}_T})|$$
$$\ge \frac{1}{2}r_0\|z_0\|_{\mathscr{R}_T} > 0. \tag{4.208}$$

Arbitrarily fix $\varepsilon \in (0, r_0)$. From (4.208) it follows that

$$\max_{1 \le j \le k} |f_j((r_0 - \varepsilon)z_0/c_0)| = \frac{r_0 - \varepsilon}{c_0} \max_{1 \le j \le k} |f_j(z_0)| = r_0 - \varepsilon < r_0.$$

This, together with (4.205), implies that

$$\frac{r_0 - \varepsilon}{c_0}z_0 \in B_{\mathscr{R}_T}. \tag{4.209}$$

By (4.209) and (4.208), we see that

$$\|z_0\|_{\mathscr{R}_T} = \frac{c_0}{r_0 - \varepsilon}\left\|\frac{r_0 - \varepsilon}{c_0}z_0\right\|_{\mathscr{R}_T} \le \frac{c_0}{r_0 - \varepsilon} = \frac{1}{r_0 - \varepsilon} \max_{1 \le j \le k} |f_j(z_0)|. \tag{4.210}$$

Passing $\varepsilon \to 0+$ in (4.210), we obtain a contradiction with (4.207). Hence, (4.206) is true.

We next show that

$$(\mathscr{R}_T)^* = \text{span}\{f_1, \dots, f_k\}. \tag{4.211}$$

To this end, we arbitrarily fix $f \in (\mathscr{R}_T)^*$. By (4.206), there exists a positive constant C_f so that

$$|f(z)| \leq C_f \max_{1 \leq j \leq k} |f_j(z)| \quad \text{for all} \ z \in \mathscr{R}_T. \tag{4.212}$$

Define a map as follows:

$$\mathscr{G}_f\big(f_1(z), \ldots, f_k(z)\big) \triangleq f(z), \quad z \in \mathscr{R}_T. \tag{4.213}$$

By (4.212), one can check that $\mathscr{G}_f$ is well defined. Moreover, $\mathscr{G}_f$ can be extended to be a linear map from $\mathbb{R}^k$ to $\mathbb{R}$. Then there exists $\{c_{f,j}\}_{j=1}^k \subseteq \mathbb{R}$ so that

$$\mathscr{G}_f\big(f_1(z), \ldots, f_k(z)\big) = \sum_{j=1}^k c_{f,j} f_j(z) \quad \text{for all} \ z \in \mathscr{R}_T.$$

This, together with (4.213), implies that

$$f(z) = \sum_{j=1}^k c_{f,j} f_j(z) \quad \text{for all} \ z \in \mathscr{R}_T,$$

which implies that

$$f = \sum_{j=1}^k c_{f,j} f_j.$$

Thus, (4.211) follows immediately.

Finally, by (4.211), we see that $(\mathscr{R}_T)^*$ is of finite dimension. Then we conclude that $\mathscr{R}_T$ is also of finite dimension.

Hence, we finish the proof of Lemma 4.6.   □

From Lemma 4.6, we see that in general, one cannot directly get the desired separation in Theorem 4.6 by the result in Remark 1.1 (after Theorem 1.11). To overcome this difficulty, we introduce the following set: For each $T > 0$,

$$\mathscr{R}_T^0 \triangleq \big\{y(T; 0, 0, u) \in Y \mid u \in L^\infty(0, T; U) \text{ with } \lim_{s \to T-} \|u\|_{L^\infty(s,T;U)} = 0\big\}. \tag{4.214}$$

By (4.214) and (4.172), we can check that $\mathscr{R}_T^0$ is a subspace of $\mathscr{R}_T$. This subspace is related to the second representation theorem in this subsection (see Theorem 4.8 below) and will be used to obtain the separation in Theorem 4.6. Before stating the second representation theorem, we make the following hypothesis:

(H1) The system (4.171) is L^∞-null controllable over $(0, T)$ for each $T > 0$, i.e., for each $T > 0$ and $y_0 \in Y$, there exists $u \in L^\infty(0, T; U)$ so that $y(T; 0, y_0, u) = 0$.

Based on the hypothesis (H1), we have the second representation theorem as follows:

Theorem 4.8 *Assume that (H1) holds. Then for each $T \in (0, +\infty)$, there is an isometry $\Psi_T : \mathcal{O}_T \mapsto (\mathcal{R}_T^0)^*$ so that for all $y_T \in \mathcal{R}_T^0$ and $f \in \mathcal{O}_T$,*

$$\langle \Psi_T(f), y_T \rangle_{(\mathcal{R}_T^0)^*, \mathcal{R}_T^0} \triangleq \langle f, y_T \rangle_{\mathcal{O}_T, \mathcal{R}_T^0} = \int_0^T \langle u(t), f(t) \rangle_U dt, \tag{4.215}$$

where u is any element in $L^\infty(0, T; U)$ satisfying $y_T = y(T; 0, 0, u)$.

Before proving Theorem 4.8, we introduce the next lemma.

Lemma 4.7 *Suppose that (H1) holds. Then the following conclusions are true:*

(i) For any $T > t > 0$, there exists a constant $C(T, t) \in (0, +\infty)$ so that for each $u \in L^\infty(0, T; U)$, there is $v_u \in L^\infty(0, T; U)$ so that

$$\begin{cases} y(T; 0, 0, \chi_{(0,t)}u) = y(T; 0, 0, \chi_{(t,T)}v_u); \\ \|v_u\|_{L^\infty(0,T;U)} \le C(T, t)\|u\|_{L^2(0,T;U)}. \end{cases} \tag{4.216}$$

(ii) Let $T \in (0, +\infty)$, $f \in L^1(0, T; U)$, and $f|_{(0,t)} \in \mathcal{O}_t$ for each $t \in (0, T)$. Then $f \in \mathcal{O}_T$. Here and throughout this subsection, $C(\cdot)$ denotes a positive constant depending on what are enclosed in the bracket.

Proof We prove the conclusions one by one.

(i) Since (H1) holds, we can apply Theorem 1.20 to find: for each $s \in (0, +\infty)$, there exists a constant $C(s) \in (0, +\infty)$ so that for each $y_0 \in Y$, there is a control $u_{y_0} \in L^\infty(0, s; U)$ so that

$$y(s; 0, y_0, u_{y_0}) = 0 \quad \text{and} \quad \|u_{y_0}\|_{L^\infty(0,s;U)} \le C(s)\|y_0\|_Y. \tag{4.217}$$

Now, for any $T > t > 0$ and $u \in L^\infty(0, T; U)$, we set

$$y_{0,u} \triangleq y(t; 0, 0, u). \tag{4.218}$$

By (4.217) (where $s = T - t$ and $y_0 = y_{0,u}$), we can find $v_{y_{0,u}} \in L^\infty(0, T - t; U)$ so that

$$y(T - t; 0, y_{0,u}, v_{y_{0,u}}) = 0 \quad \text{and} \quad \|v_{y_{0,u}}\|_{L^\infty(0,T-t;U)} \le C(T - t)\|y_{0,u}\|_Y.$$

These, together with (4.218), imply that

$$\|v_{y_{0,u}}\|_{L^\infty(0,T-t;U)} \leq C(T-t) \left\| \int_0^t e^{A(t-\tau)} Bu(\tau) d\tau \right\|_Y$$
$$\leq C(T,t) \max_{0 \leq \tau \leq t} \|e^{A\tau}\|_{\mathscr{L}(Y;Y)} \|B\|_{\mathscr{L}(U;Y)} \|u\|_{L^2(0,t;U)} \tag{4.219}$$

and that

$$y(T; 0, 0, \chi_{(0,t)}u) = e^{A(T-t)} y(t; 0, 0, u) \tag{4.220}$$
$$= e^{A(T-t)} y_{0,u} = y(T-t; 0, 0, -v_{y_{0,u}}).$$

Define

$$v_u(\tau) \triangleq \begin{cases} 0, & \tau \in (0, t], \\ -v_{y_{0,u}}(\tau - t), & \tau \in (t, T). \end{cases}$$

By (4.219) and (4.220), one can directly check that (4.216) holds for the above u and v_u.

(ii) Arbitrarily fix $u \in L^\infty(0, T; U)$ satisfying that

$$y(T; 0, 0, u) = 0. \tag{4.221}$$

By Lemma 4.5, we see that it suffices to show

$$\int_0^T \langle f(t), u(t) \rangle_U dt = 0. \tag{4.222}$$

To prove (4.222), we arbitrarily fix $\varepsilon \in (0, +\infty)$. Then choose $\delta_\varepsilon \in (0, \min\{\varepsilon, T/2\})$ so that

$$\|y(\delta_\varepsilon; 0, 0, u)\|_Y < \varepsilon. \tag{4.223}$$

Set

$$z_\varepsilon \triangleq y(\delta_\varepsilon; 0, 0, u). \tag{4.224}$$

By (4.217), there exists $v_u^\varepsilon \in L^\infty(0, T/2; U)$ so that

$$y(T/2; 0, z_\varepsilon, v_u^\varepsilon) = 0 \quad \text{and} \quad \|v_u^\varepsilon\|_{L^\infty(0,T/2;U)} \leq C(T/2) \|z_\varepsilon\|_Y, \tag{4.225}$$

where $C(T/2)$ is given in (4.217). Define a new control

$$v_\varepsilon(t) \triangleq \begin{cases} u(t + \delta_\varepsilon) - v_u^\varepsilon(t), & t \in (0, T/2), \\ u(t + \delta_\varepsilon), & t \in (T/2, T - \delta_\varepsilon). \end{cases} \tag{4.226}$$

From (4.225) and (4.226), we find that

$$y(T - \delta_\varepsilon; 0, 0, v_\varepsilon) = \int_0^{T-\delta_\varepsilon} e^{A(T-\delta_\varepsilon-t)} Bu(t + \delta_\varepsilon) dt$$

$$- \int_0^{T/2} e^{A(T-\delta_\varepsilon-t)} B v_u^\varepsilon(t) dt$$

$$= \int_{\delta_\varepsilon}^T e^{A(T-t)} Bu(t) dt + e^{A(T/2-\delta_\varepsilon)} e^{AT/2} z_\varepsilon,$$

which, combined with (4.224) and (4.221), indicates that

$$y(T - \delta_\varepsilon; 0, 0, v_\varepsilon) = \int_0^T e^{A(T-t)} Bu(t) dt = 0. \tag{4.227}$$

Since $f|_{(0,T-\delta_\varepsilon)} \in \mathscr{O}_{T-\delta_\varepsilon}$, by (4.227) and Lemma 4.5, we get that

$$\int_0^{T-\delta_\varepsilon} \langle f(t), v_\varepsilon(t) \rangle_U dt = 0.$$

This, along with (4.226), (4.225), (4.224), and (4.223), yields that

$$\left| \int_0^{T-\delta_\varepsilon} \langle f(t), u(t + \delta_\varepsilon) \rangle_U dt \right|$$

$$= \left| \int_0^{T/2} \langle f(t), v_u^\varepsilon(t) \rangle_U dt \right| \le C(T/2) \| f \|_{L^1(0,T/2;U)} \varepsilon.$$

Since $\varepsilon > 0$ was arbitrarily taken and $\delta_\varepsilon < \varepsilon$, the above yields that

$$\int_0^{T-\delta_\varepsilon} \langle f(t), u(t + \delta_\varepsilon) \rangle_U dt \to 0, \quad \text{as} \quad \varepsilon \to 0. \tag{4.228}$$

Now, by the density of $C([0, T]; U)$ in $L^1(0, T; U)$, we obtain (4.222) from (4.228).

Hence, we complete the proof of Lemma 4.7. □

We now are on the position to prove Theorem 4.8.

Proof Let $T > 0$. According to Theorem 4.7, each $g \in \mathscr{O}_T$ induces a linear and bounded functional $\widehat{\mathscr{F}}_g$ over $\mathscr{R}_T^0$, via

$$\widehat{\mathscr{F}}_g(y_T) \triangleq \langle y_T, g \rangle_{\mathscr{R}_T, \mathscr{O}_T}, \quad y_T \in \mathscr{R}_T^0, \tag{4.229}$$

where $\langle \cdot, \cdot \rangle_{\mathscr{R}_T, \mathscr{O}_T}$ is given by (4.188). Thus, we obtain a map Ψ_T from $\mathscr{O}_T$ to $(\mathscr{R}_T^0)^*$ defined by

$$\Psi_T(g) \triangleq \widehat{\mathscr{F}}_g \quad \text{for all} \quad g \in \mathscr{O}_T. \tag{4.230}$$

One can easily check that $\Psi_T : \mathscr{O}_T \mapsto (\mathscr{R}_T^0)^*$ is linear. The rest of the proof is organized by the following three steps.

Step 1. We show that

$$\|g\|_{\mathscr{O}_T} = \|\Psi_T(g)\|_{(\mathscr{R}_T^0)^*} \quad \text{for all} \ g \in \mathscr{O}_T. \tag{4.231}$$

Arbitrarily fix $g \in \mathscr{O}_T$. On one hand, it follows from (4.229) that

$$\|\widehat{\mathscr{F}}_g\|_{(\mathscr{R}_T^0)^*} = \sup_{y_T \in B_{\mathscr{R}_T^0}(0,1)} \langle y_T, g \rangle_{\mathscr{R}_T, \mathscr{O}_T} \leq \|g\|_{\mathscr{O}_T}, \tag{4.232}$$

where $B_{\mathscr{R}_T^0}(0,1)$ is the closed unit ball in $\mathscr{R}_T^0$.

On the other hand, we arbitrarily fix $t_0 \in (0, T)$. Then according to the Hahn-Banach Theorem and the Riesz Representation Theorem (see Theorems 1.5 and 1.4), there is a control $u_{t_0} \in L^\infty(0, t_0; U)$ so that

$$\|g\|_{L^1(0,t_0;U)} = \langle u_{t_0}, g \rangle_{L^\infty(0,t_0;U), L^1(0,t_0;U)} \quad \text{and} \quad \|u_{t_0}\|_{L^\infty(0,t_0;U)} = 1. \tag{4.233}$$

Write $\widetilde{u}_{t_0}$ for the zero extension of u_{t_0} over $(0, T)$. Then it follows from (4.214) that $y(T; 0, 0, \widetilde{u}_{t_0}) \in \mathscr{R}_T^0$. Now, by (4.233), (4.188), (4.229), and (4.173), one can directly check that

$$\begin{aligned}
\|g\|_{L^1(0,t_0;U)} &= \langle \widetilde{u}_{t_0}, g \rangle_{L^\infty(0,T;U), L^1(0,T;U)} = \langle y(T; 0, 0, \widetilde{u}_{t_0}), g \rangle_{\mathscr{R}_T, \mathscr{O}_T} \\
&= \widehat{\mathscr{F}}_g\big(y(T; 0, 0, \widetilde{u}_{t_0})\big) \leq \|\widehat{\mathscr{F}}_g\|_{(\mathscr{R}_T^0)^*} \|y(T; 0, 0, \widetilde{u}_{t_0})\|_{\mathscr{R}_T} \\
&\leq \|\widehat{\mathscr{F}}_g\|_{(\mathscr{R}_T^0)^*} \|\widetilde{u}_{t_0}\|_{L^\infty(0,T;U)} = \|\widehat{\mathscr{F}}_g\|_{(\mathscr{R}_T^0)^*}.
\end{aligned}$$

Since t_0 was arbitrarily taken from $(0, T)$, the above yields that

$$\|g\|_{\mathscr{O}_T} = \|g\|_{L^1(0,T;U)} \leq \|\widehat{\mathscr{F}}_g\|_{(\mathscr{R}_T^0)^*}.$$

This, together with (4.232), implies (4.231).

Step 2. We claim that

$$\Psi_T : \mathscr{O}_T \mapsto (\mathscr{R}_T^0)^* \quad \text{is surjective.}$$

Let $\widehat{f} \in (\mathscr{R}_T^0)^*$. We aim to find $\widehat{g} \in \mathscr{O}_T$ so that

$$\widehat{f} = \Psi_T(\widehat{g}) \quad \text{in} \ (\mathscr{R}_T^0)^*. \tag{4.234}$$

In what follows, for each $u \in L^\infty(0, t_0; U)$, with $t_0 \in (0, T)$, we denote by $\widetilde{u}$ the zero extension of u over $(0, T)$. Then it follows from (4.214) that $y(T; 0, 0, \widetilde{u}) \in \mathscr{R}_T^0$. We define, for each $t_0 \in (0, T)$, a map $\mathscr{G}_{\widehat{f}, t_0}$ from $L^\infty(0, t_0; U)$ to $\mathbb{R}$ by setting

$$\mathscr{G}_{\widehat{f}, t_0}(u) \triangleq \langle \widehat{f}, y(T; 0, 0, \widetilde{u}) \rangle_{(\mathscr{R}_T^0)^*, \mathscr{R}_T^0}, \quad u \in L^\infty(0, t_0; U). \tag{4.235}$$

From (4.235), we see that for each $t_0 \in (0, T)$,

$$|\mathscr{G}_{\widehat{f},t_0}(u)| \leq \|\widehat{f}\|_{(\mathscr{R}_T^0)^*}\|y(T;0,0,\widetilde{u})\|_{\mathscr{R}_T} \quad \text{for all} \ \ u \in L^\infty(0,t_0;U). \tag{4.236}$$

Arbitrarily fix $t_0 \in (0,T)$. By (i) of Lemma 4.7, there exists a constant $C(T,t_0) > 0$ so that for each $u \in L^\infty(0,t_0;U)$, there is a control $v_u \in L^\infty(0,T;U)$ satisfying that

$$y(T;0,0,\widetilde{u}) = y(T;0,0,\chi_{(t_0,T)}v_u); \quad \|v_u\|_{L^\infty(0,T;U)} \leq C(T,t_0)\|\widetilde{u}\|_{L^2(0,T;U)}. \tag{4.237}$$

From the equality in (4.237) and (4.173), we find that

$$\|y(T;0,0,\widetilde{u})\|_{\mathscr{R}_T} \leq \|v_u\|_{L^\infty(0,T;U)},$$

which, combined with the inequality in (4.237), indicates that

$$\|y(T;0,0,\widetilde{u})\|_{\mathscr{R}_T} \leq C(T,t_0)\|u\|_{L^2(0,t_0;U)} \quad \text{for all} \ \ u \in L^\infty(0,t_0;U).$$

This, together with (4.236), implies that for each $t_0 \in (0,T)$,

$$|\mathscr{G}_{\widehat{f},t_0}(u)| \leq C(T,t_0)\|\widehat{f}\|_{(\mathscr{R}_T^0)^*}\|u\|_{L^2(0,t_0;U)} \quad \text{for all} \ \ u \in L^\infty(0,t_0;U). \tag{4.238}$$

By (4.238) and the Hahn-Banach Theorem (see Theorem 1.5), we can uniquely extend $\mathscr{G}_{\widehat{f},t_0}$ to be an element in $(L^2(0,t_0;U))^*$, denoted in the same manner, so that

$$|\mathscr{G}_{\widehat{f},t_0}(u)| \leq C(T,t_0)\|\widehat{f}\|_{(\mathscr{R}_T^0)^*}\|u\|_{L^2(0,t_0;U)} \quad \text{for all} \ \ u \in L^2(0,t_0;U). \tag{4.239}$$

By (4.239), using the Riesz Representation Theorem (see Theorem 1.4), we find that for each $t_0 \in (0,T)$, there exists a $g_{t_0} \in L^2(0,t_0;U)$ so that

$$\mathscr{G}_{\widehat{f},t_0}(u) = \int_0^{t_0} \langle g_{t_0}(t), u(t)\rangle_U\,dt \quad \text{for all} \ \ u \in L^2(0,t_0;U). \tag{4.240}$$

We now take $v \in L^\infty(0,t_0;U)$ so that $y(T;0,0,\widetilde{v}) = 0$. (Here, $\widetilde{v}$ is the zero extension of v over $(0,T)$.) By (4.240) and (4.235), we see that

$$\int_0^{t_0} \langle g_{t_0}(t), v(t)\rangle_U\,dt = \mathscr{G}_{\widehat{f},t_0}(v) = 0.$$

This, together with Lemma 4.5, implies that

$$g_{t_0} \in \mathscr{O}_{t_0} \quad \text{for each} \ \ t_0 \in (0,T). \tag{4.241}$$

Meanwhile, by (4.240), (4.236), and (4.173), one can easily check that for each $u \in L^\infty(0, t_0; U)$,

$$\int_0^{t_0} \langle g_{t_0}(t), u(t) \rangle_U \, dt \leq \|\widehat{f}\|_{(\mathscr{R}_T^0)^*} \|y(T; 0, 0, \widetilde{u})\|_{\mathscr{R}_T} \leq \|\widehat{f}\|_{(\mathscr{R}_T^0)^*} \|u\|_{L^\infty(0, t_0; U)}.$$

This, together with (4.241), implies that

$$\|g_{t_0}\|_{\mathscr{O}_{t_0}} = \|g_{t_0}\|_{L^1(0, t_0; U)} \leq \|\widehat{f}\|_{(\mathscr{R}_T^0)^*} \text{ for all } t_0 \in (0, T). \qquad (4.242)$$

We next define a function $\widehat{g} : (0, T) \mapsto U$ in the following manner: For each $t_0 \in (0, T)$,

$$\widehat{g}(t) \triangleq g_{t_0}(t) \text{ for all } t \in (0, t_0). \qquad (4.243)$$

Then, the map $\widehat{g}$ is well defined on $(0, T)$. In fact, when $0 < t_1 < t_2 < T$, it follows from (4.240) and (4.235) that for each $u \in L^\infty(0, t_1; U)$,

$$\int_0^{t_1} \langle g_{t_1}(t), u(t) \rangle_U \, dt = \mathscr{G}_{\widehat{f}, t_1}(u) = \langle \widehat{f}, y(T; 0, 0, \widetilde{u}) \rangle_{(\mathscr{R}_T^0)^*, \mathscr{R}_T^0} = \mathscr{G}_{\widehat{f}, t_2}(\widetilde{u}|_{(0, t_2)})$$

$$= \int_0^{t_2} \langle g_{t_2}(t), \widetilde{u}(t) \rangle_U \, dt = \int_0^{t_1} \langle g_{t_2}(t), u(t) \rangle_U \, dt,$$

which indicates that $g_{t_1}(\cdot) = g_{t_2}(\cdot)$ over $(0, t_1)$. So we can check from (4.243) that the function $\widehat{g}$ is well defined. By (4.243) and (4.242), we see that

$$\|\widehat{g}\|_{L^1(0, T; U)} \leq \|\widehat{f}\|_{(\mathscr{R}_T^0)^*}. \qquad (4.244)$$

From (H1), (4.244), (4.243), (4.241), and (ii) of Lemma 4.7, it follows that

$$\widehat{g} \in \mathscr{O}_T \text{ and } \|\widehat{g}\|_{\mathscr{O}_T} \leq \|\widehat{f}\|_{(\mathscr{R}_T^0)^*}. \qquad (4.245)$$

Moreover, by (4.235), (4.240), and (4.243), we deduce that for each $t_0 \in (0, T)$,

$$\langle \widehat{f}, y(T; 0, 0, \widetilde{u}) \rangle_{(\mathscr{R}_T^0)^*, \mathscr{R}_T^0} = \int_0^T \langle \widehat{g}(t), \widetilde{u}(t) \rangle_U \, dt \text{ for all } u \in L^\infty(0, t_0; U).$$

$$(4.246)$$

Finally, for each $y_T \in \mathscr{R}_T^0$, it follows from (4.214) that there is $u_{y_T} \in L^\infty(0, T; U)$ so that

$$y_T = y(T; 0, 0, u_{y_T}) \text{ and } \lim_{s \to T} \|u_{y_T}\|_{L^\infty(s, T; U)} = 0.$$

By these and (4.173), we can check that when $s \to T$,

$$\|y(T; 0, 0, \chi_{(0,s)}u_{y_T}) - y_T\|_{\mathscr{R}_T} = \|y(T; 0, 0, \chi_{(s,T)}u_{y_T})\|_{\mathscr{R}_T}$$

$$\leq \|u_{y_T}\|_{L^\infty(s,T;U)} \to 0,$$

which indicates that

$$y(T; 0, 0, \chi_{(0,s)}u_{y_T}) \to y_T \quad \text{in} \quad \mathscr{R}_T. \tag{4.247}$$

Notice that $y(T; 0, 0, \chi_{(0,s)}u_{y_T}) \in \mathscr{R}_T^0$ and $\widehat{g} \in \mathscr{O}_T$ (see (4.245)). Thus, by (4.247), (4.246), (4.188), and the Lebesgue Dominated Convergence Theorem, we find that for each $y_T \in \mathscr{R}_T^0$,

$$\langle \widehat{f}, y_T \rangle_{(\mathscr{R}_T^0)^*, \mathscr{R}_T^0} = \lim_{s \to T} \langle \widehat{f}, y(T; 0, 0, \chi_{(0,s)}u_{y_T}) \rangle_{(\mathscr{R}_T^0)^*, \mathscr{R}_T^0}$$

$$= \lim_{s \to T} \int_0^T \langle \widehat{g}(t), \chi_{(0,s)}(t)u_{y_T}(t) \rangle_U \, dt$$

$$= \int_0^T \langle \widehat{g}(t), u_{y_T}(t) \rangle_U \, dt = \langle y_T, \widehat{g} \rangle_{\mathscr{R}_T, \mathscr{O}_T}.$$

This, together with (4.229), implies that

$$\langle \widehat{f}, y_T \rangle_{(\mathscr{R}_T^0)^*, \mathscr{R}_T^0} = \widehat{\mathscr{F}}_{\widehat{g}}(y_T) \quad \text{for all} \quad y_T \in \mathscr{R}_T^0, \quad \text{i.e.,} \quad \widehat{f} = \widehat{\mathscr{F}}_{\widehat{g}} \quad \text{in} \quad (\mathscr{R}_T^0)^*.$$

Hence, (4.234) follows immediately from the latter and (4.230).

Step 3. We show the second equality in (4.215).

The second equality in (4.215) follows from (4.188).

Hence, we finish the proof of Theorem 4.8. □

Remark 4.10 The assumption (H1) in Theorem 4.8 can be replaced by the following weaker condition (see [19, Theorem 2.6 and Proposition 9]).
(H2) There is $p_0 \in [2, +\infty)$ so that $\mathscr{A}_{p_0}(T, \widehat{t}) \subseteq \mathscr{A}_\infty(T, \widehat{t})$ for all $T, \widehat{t}$, with $0 < \widehat{t} < T < +\infty$, where

$$\mathscr{A}_{p_0}(T, \widehat{t}) \triangleq \{y(T; 0, 0, u) \mid u \in L^{p_0}(0, T; U) \text{ with } u|_{(\widehat{t}, T)} = 0\};$$

$$\mathscr{A}_\infty(T, \widehat{t}) \triangleq \{y(T; 0, 0, v) \mid v \in L^\infty(0, T; U) \text{ with } v|_{(0, \widehat{t})} = 0\}.$$

It deserves to mention that (H2) is essentially weaker than (H1), since the next two facts are proved in [19]. Fact one: (H1) implies (H2), while (H1) does not hold when (H2) is true for many cases. Fact two: In finite dimensional cases, any pair of matrices (A, B) in $\mathbb{R}^{n \times n} \times \mathbb{R}^{n \times m}$ satisfies (H2).

4.3.3   Conditions on Separation in Weak Maximum Principle

The aim of this subsection is to give a condition that ensures the separation in Theorem 4.6.

Theorem 4.9 *Suppose that (H1) is true. Then the target $\{0\}$ and $Y_R(t_E^*)$ are separable in $\mathscr{R}_{t_E^*}$ by a regular vector $\psi \in (\mathscr{R}_{t_E^*})^* \setminus \{0\}$.*

The proof of Theorem 4.9 needs the next lemma.

Lemma 4.8 *For each $T > 0$, it holds that $B_{\mathscr{R}_T} = \overline{B}_{\mathscr{R}_T^0}^{\sigma(\mathscr{R}_T, \mathscr{O}_T)}$. Here, $B_{\mathscr{R}_T}$ and $B_{\mathscr{R}_T^0}$ are the closed unit balls in $\mathscr{R}_T$ and $\mathscr{R}_T^0$, respectively, and the set $\overline{B}_{\mathscr{R}_T^0}^{\sigma(\mathscr{R}_T, \mathscr{O}_T)}$ is the closure of $B_{\mathscr{R}_T^0}$ in the space $\mathscr{R}_T$, under the weak star topology $\sigma(\mathscr{R}_T, \mathscr{O}_T)$.*

Proof Let $T > 0$. We first prove that

$$B_{\mathscr{R}_T} \subseteq \overline{B}_{\mathscr{R}_T^0}^{\sigma(\mathscr{R}_T, \mathscr{O}_T)}. \tag{4.248}$$

To this end, let $y_T \in B_{\mathscr{R}_T}$. By (4.172) and (4.173), there exists a sequence $\{v_k\}_{k \geq 1} \subseteq L^\infty(0, T; U)$ so that for all $k \geq 1$,

$$y_T = y(T; 0, 0, v_k) \quad \text{and} \quad \|y_T\|_{\mathscr{R}_T} \leq \|v_k\|_{L^\infty(0,T;U)} \leq \|y_T\|_{\mathscr{R}_T} + T/2k. \tag{4.249}$$

For each $k \geq 1$, we set

$$\lambda_k \triangleq \frac{\|y_T\|_{\mathscr{R}_T}}{\|y_T\|_{\mathscr{R}_T} + T/2k} \quad \text{and} \quad u_k \triangleq \chi_{(0, T - T/2k)} \lambda_k v_k. \tag{4.250}$$

On one hand, it follows from (4.250) and the second inequality in (4.249) that

$$\|u_k\|_{L^\infty(0,T;U)} \leq \|y_T\|_{\mathscr{R}_T} \leq 1 \quad \text{for all } k \geq 1. \tag{4.251}$$

This, together with (4.214), (4.250), and (4.173), implies that

$$y(T; 0, 0, u_k) \in B_{\mathscr{R}_T^0} \quad \text{for all } k \geq 1. \tag{4.252}$$

On the other hand, by the equality in (4.249), (4.188), and (4.250), we find that for each $f \in \mathscr{O}_T$,

$$\langle y(T; 0, 0, u_k) - y_T, f \rangle_{\mathscr{R}_T, \mathscr{O}_T} = \langle y(T; 0, 0, u_k - v_k), f \rangle_{\mathscr{R}_T, \mathscr{O}_T}$$

$$= \int_0^T \langle u_k(t) - v_k(t), f(t) \rangle_U \, dt \to 0 \quad \text{as } k \to +\infty.$$

This, together with (4.252), implies that $y_T \in \overline{B}_{\mathscr{R}_T^0}^{\sigma(\mathscr{R}_T, \mathscr{O}_T)}$. Hence, (4.248) follows.

We next show that

$$B_{\mathscr{R}_T} \supseteq \overline{B}_{\mathscr{R}_T^0}^{\sigma(\mathscr{R}_T, \mathscr{O}_T)}.$$ (4.253)

For this purpose, we let $y_T \in \mathscr{R}_T$ and $\{y_k\}_{k \geq 1} \subseteq B_{\mathscr{R}_T^0}$ so that

$$y_k \to y_T \text{ in the topology } \sigma(\mathscr{R}_T, \mathscr{O}_T) \text{ as } k \to +\infty.$$

Since $\mathscr{R}_T = \mathscr{O}_T^*$ (see Theorem 4.7), we have that

$$y_k \to y_T \text{ in the weak star topology as } k \to +\infty.$$

Hence,

$$\|y_T\|_{\mathscr{R}_T} \leq \liminf_{k \to +\infty} \|y_k\|_{\mathscr{R}_T} \leq 1,$$

which indicates that $y_T \in B_{\mathscr{R}_T}$. Then (4.253) holds.

Finally, it follows from (4.248) and (4.253) that $B_{\mathscr{R}_T} = \overline{B}_{\mathscr{R}_T^0}^{\sigma(\mathscr{R}_T, \mathscr{O}_T)}$. This completes the proof of Lemma 4.8. $\square$

We now are in a position to prove Theorem 4.9.

Proof (Proof of Theorem 4.9) The proof will be carried out by the following three stages.

Stage 1. We show that

$$e^{At_E^*} \widehat{y}_0 \in \mathscr{R}_{t_E^*}^0 \text{ and } \|e^{At_E^*} \widehat{y}_0\|_{\mathscr{R}_{t_E^*}} = M.$$ (4.254)

Its proof will be divided into the following two steps:
Step 1. We claim that

$$e^{At_E^*} \widehat{y}_0 \in \mathscr{R}_{t_E^*}^0.$$ (4.255)

In fact, it follows by (H1) that there exists $u_1 \in L^\infty(0, t_E^*/2; U)$ so that

$$y(t_E^*/2; 0, \widehat{y}_0, u_1) = 0.$$ (4.256)

Write $\widetilde{u}_1$ for the zero extension of u_1 over $(0, +\infty)$. Then it follows from (4.256) that

$$y(t_E^*; 0, \widehat{y}_0, \widetilde{u}_1) = e^{At_E^*/2} y(t_E^*/2; 0, \widehat{y}_0, u_1) = 0,$$ (4.257)

which indicates that

$$e^{At_E^*} \widehat{y}_0 = -y(t_E^*; 0, 0, \widetilde{u}_1).$$

Since $\widetilde{u}_1 = 0$ over $(t_E^*/2, t_E^*)$, by the above equality and the definition of $\mathscr{R}_{t_E^*}^0$ (see (4.214)), we obtain (4.255).

Step 2. We show that

$$\|e^{At_E^*}\widehat{y}_0\|_{\mathcal{R}_{t_E^*}^0} = M. \tag{4.258}$$

To this end, we let $(0, \widehat{y}_0, t_E^*, u^*)$ be an optimal tetrad to the problem $(TP)_{min}^{Q_S, Q_E}$. Then we have that

$$y(t_E^*; 0, \widehat{y}_0, u^*) = 0 \quad \text{and} \quad \|u^*\|_{L^\infty(0, t_E^*; U)} \leq M.$$

These yield that

$$e^{At_E^*}\widehat{y}_0 = -y(t_E^*; 0, 0, u^*) \in \mathcal{R}_{t_E^*}$$

and

$$\|e^{At_E^*}\widehat{y}_0\|_{\mathcal{R}_{t_E^*}} \leq M. \tag{4.259}$$

Now, by contradiction, we suppose that (4.258) was not true. Then by (4.259), we would find that

$$\|e^{At_E^*}\widehat{y}_0\|_{\mathcal{R}_{t_E^*}} < M.$$

This, together with (4.172) and (4.173), implies that there is a control $u_2 \in L^\infty(0, t_E^*; U)$ so that

$$e^{At_E^*}\widehat{y}_0 = y(t_E^*; 0, 0, u_2) \quad \text{and} \quad \|u_2\|_{L^\infty(0, t_E^*; U)} < M. \tag{4.260}$$

Choose $\varepsilon > 0$ small enough so that

$$0 < \varepsilon \leq M - \|u_2\|_{L^\infty(0, t_E^*; U)}. \tag{4.261}$$

Then, by (H1) we can apply Theorem 1.20 to find $\delta \in (0, t_E^*/2)$ and $u_\delta \in L^\infty(0, +\infty; U)$ so that

$$y(t_E^*/2; 0, z_\delta, u_\delta) = 0 \quad \text{and} \quad \|u_\delta\|_{L^\infty(0, +\infty; U)} \leq \varepsilon, \tag{4.262}$$

where

$$z_\delta \triangleq \widehat{y}_0 - y(\delta; 0, \widehat{y}_0, -u_2). \tag{4.263}$$

We define a new control in the following manner:

$$u_3(t) \triangleq \chi_{(0, t_E^*/2)} u_\delta(t) - u_2(t + \delta) \quad \text{for a.e. } t \in (0, t_E^* - \delta). \tag{4.264}$$

By (4.263) and (4.264), we have that

$$y(t_E^* - \delta; 0, \widehat{y}_0, u_3)$$

$$= e^{A(t_E^* - \delta)}\widehat{y}_0 + \int_0^{t_E^* - \delta} e^{A(t_E^* - \delta - t)} B u_3(t)dt$$

$$= e^{A(t_E^* - \delta)}[z_\delta + y(\delta; 0, \widehat{y}_0, -u_2)]$$

$$+ \int_0^{t_E^* - \delta} e^{A(t_E^* - \delta - t)} B[\chi_{(0,t_E^*/2)} u_\delta(t) - u_2(t + \delta)]dt,$$

which, combined with the equalities in (4.260) and (4.262), indicates that

$$y(t_E^* - \delta; 0, \widehat{y}_0, u_3)$$

$$= e^{A(t_E^* - \delta)} z_\delta + \int_0^{t_E^* - \delta} e^{A(t_E^* - \delta - t)} B \chi_{(0,t_E^*/2)} u_\delta(t)dt$$

$$+ e^{A(t_E^* - \delta)} \left(e^{A\delta}\widehat{y}_0 - \int_0^\delta e^{A(\delta - t)} B u_2(t)dt \right) - \int_\delta^{t_E^*} e^{A(t_E^* - t)} B u_2(t)dt \qquad (4.265)$$

$$= y(t_E^* - \delta; 0, z_\delta, \chi_{(0,t_E^*/2)} u_\delta) + y(t_E^*; 0, \widehat{y}_0, -u_2) = 0.$$

Moreover, it follows from (4.264), (4.262), and (4.261) that

$$\|u_3\|_{L^\infty(0,t_E^* - \delta; U)} \leq \|u_\delta\|_{L^\infty(0,+\infty; U)} + \|u_2\|_{L^\infty(0,t_E^*; U)} \leq M.$$

This, together with (4.265), contradicts the optimality of t_E^*. Hence, (4.258) holds. Finally, (4.254) follows from (4.255) and (4.258) immediately.

Stage 2. We claim that there exists a regular vector $\varphi^* \in (\mathcal{R}_{t_E^*})^* \setminus \{0\}$ so that

$$\sup_{z \in \mathscr{A}_M} \langle \varphi^*, z \rangle_{(\mathcal{R}_{t_E^*})^*, \mathcal{R}_{t_E^*}} \leq \langle \varphi^*, -e^{At_E^*}\widehat{y}_0 \rangle_{(\mathcal{R}_{t_E^*})^*, \mathcal{R}_{t_E^*}}. \qquad (4.266)$$

Here

$$\mathscr{A}_M \triangleq \{z \in \mathcal{R}_{t_E^*}^0 \mid \|z\|_{\mathcal{R}_{t_E^*}} \leq M\}. \qquad (4.267)$$

By (4.267), (4.254), and Theorem 1.11, there exists $\widehat{\varphi} \in (\mathcal{R}_{t_E^*}^0)^* \setminus \{0\}$ so that

$$\sup_{z \in \mathscr{A}_M} \langle \widehat{\varphi}, z \rangle_{(\mathcal{R}_{t_E^*}^0)^*, \mathcal{R}_{t_E^*}^0} \leq \langle \widehat{\varphi}, -e^{At_E^*}\widehat{y}_0 \rangle_{(\mathcal{R}_{t_E^*}^0)^*, \mathcal{R}_{t_E^*}^0}. \qquad (4.268)$$

Meanwhile, by Theorem 4.8, there exists $f_{\widehat{\varphi}} \in \mathcal{O}_{t_E^*} \setminus \{0\}$ so that

$$\|f_{\widehat{\varphi}}\|_{\mathcal{O}_{t_E^*}} = \|\widehat{\varphi}\|_{(\mathcal{R}_{t_E^*}^0)^*} \text{ and } \langle f_{\widehat{\varphi}}, z \rangle_{\mathcal{O}_{t_E^*}, \mathcal{R}_{t_E^*}^0} = \langle \widehat{\varphi}, z \rangle_{(\mathcal{R}_{t_E^*}^0)^*, \mathcal{R}_{t_E^*}^0} \text{ for all } z \in \mathcal{R}_{t_E^*}^0.$$

These, together with (4.268), imply that

$$\sup_{z \in \mathscr{A}_M} \langle f_{\widehat{\varphi}}, z \rangle_{\mathscr{O}_{t_E^*}, \mathscr{R}_{t_E^*}^0} \leq \langle f_{\widehat{\varphi}}, -e^{At_E^*} \widehat{y}_0 \rangle_{\mathscr{O}_{t_E^*}, \mathscr{R}_{t_E^*}^0}. \tag{4.269}$$

Two observations are given in order: First, it follows from Theorem 4.8 and Theorem 4.7, that

$$\langle f_{\widehat{\varphi}}, z \rangle_{\mathscr{O}_{t_E^*}, \mathscr{R}_{t_E^*}^0} = \langle z, f_{\widehat{\varphi}} \rangle_{\mathscr{R}_{t_E^*}, \mathscr{O}_{t_E^*}} \quad \text{for all } z \in \mathscr{R}_{t_E^*}^0.$$

Second, it follows from Lemma 4.8 and (4.267) that

$$\sup_{z \in \mathscr{A}_M} \langle z, f_{\widehat{\varphi}} \rangle_{\mathscr{R}_{t_E^*}, \mathscr{O}_{t_E^*}} = \sup_{z \in \mathscr{R}_{t_E^*}, \|z\|_{\mathscr{R}_{t_E^*}} \leq M} \langle z, f_{\widehat{\varphi}} \rangle_{\mathscr{R}_{t_E^*}, \mathscr{O}_{t_E^*}}.$$

These two observations, together with (4.269), imply that

$$\begin{aligned}\sup_{z \in \mathscr{R}_{t_E^*}, \|z\|_{\mathscr{R}_{t_E^*}} \leq M} \langle z, f_{\widehat{\varphi}} \rangle_{\mathscr{R}_{t_E^*}, \mathscr{O}_{t_E^*}} &= \sup_{z \in \mathscr{A}_M} \langle f_{\widehat{\varphi}}, z \rangle_{\mathscr{O}_{t_E^*}, \mathscr{R}_{t_E^*}^0} \\ &\leq \langle f_{\widehat{\varphi}}, -e^{At_E^*} \widehat{y}_0 \rangle_{\mathscr{O}_{t_E^*}, \mathscr{R}_{t_E^*}^0} = \langle -e^{At_E^*} \widehat{y}_0, f_{\widehat{\varphi}} \rangle_{\mathscr{R}_{t_E^*}, \mathscr{O}_{t_E^*}}.\end{aligned} \tag{4.270}$$

Meanwhile, according to Theorem 4.7, there exists $\widetilde{\varphi} \in (\mathscr{R}_{t_E^*})^*$ so that

$$\langle z, f_{\widehat{\varphi}} \rangle_{\mathscr{R}_{t_E^*}, \mathscr{O}_{t_E^*}} = \langle \widetilde{\varphi}, z \rangle_{(\mathscr{R}_{t_E^*})^*, \mathscr{R}_{t_E^*}} \quad \text{for all } z \in \mathscr{R}_{t_E^*}; \quad \text{and } \|f_{\widehat{\varphi}}\|_{\mathscr{O}_{t_E^*}} = \|\widetilde{\varphi}\|_{(\mathscr{R}_{t_E^*})^*}.$$

This, along with (4.270), Definition 4.6, and Theorem 4.7, leads to (4.266), with $\varphi^* = \widetilde{\varphi}$.

Stage 3. We show the conclusions in Theorem 4.9.

First, we claim that

$$Y_R(t_E^*) = \{e^{At_E^*} \widehat{y}_0\} + \{z \in \mathscr{R}_{t_E^*} \mid \|z\|_{\mathscr{R}_{t_E^*}} \leq M\}. \tag{4.271}$$

To this end, we arbitrarily fix $y_1 \in Y_R(t_E^*)$. By the definition of $Y_R(t_E^*)$ (see (4.174)), there exists $u \in L^\infty(0, t_E^*; U)$ so that

$$y_1 = y(t_E^*; 0, \widehat{y}_0, u) \quad \text{and} \quad \|u\|_{L^\infty(0, t_E^*; U)} \leq M.$$

It follows from the latter, and definitions of $\mathscr{R}_{t_E^*}$ and $\|\cdot\|_{\mathscr{R}_{t_E^*}}$ (see (4.172) and (4.173)) that

$$y_1 = e^{At_E^*} \widehat{y}_0 + y(t_E^*; 0, 0, u) \in \{e^{At_E^*} \widehat{y}_0\} + \{z \in \mathscr{R}_{t_E^*} \mid \|z\|_{\mathscr{R}_{t_E^*}} \leq M\},$$

which indicates that

$$Y_R(t_E^*) \subseteq \{e^{At_E^*} \widehat{y}_0\} + \{z \in \mathscr{R}_{t_E^*} \mid \|z\|_{\mathscr{R}_{t_E^*}} \leq M\}. \tag{4.272}$$

On the other hand, for any $z \in \mathscr{R}_{t_E^*}$ with $\|z\|_{\mathscr{R}_{t_E^*}} \leq M$, by definitions of $\mathscr{R}_{t_E^*}$ and $\|\cdot\|_{\mathscr{R}_{t_E^*}}$ (see (4.172) and (4.173)), there exists $v \in L^\infty(0, t_E^*; U)$ so that

$$z = y(t_E^*; 0, 0, v) \quad \text{and} \quad \|v\|_{L^\infty(0,t_E^*;U)} \leq M.$$

These, together with (4.174), imply that

$$e^{A t_E^*}\widehat{y}_0 + z = y(t_E^*; 0, \widehat{y}_0, v) \in Y_R(t_E^*),$$

which indicates that

$$\{e^{A t_E^*}\widehat{y}_0\} + \{z \in \mathscr{R}_{t_E^*} \mid \|z\|_{\mathscr{R}_{t_E^*}} \leq M\} \subseteq Y_R(t_E^*). \tag{4.273}$$

Hence, (4.271) follows from (4.272) and (4.273) immediately.

Next, by (4.266) and (4.267), we obtain that

$$\langle \varphi^*, e^{A t_E^*}\widehat{y}_0 + z \rangle_{(\mathscr{R}_{t_E^*})^*, \mathscr{R}_{t_E^*}} \leq 0 \quad \text{for all} \ z \in \mathscr{R}_{t_E^*}^0 \ \text{with} \ \|z\|_{\mathscr{R}_{t_E^*}} \leq M. \tag{4.274}$$

We claim that

$$\langle \varphi^*, e^{A t_E^*}\widehat{y}_0 + z \rangle_{(\mathscr{R}_{t_E^*})^*, \mathscr{R}_{t_E^*}} \leq 0 \quad \text{for all} \ z \in \mathscr{R}_{t_E^*} \ \text{with} \ \|z\|_{\mathscr{R}_{t_E^*}} \leq M. \tag{4.275}$$

Indeed, on one hand, by Lemma 4.8, for any fixed $z \in \mathscr{R}_{t_E^*}$ with $\|z\|_{\mathscr{R}_{t_E^*}} \leq M$, there exists a sequence $\{z_k\}_{k \geq 1} \subseteq \mathscr{R}_{t_E^*}^0$ with $\|z_k\|_{\mathscr{R}_{t_E^*}} \leq M$ so that

$$\langle z_k, \psi \rangle_{\mathscr{R}_{t_E^*}, \mathscr{O}_{t_E^*}} \to \langle z, \psi \rangle_{\mathscr{R}_{t_E^*}, \mathscr{O}_{t_E^*}} \quad \text{for all} \ \psi \in \mathscr{O}_{t_E^*}. \tag{4.276}$$

On the other hand, since $\varphi^* \in (\mathscr{R}_{t_E^*})^* \setminus \{0\}$ is regular, by Definition 4.6, Corollary 4.3, and Theorem 4.7, we obtain that there exists $f_{\varphi^*} \in \mathscr{O}_{t_E^*}$ so that for all $u \in L^\infty(0, t_E^*; U)$,

$$\langle \varphi^*, y(t_E^*; 0, 0, u) \rangle_{(\mathscr{R}_{t_E^*})^*, \mathscr{R}_{t_E^*}} = \langle y(t_E^*; 0, 0, u), f_{\varphi^*} \rangle_{\mathscr{R}_{t_E^*}, \mathscr{O}_{t_E^*}}. \tag{4.277}$$

It follows from (4.277) and (4.276) that

$$\langle \varphi^*, z \rangle_{(\mathscr{R}_{t_E^*})^*, \mathscr{R}_{t_E^*}} = \langle z, f_{\varphi^*} \rangle_{\mathscr{R}_{t_E^*}, \mathscr{O}_{t_E^*}}$$
$$= \lim_{k \to +\infty} \langle z_k, f_{\varphi^*} \rangle_{\mathscr{R}_{t_E^*}, \mathscr{O}_{t_E^*}} = \lim_{k \to +\infty} \langle \varphi^*, z_k \rangle_{(\mathscr{R}_{t_E^*})^*, \mathscr{R}_{t_E^*}},$$

which, combined with (4.274), indicates (4.275).

By (4.275) and (4.271), we see that

$$\sup_{z \in Y_R(t_E^*)} \langle \varphi^*, z \rangle_{(\mathscr{R}_{t_E^*})^*, \mathscr{R}_{t_E^*}} \leq 0.$$

From the above inequality and Definitions 4.5 and 4.6, it follows that $Y_R(t_E^*)$ and the target $\{0\}$ are separable in $\mathscr{R}_{t_E^*}$ by a regular vector φ^*.

Hence, we finish the proof of Theorem 4.9. $\square$

At the end of this subsection, we give two examples, which may help us to understand Theorem 4.9 and Theorem 4.6 better.

Example 4.19 Consider the problem $(TP)_{min}^{Q_S, Q_E}$ in Example 4.13. One can easily check two facts as follows: First, this problem can be put into the framework $(\mathscr{H}_{TP})$ (given at the beginning of Section 4.3); Second, (H1) (given before Theorem 4.8) is true. (Here we used Theorem 1.23.) We further assume that $(TP)_{min}^{Q_S, Q_E}$ has optimal controls.

According to Theorem 4.9, the target $\{0\}$ and the set $Y_R(t_E^*)$ are separable in $\mathscr{R}_{t_E^*}$ by a regular vector in $(\mathscr{R}_{t_E^*})^* \setminus \{0\}$. Hence, by Theorem 4.6, $(TP)_{min}^{Q_S, Q_E}$ satisfies the weak Pontryagin Maximum Principle. We end Example 4.19.

Example 4.20 Consider the problem $(TP)_{min}^{Q_S, Q_E}$ in Example 4.14. One can easily check that it can be put into the framework $(\mathscr{H}_{TP})$ (given at the beginning of Section 4.3) and that (H1) (given before Theorem 4.8) holds. (Here we used Theorems 1.22 and 1.21.) Moreover, $(TP)_{min}^{Q_S, Q_E}$ has optimal controls. Then according to Theorem 4.9, the target $\{0\}$ and the set $Y_R(t_E^*)$ are separable in $\mathscr{R}_{t_E^*}$ by a regular vector in $(\mathscr{R}_{t_E^*})^* \setminus \{0\}$. Hence, by Theorem 4.6, $(TP)_{min}^{Q_S, Q_E}$ satisfies the weak Pontryagin Maximum Principle. We end Example 4.20.

4.4   Maximum Principle for Maximal Time Controls

This section studies the maximal time control problem $(TP)_{max}^{Q_S, Q_E}$ under the following framework $(\mathscr{A}_{TP}^{max})$:

(i) The state space Y and the control space U are real separable Hilbert spaces.
(ii) Let $T > 0$. Let $A : \mathscr{D}(A) \subseteq Y \mapsto Y$ generate a C_0 semigroup $\{e^{At}\}_{t \geq 0}$ on Y. Let $D(\cdot) \in L^\infty(0, T; \mathscr{L}(Y))$ and $B(\cdot) \in L^\infty(0, T; \mathscr{L}(U, Y))$. The controlled system is as:

$$\dot{y}(t) = Ay(t) + D(t)y(t) + B(t)u(t), \quad t \in (0, T). \tag{4.278}$$

Let $\{\Phi(t, s) : t \geq s \geq 0\}$ be the evolution system generated by $A + D(\cdot)$ over Y. Given $\tau \in [0, T)$, $y_0 \in Y$ and $u \in L^\infty(\tau, T; U)$, write $y(\cdot; \tau, y_0, u)$ for the solution of (4.278) with the initial condition that $y(\tau) = y_0$.

(iii) The control constraint set $\mathbb{U}$ is a nonempty, bounded, convex, and closed subset in U.

(iv) Let $Q_S = [0, T) \times Y_S$ and $Q_E = \{T\} \times Y_E$, where Y_S and Y_E are two nonempty, bounded, convex, and closed subsets in Y so that $Y_S \cap Y_E = \emptyset$.

As what we did in the previous sections, we always assume that the problem $(TP)_{max}^{Q_S, Q_E}$ *has an optimal control.* The aim of this section is to derive the classical and the local Pontryagin Maximum Principles for $(TP)_{max}^{Q_S, Q_E}$, through using the similar methods to those used in Sections 4.1 and 4.2.

Recall that $(TP)_{max}^{Q_S, Q_E}$ is defined as follows:

$$\sup\{t_S \mid t_S \in [0, T), \ y_0 \in Y_S, u \in L(t_S, T; \mathbb{U}) \ \text{and} \ y(T; t_S, y_0, u) \in Y_E\}.$$

When (t_S^*, y_0^*, T, u^*) is an optimal tetrad for $(TP)_{max}^{Q_S, Q_E}$, u^* is called an optimal control and t_S^* is called the maximal time. (Throughout this section, we use t_S^* to denote the maximal time of $(TP)_{max}^{Q_S, Q_E}$.) We write $y^*(\cdot)$ for $y(\cdot; t_S^*, y_0^*, u^*)$, and call it as the corresponding optimal trajectory (or the optimal state).

4.4.1 Classical Maximum Principle for Maximal Time Controls

We first give the definition of classical Pontryagin Maximum Principle for the problem $(TP)_{max}^{Q_S, Q_E}$.

Definition 4.8 Several definitions are given in order.

(i) An optimal control u^* (associated with an optimal tetrad (t_S^*, y_0^*, T, u^*)) to $(TP)_{max}^{Q_S, Q_E}$ is said to satisfy the classical Pontryagin Maximum Principle, if there exists $\varphi(\cdot) \in C([t_S^*, T]; Y)$, with $\varphi(\cdot) \neq 0$ (i.e., φ is not a zero function), so that

$$\dot{\varphi}(t) = -A^*\varphi(t) - D(t)^*\varphi(t) \ \text{for a.e.} \ t \in (t_S^*, T); \qquad (4.279)$$

$$H(t, y^*(t), u^*(t), \varphi(t)) = \max_{u \in \mathbb{U}} H(t, y^*(t), u, \varphi(t)) \ \text{for a.e.} \ t \in (t_S^*, T),$$
$$\qquad (4.280)$$

where $y^* \in C([t_S^*, T]; Y)$ is the corresponding optimal trajectory, and where

$$H(t, y, u, \varphi) \triangleq \langle \varphi, D(t)y + B(t)u \rangle_Y \qquad (4.281)$$

for a.e. $t \in (t_S^*, T)$ and all $(y, u, \varphi) \in Y \times U \times Y$;

$$\langle \varphi(t_S^*), y_0 - y_0^* \rangle_Y \leq 0 \ \text{for each} \ y_0 \in Y_S; \qquad (4.282)$$

and

$$\langle \varphi(T), z - y^*(T) \rangle_Y \geq 0 \ \text{ for each } \ z \in Y_E. \tag{4.283}$$

Here, (4.279) is called the dual (or adjoint) equation; (4.280) is called the maximum condition; the function $H(\cdot)$ defined by (4.281) is called the Hamiltonian (associated with $(TP)_{max}^{Qs,QE}$); (4.282) and (4.283) are called the transversality conditions.

(ii) The problem $(TP)_{max}^{Qs,QE}$ is said to satisfy the classical Pontryagin Maximum Principle, if any optimal control satisfies the classical Pontryagin Maximum Principle.

Before presenting main results of this subsection, we define, for any $0 \leq t_1 < t_2 \leq T$, the following two sets:

$$Y_R^{max}(t_2, t_1) \triangleq \left\{ y(t_2; t_1, y_0, u) \mid y_0 \in Y_S, \ u \in L(t_1, t_2; \mathbb{U}) \right\} \tag{4.284}$$

and

$$Y_C^{max}(t_1) \triangleq \left\{ y_0 \in Y \mid \exists u \in L(t_1, T; \mathbb{U}) \ \text{ so that } \ y(T; t_1, y_0, u) \in Y_E \right\}. \tag{4.285}$$

We call $Y_R^{max}(t_2, t_1)$ the reachable set at the time t_2 from the time t_1 of the Equation (4.278). We call $Y_C^{max}(t_1)$ the controllable set at the time t_1 for the Equation (4.278).

We now present the first main result of this subsection.

Theorem 4.10 *The problem* $(TP)_{max}^{Qs,QE}$ *satisfies the classical Pontryagin Maximum Principle if and only if* $Y_R^{max}(T, t_S^*)$ *and* Y_E *are separable in* Y.

Proof We first show the sufficiency. Assume that $Y_R^{max}(T, t_S^*)$ and Y_E are separable in Y. Let (t_S^*, y_0^*, T, u^*) be an optimal tetrad to $(TP)_{max}^{Qs,QE}$. By the similar arguments to those used to show Theorem 4.1 (where $Y_R(t_E^*)$ is replaced by $Y_R(T, t_S^*)$), we can prove that there exists $\varphi(\cdot) \in C([t_S^*, T]; Y)$ with $\varphi(\cdot) \neq 0$ so that (4.279)–(4.283) hold. Then by Definition 4.8, we find that (t_S^*, y_0^*, T, u^*) satisfies the classical Pontryagin Maximum Principle, consequently, so does $(TP)_{max}^{Qs,QE}$. Hence, we have proved the sufficiency.

We next show the necessity. Suppose that $(TP)_{max}^{Qs,QE}$ satisfies the classical Pontryagin Maximum Principle. Let (t_S^*, y_0^*, T, u^*) be an optimal tetrad to $(TP)_{max}^{Qs,QE}$. By Definition 4.8, there exists $\varphi(\cdot) \in C([t_S^*, T]; Y)$ with $\varphi(\cdot) \neq 0$ so that (4.279)–(4.283) hold. Then using the similar arguments as those used in the proof of Theorem 4.1, we can prove that $Y_R^{max}(T, t_S^*)$ and Y_E are separable in Y. Hence, the necessity has been shown.

Thus, we complete the proof of Theorem 4.10. □

We next give the second main result of this subsection.

Theorem 4.11 *Assume that $A : \mathscr{D}(A) \subseteq Y \mapsto Y$ generates a C_0 group. Then the problem $(TP)_{max}^{Q_S,Q_E}$ satisfies the classical Pontryagin Maximum Principle if and only if $Y_C^{max}(t_S^*)$ and Y_S are separable in Y.*

Proof We define a minimal time optimal control problem $(TP)_{min}^{\widehat{Q}_S,\widehat{Q}_E}$, where the controlled system is

$$\dot{z}(t) = \begin{cases} -Az(t) - D(T-t)z(t) - B(T-t)v(t), & t \in (0,T), \\ 0, & t \in [T, +\infty), \end{cases} \quad (4.286)$$

and where

$$\widehat{Q}_S \triangleq \{0\} \times Y_E \quad \text{and} \quad \widehat{Q}_E \triangleq (0,T] \times Y_S. \quad (4.287)$$

Write $z(\cdot; \tau, z_0, v)$ for the solution of (4.286) with the initial condition that $z(\tau) = z_0$, where $\tau \in [0, +\infty)$ and $z_0 \in Y$ are arbitrarily fixed.

We now show the sufficiency. Assume that $Y_C^{max}(t_S^*)$ and Y_S are separable in Y. Let (t_S^*, y_0^*, T, u^*) be an optimal tetrad to $(TP)_{max}^{Q_S,Q_E}$. Then it follows by Theorem 2.2 and Definition 2.2 that

$$(0, y(T; t_S^*, y_0^*, u^*), T - t_S^*, u^*(T - \cdot)) \quad \text{is an optimal tetrad to} \quad (TP)_{min}^{\widehat{Q}_S,\widehat{Q}_E}. \quad (4.288)$$

Meanwhile, by (4.285), we can easily check that

$$Y_C^{max}(t_S^*) = \left\{ z(T - t_S^*; 0, z_0, v) \mid z_0 \in Y_E, v \in L(0, T - t_S^*; \mathbb{U}) \right\}. \quad (4.289)$$

Since $Y_C^{max}(t_S^*)$ is the reachable set at $T - t_S^*$ of the system (4.286) (see (4.289)) and Y_S is the target set of $(TP)_{min}^{\widehat{Q}_S,\widehat{Q}_E}$ (see (4.287)), we can apply Theorem 4.1 to find $\varphi(\cdot) \in C([t_S^*, T]; Y)$ with $\varphi(\cdot) \neq 0$ so that (4.279)–(4.283) hold. From this and Definition 4.8, we see that (t_S^*, y_0^*, T, u^*) satisfies the classical Pontryagin Maximum Principle, consequently, so does $(TP)_{max}^{Q_S,Q_E}$. Hence, the sufficiency has been proved.

We next show the necessity. Suppose that $(TP)_{max}^{Q_S,Q_E}$ satisfies the classical Pontryagin Maximum Principle. Let (t_S^*, y_0^*, T, u^*) be an optimal tetrad to $(TP)_{max}^{Q_S,Q_E}$. Then by Definition 4.8, there exists $\varphi(\cdot) \in C([t_S^*, T]; Y)$ with $\varphi(\cdot) \neq 0$ so that (4.279)–(4.283) hold. Define the following function:

$$\psi(t) \triangleq -\varphi(T - t), \quad t \in [0, T - t_S^*].$$

Then $\psi(\cdot) \neq 0$, and it follows by (4.279)–(4.283) that

$$\dot{\psi}(t) = A^*\psi(t) + D(T-t)^*\varphi(t) \quad \text{a.e.} \ t \in (0, T - t_S^*), \quad (4.290)$$

$$\langle -B(T-t)^*\psi(t), u^*(T-t)\rangle_U$$
$$= \max_{u\in\mathbb{U}}\langle -B(T-t)^*\psi(t), u\rangle_U \quad \text{a.e.} \ t \in (0, T-t_S^*), \tag{4.291}$$

$$\langle \psi(0), z - y(T; t_S^*, y_0^*, u^*)\rangle_Y \le 0 \quad \text{for all} \ z \in Y_E, \tag{4.292}$$

and for all $y_0 \in Y_S$,

$$\langle \psi(T-t_S^*), y_0 - z(T-t_S^*; 0, y(T; t_S^*, y_0^*, u^*), u^*(T-\cdot))\rangle_Y \ge 0. \tag{4.293}$$

Meanwhile, by Theorem 2.2 and Definition 2.2, we obtain that

$$(0, y(T; t_S^*, y_0^*, u^*), T-t_S^*, u^*(T-\cdot)) \ \text{is an optimal tetrad to} \ (TP)_{min}^{\widehat{Q}_S, \widehat{Q}_E}. \tag{4.294}$$

By (4.290)–(4.293), using the similar arguments to those used in the proof of Theorem 4.1 (replacing respectively $Y_R(t_E^*)$ and Y_E by $Y_C^{max}(t_S^*)$ and Y_S), we can prove that $Y_C^{max}(t_S^*)$ and Y_S are separable in Y. Thus, the necessity is true.

Hence, we finish the proof of Theorem 4.11.   □

The next two results of this subsection are comparable with Theorem 4.2.

Theorem 4.12 *For the problem* $(TP)_{max}^{Q_S, Q_E}$, *the set* $Y_R^{max}(T, t_S^*)$ *and the target* Y_E *are separable in* Y, *provided that one of the following conditions holds:*

(i) *The space* Y *is of finite dimension.*
(ii) *The space* Y *is of infinite dimension and* $Int(Y_R^{max}(T, t_S^*) - Y_E) \ne \emptyset.$

Proof Its proof is the same as that of Theorem 4.2. We omit it here.   □

Theorem 4.13 *Assume that* $A : \mathscr{D}(A) \subseteq Y \mapsto Y$ *generates a* C_0 *group. Then for the problem* $(TP)_{max}^{Q_S, Q_E}$, *the set* $Y_C^{max}(t_S^*)$ *and* Y_S *are separable in* Y, *if one of the following conditions holds:*

(i) *The space* Y *is of finite dimension.*
(ii) *The space* Y *is of infinite dimension and* $Int(Y_C^{max}(t_S^*) - Y_S) \ne \emptyset.$

Proof Let $(TP)_{min}^{\widehat{Q}_S, \widehat{Q}_E}$ be defined in the proof of Theorem 4.11. As we pointed out there, $T - t_S^*$ is the optimal time of $(TP)_{min}^{\widehat{Q}_S, \widehat{Q}_E}$ (see (4.288)), $Y_C^{max}(t_S^*)$ is the reachable set at $T - t_S^*$ of the system (4.286) and Y_S is the target set of $(TP)_{min}^{\widehat{Q}_S, \widehat{Q}_E}$. By Theorem 4.2, we see that if either (i) or (ii) holds, then $Y_C^{max}(t_S^*)$ and Y_S are separable in Y. This completes the proof.   □

At the end of this subsection, we give some examples, which may help us to understand the main results of this subsection.

Example 4.21 Let $A(\cdot) = (a_{ij}(\cdot))_{n\times n}$ and $B(\cdot) = (b_{ij}(\cdot))_{n\times m}$, with $a_{ij}(\cdot) \in L^\infty(0, T)$ $(1 \le i, j \le n)$ and $b_{ij}(\cdot) \in L^\infty(0, T)$ $(1 \le i \le n, 1 \le j \le m)$. Let

y_0, $y_1 \in \mathbb{R}^n$, with $y_0 \neq y_1$. Consider the maximal time control problem $(TP)_{max}^{Q_S, Q_E}$, where the controlled system is as:

$$\dot{y}(t) = A(t)y(t) + B(t)u(t), \quad t \in (0, T), \quad \text{with } y(t) \in \mathbb{R}^n, \ u(t) \in \mathbb{R}^m, \tag{4.295}$$

and where

$$Q_S = [0, T) \times \{y_0\}; \quad Q_E = \{T\} \times \{y_1\}; \quad \mathbb{U} = B_\rho(0) \text{ with } \rho > 0.$$

One can easily check that the above $(TP)_{max}^{Q_S, Q_E}$ can be put into the framework $(\mathscr{A}_{TP}^{max})$ (given at the beginning of Section 4.4).

We further assume that the above $(TP)_{max}^{Q_S, Q_E}$ has an optimal control. Then according to Theorem 4.12 and Theorem 4.10 (or Theorem 4.13 and Theorem 4.11), the problem $(TP)_{max}^{Q_S, Q_E}$ satisfies the classical Pontryagin Maximum Principle. We end Example 4.21.

Example 4.22 Let $\Omega \subseteq \mathbb{R}^d$ (with $d \geq 1$) be a bounded domain with a C^2-boundary $\partial\Omega$. Let $\omega \subseteq \Omega$ be a nonempty and open subset with its characteristic function χ_ω. Consider the following controlled heat equation:

$$\begin{cases} \partial_t y - \Delta y + a(x, t)y = \chi_\omega u & \text{in } \Omega \times (0, T), \\ y = 0 & \text{on } \partial\Omega \times (0, T), \end{cases}$$

where $a(\cdot, \cdot) \in C(\overline{\Omega} \times [0, T])$. By setting $Y = U = L^2(\Omega)$, we can easily put the above equation into the framework (4.278).

We now consider a time optimal control problem $(TP)_{max}^{Q_S, Q_E}$, where the controlled system is the above heat equation, and where

$$Q_S = [0, T) \times \{y_0\} \text{ with } y_0 \in L^2(\Omega) \setminus B_r(0) \text{ and } r > 0;$$

$$Q_E = \{T\} \times B_r(0); \quad \mathbb{U} = B_\rho(0) \text{ with } \rho > 0.$$

One can easily check that $(TP)_{max}^{Q_S, Q_E}$ can be put into the framework $(\mathscr{A}_{TP}^{max})$ (given at the beginning of Section 4.4). We further assume that $(TP)_{max}^{Q_S, Q_E}$ has an optimal control. Notice that when $\rho > 0$ is large enough, this problem has an optimal control (see Theorem 3.7).

According to Theorem 4.12 and Theorem 4.10, the problem $(TP)_{max}^{Q_S, Q_E}$ satisfies the classical Pontryagin Maximum Principle. We end Example 4.22.

4.4.2   Local Maximum Principle for Maximal Time Controls

We first give the definition of the local Pontryagin Maximum Principle for the problem $(TP)_{max}^{Q_S, Q_E}$.

Definition 4.9 Several definitions are given in order.

(i) An optimal control u^* (associated with an optimal tetrad (t_S^*, y_0^*, T, u^*)) to $(TP)_{max}^{Q_S, Q_E}$ is said to satisfy the local Pontryagin Maximum Principle, if for any $\widetilde{T} \in (t_S^*, T)$, there exists $\varphi_{\widetilde{T}}(\cdot) \in C([t_S^*, \widetilde{T}]; Y)$, with $\varphi_{\widetilde{T}}(\cdot) \neq 0$ (i.e., $\varphi_{\widetilde{T}}$ is not a zero function), so that

$$\dot{\varphi}_{\widetilde{T}}(t) = -A^* \varphi_{\widetilde{T}}(t) - D(t)^* \varphi_{\widetilde{T}}(t) \quad \text{for a.e. } t \in (t_S^*, \widetilde{T}); \tag{4.296}$$

$$\begin{aligned} &H(t, y^*(t), u^*(t), \varphi_{\widetilde{T}}(t)) \\ &= \max_{u \in \mathbb{U}} H(t, y^*(t), u, \varphi_{\widetilde{T}}(t)) \quad \text{for a.e. } t \in (t_S^*, \widetilde{T}), \end{aligned} \tag{4.297}$$

where $y^* \in C([t_S^*, T]; Y)$ is the corresponding optimal trajectory and

$$H(t, y, u, \varphi) \triangleq \langle \varphi, D(t)y + B(t)u \rangle_Y \tag{4.298}$$

for a.e. $t \in (t_S^*, T)$ and all $(y, u, \varphi) \in Y \times U \times Y$. Besides,

$$\langle \varphi_{\widetilde{T}}(t_S^*), y_0 - y_0^* \rangle_Y \leq 0 \quad \text{for each } y_0 \in Y_S, \tag{4.299}$$

and

$$\langle \varphi_{\widetilde{T}}(\widetilde{T}), z - y^*(\widetilde{T}) \rangle_Y \geq 0 \quad \text{for each } z \in Y_C^{max}(\widetilde{T}). \tag{4.300}$$

Here, (4.296) is called the dual (or adjoint) equation; (4.297) is called the maximum condition; the function $H(\cdot)$ defined by (4.298) is called the Hamiltonian (associated with $(TP)_{max}^{Q_S, Q_E}$); (4.299) and (4.300) are called the transversality condition.

(ii) The problem $(TP)_{max}^{Q_S, Q_E}$ is said to satisfy the local Pontryagin Maximum Principle, if any optimal control satisfies the local Pontryagin Maximum Principle.

The main result of this subsection is as follows.

Theorem 4.14 *The problem $(TP)_{max}^{Q_S, Q_E}$ satisfies the local Pontryagin Maximum Principle if and only if $Y_R^{max}(t, t_S^*)$ and $Y_C^{max}(t)$ are separable in Y for any $t \in (t_S^*, T)$.*

Proof We first show the sufficiency. Assume that $Y_R^{max}(t, t_S^*)$ and $Y_C^{max}(t)$ are separable in Y for any $t \in (t_S^*, T)$. Let (t_S^*, y_0^*, T, u^*) be an optimal tetrad to $(TP)_{max}^{Q_S, Q_E}$. For any $\widetilde{T} \in (t_S^*, T)$, by the similar arguments to those used in the proof of Theorem 4.1 (replacing respectively Y_E and $Y_R(t_E^*)$ by $Y_C^{max}(\widetilde{T})$ and $Y_R^{max}(\widetilde{T}, t_S^*)$), we can show that there exists $\varphi_{\widetilde{T}}(\cdot) \in C([t_S^*, \widetilde{T}]; Y)$ with $\varphi_{\widetilde{T}}(\cdot) \neq 0$ so that (4.296)–(4.300) hold. From this and Definition 4.9, we see that (t_S^*, y_0^*, T, u^*) satisfies the local Pontryagin Maximum Principle, consequently, so does $(TP)_{max}^{Q_S, Q_E}$. Hence, we have proved the sufficiency.

We next show the necessity. Suppose that $(TP)_{max}^{Q_S,Q_E}$ satisfies the local Pontryagin Maximum Principle. Let (t_S^*, y_0^*, T, u^*) be an optimal tetrad to $(TP)_{max}^{Q_S,Q_E}$. Then by Definition 4.9, for any $\widetilde{T} \in (t_S^*, T)$, there exists $\varphi_{\widetilde{T}}(\cdot) \in C([t_S^*, \widetilde{T}]; Y)$ with $\varphi_{\widetilde{T}}(\cdot) \neq 0$ so that (4.296)–(4.300) hold. Next, by the similar arguments as those used in the proof of Theorem 4.1, we can show that $Y_R^{max}(\widetilde{T}, t_S^*)$ and $Y_C^{max}(\widetilde{T})$ are separable in Y. Hence, we have proved the necessity.

Thus, we complete the proof.   □

The next result is concerned with relations between the classical Pontryagin Maximum Principle and the local Pontryagin Maximum Principle for the problem $(TP)_{max}^{Q_S,Q_E}$. We start with the following assumption:

(H_C) $\varphi(\cdot) = 0$ in (t_S^*, T), provided that $\varphi(t_0) = 0$ for some $t_0 \in (t_S^*, T]$ and $\varphi(\cdot)$ solves the equation:

$$\begin{cases} \dot{\varphi}(t) = -[A^* + D(t)^*]\varphi(t), & t \in (t_S^*, T), \\ \varphi(T) \in Y. \end{cases}$$

Corollary 4.4 *Assume that the problem $(TP)_{max}^{Q_S,Q_E}$ satisfies the classical Pontryagin Maximum Principle. Suppose that (H_C) is true. Then $(TP)_{max}^{Q_S,Q_E}$ also satisfies the local Pontryagin Maximum Principle.*

Proof The proof is the same as that of Corollary 4.2. We omit it here.   □

Example 4.23 Consider the maximal time control problems in Examples 4.21 and 4.22. In those examples, we have seen that these problems satisfy the classical Pontryagin Maximum Principle. Moreover, one can easily check that the assumption (H_C) holds for these two cases. (For the case of ODEs, we use the theory of existence and uniqueness for ODEs; for the case of the heat equation, we use (i) of Remark 1.5 (after Theorem 1.22).) Then by Corollary 4.4, these problems satisfy the local Pontryagin Maximum Principle.

Miscellaneous Notes

The studies on the Pontryagin Maximum Principle can be dated back to 1950s (see, for instance, [5, 6] and [14]). Due to the importance of the Pontryagin Maximum Principle, many mathematicians have worked on it. Here, we mention literatures [2–4, 7, 8, 10, 12, 13, 15], and [11]. In particular, we would like to recommend the book [9]. In this book, the author studied the Pontryagin Maximum Principle for infinite dimensional evolution equations in an abstract setting; introduced two different maximum principles: the strong maximum principle and the weak maximum principle. (Notice that the weak maximum principle in [9] differs from that introduced in our monograph.)

About three kinds of maximum principles introduced in this chapter, several notes are given in order:

- Definitions on these maximum principles are summarized from [9, 19] and [18].
- Connections among these maximum principles are as follows: (i) Some problems do not hold the classical maximum principle, but have the local one (see Example 4.9); (ii) The classical one cannot imply the local one (see Remark 4.4); (iii) If the adjoint equation has some unique continuation property, then the classical one implies the local one (see Corollary 4.2); (iv) Some problems hold neither the classical one nor the local one, but have the weak one (see Example 4.17).
- Our methods to derive these maximum principles are based on the separability results and the representation formulas introduced in Sections 1.1.2 and 4.3.2. The two representation formulas in Section 4.3.2 are taken from [19], where the first representation formula is originally from [18].

Materials of this chapter come from two parts: First, some of them are taken from papers [19] and [18]; Second, some of them, including several examples, are given specially to this monograph by us.

Next, we would like to stress the following fact about our control operator: In this chapter, though it is assumed that $B(\cdot) \in L^{\infty}(0, +\infty; \mathscr{L}(U, Y))$ or $B(\cdot) \in L^{\infty}(0, T; \mathscr{L}(U, Y))$, many results are still true for the case that $B(\cdot) \in L^{\infty}(0, +\infty; \mathscr{L}(U, Y_{-1}))$ or $B(\cdot) \in L^{\infty}(0, T; \mathscr{L}(U, Y_{-1}))$ is admissible for the semigroup generated by A. (Here, $Y_{-1} \triangleq D(A^*)'$ is the dual space of $D(A^*)$.) When $B(\cdot) \equiv B$, its admissibility has been defined in [17]. This definition can be easily extended to the case that $B(\cdot)$ is time varying. The case that $B(\cdot)$ is admissible covers some time optimal control problems governed by boundary controlled PDEs. With regard to the maximum principles for the case when $B(\cdot) \equiv B$ is admissible, we refer the readers to [19].

Finally, we give several open problems that might interest readers:

- Can we have weak maximum principles for time-invariant systems with general control constraints?
- Can we have weak maximum principles for general time-varying systems?
- For heat equations, what is the best regularity for the multiplier in the local/weak maximum principle?

References

1. A.A. Agrachev, Y.L. Sachkov, *Control Theory from the Geometric Viewpoint*. Encyclopaedia of Mathematical Sciences, Control Theory and Optimization, II, vol. 87 (Springer, Berlin, 2004)
2. N. Arada, J.P. Raymond, Time optimal problems with Dirichlet boundary controls Discrete Contin. Dyn. Syst. **9**, 1549–1570 (2003)
3. V. Barbu, *Optimal Control of Variational Inequalities*. Research Notes in Mathematics, vol. 100 (Pitman (Advanced Publishing Program), Boston, MA, 1984)

4. V. Barbu, Analysis and control of nonlinear infinite-dimensional systems, in *Mathematics in Science and Engineering*, vol. 190 (Academic, Boston, MA, 1993)
5. R. Bellman, I. Glicksberg, O. Gross, On the "bang-bang" control problem. Q. Appl. Math. **14**, 11–18 (1956)
6. V.G. Boltyanskiĭ, R.V. Gamkrelidz, L.S. Pontryagin, On the theory of optimal processes (Russian), Dokl. Akad. Nauk. SSSR **110**, 7–10 (1956)
7. R. Conti, Time-optimal solution of a linear evolution equation in Banach spaces. J. Optim. Theory Appl. **2**, 277–284 (1968)
8. H.O. Fattorini, The time optimal control problems in Banach spaces. Appl. Math. Optim. **1**, 163–188 (1974)
9. H.O. Fattorini, *Infinite Dimensional Linear Control Systems, the Time Optimal and Norm Optimal Problems*. North-Holland Mathematics Studies, vol. 201 (Elsevier Science B.V., Amsterdam, 2005)
10. K. Kunisch, L. Wang, Time optimal controls of the linear Fitzhugh-Nagumo equation with pointwise control constraints. J. Math. Anal. Appl. **395**, 114–130 (2012)
11. K. Kunisch, L. Wang, Time optimal control of the heat equation with pointwise control constraints. ESAIM Control Optim. Calc. Var. **19**, 460–485 (2013)
12. X. Li, J. Yong, *Optimal Control Theory for Infinite-Dimensional Systems*. Systems & Control: Foundations & Applications (Birkhäuser Boston, Boston, MA, 1995)
13. J. Lohéac, M. Tucsnak, Maximum principle and bang-bang property of time optimal controls for Schrodinger-type systems. SIAM J. Control Optim. **51**, 4016–4038 (2013)
14. L.S. Pontryagin, V.G. Boltyanskii, R.V. Gamkrelidze, E.F. Mischenko, *The Mathematical Theory of Optimal Processes*, ed. by L.W. Neustadt, (Interscience Publishers Wiley, New York, London, 1962) [Translated from the Russian by K. N. Trirogoff]
15. J.P. Raymond, H. Zidani, Pontryagin's principle for time-optimal problems. J. Optim. Theory Appl. **101**, 375–402 (1999)
16. E. Roxin, A geometric interpretation of Pontryagin's maximum principle, in *International Symposium on Nonlinear Differential Equations and Nonlinear Mechanics* (Academic, New York, 1963), pp. 303–324
17. M. Tucsnak, G. Weiss, Observation and control for operator semigroups, (Birkhäuser Verlag, Basel, 2009)
18. G. Wang, Y. Xu, Y. Zhang, Attainable subspaces and the bang-bang property of time optimal controls for heat equations. SIAM J. Control Optim. **53**, 592–621 (2015)
19. G. Wang, Y. Zhang, Decompositions and bang-bang problems. Math. Control Relat. Fields **7**, 73–170 (2017)

Chapter 5
Equivalence of Several Kinds of Optimal Controls

Connections of time optimal control problems with other kinds of optimal control problems can allow us to get properties for one kind of optimal control problems through studying another kind of optimal control problems. In [4] (see also [5]), the author derived the Pontryagin Maximum Principle for minimal time controls (or minimal norm controls) via the Pontryagin Maximum Principle for minimal norm controls (or minimal time controls). Such connections can also help us in the following subjects: (i) To study the regularity of the Bellman functions associated with minimal time control problems (see for instance, [1, 6] and [3]); (ii) To study the behaviors of optimal time and an optimal control, when a controlled system has a small perturbation (see [19] and [13]); (iii) To get necessary and sufficient conditions on both optimal time and optimal controls for some time optimal control problems (see [17] and [15]); (iv) To find some iterative algorithms for solutions of time optimal control problems (see [15]).

To our best knowledge, the terminology "equivalence" on optimal controls arose first in [17]. It means that several optimal control problems share the same optimal control in certain senses. In this chapter, we first present three equivalence theorems for minimal time optimal control problems and minimal norm control problems. Among these equivalent theorems, the first one is concerned with time invariant systems with terminal singleton constraints; the second one is related to time-varying systems with terminal non-singleton constraints; the third one is about time-varying systems with terminal singleton constraints. We then present the equivalence among maximal time optimal control problems, minimal norm control problems, and optimal target control problems.

© Springer International Publishing AG, part of Springer Nature 2018
G. Wang et al., *Time Optimal Control of Evolution Equations*, Progress
in Nonlinear Differential Equations and Their Applications 92,
https://doi.org/10.1007/978-3-319-95363-2_5

5.1 Minimal Time Controls, Minimal Norm Controls, and Time Invariant Systems

In this section, we will study the equivalence of the problem $(TP)_{min}^{Q_S,Q_E}$ and the corresponding minimal norm control problem under the following framework $(\mathscr{A}_1)$:

(i) The state space Y and the control space U are real separable Hilbert spaces.
(ii) The controlled system is as:

$$\dot{y}(t) = Ay(t) + Bu(t), \quad t \in (0, +\infty), \tag{5.1}$$

where $A : \mathscr{D}(A) \subseteq Y \mapsto Y$ generates a C_0 semigroup $\{e^{At}\}_{t \geq 0}$ on Y and $B \in \mathscr{L}(U, Y)$. Given $\tau \in [0, +\infty)$, $y_0 \in Y$ and $u \in L^\infty(\tau, +\infty; U)$, we write $y(\cdot; \tau, y_0, u)$ for the solution of (5.1) over $[\tau, +\infty)$, with the initial condition that $y(\tau) = y_0$.
(iii) Let $\mathbb{U} = B_M(0)$, where $B_M(0)$ is the closed ball in U, centered at 0 and of radius $M \geq 0$. (When $M = 0$, we agree that $B_M(0) \triangleq \{0\}$.)
(iv) Let $Q_S = \{0\} \times \{\widehat{y}_0\}$ and $Q_E = (0, +\infty) \times \{0\}$, where $\widehat{y}_0 \in Y \setminus \{0\}$ is arbitrarily fixed.
(v) The system (5.1) is L^∞-null controllable over $[0, T]$ for each $T \in (0, +\infty)$ (i.e., for each $T \in (0, +\infty)$ and $y_0 \in Y$, there exists $u \in L^\infty(0, T; U)$ so that $y(T; 0, y_0, u) = 0$).
(vi) The semigroup $\{e^{At}\}_{t \geq 0}$ holds the backward uniqueness property (i.e., if $e^{At}y_0 = 0$ for some $y_0 \in Y$ and $t \geq 0$, then $y_0 = 0$).

In this section, we simply write $(TP)_{min}^{Q_S,Q_E}$ for $(TP)_M^{\widehat{y}_0}$ which reads as:

$$(TP)_M^{\widehat{y}_0} \qquad \mathbf{T}(M, \widehat{y}_0) \triangleq \inf\{T \in (0, +\infty) \mid y(T; 0, \widehat{y}_0, u) = 0, \, u \in L(0, T; \mathbb{U})\}. \tag{5.2}$$

We consider the following problem:

$$(\widetilde{TP})_M^{\widehat{y}_0} \qquad \widetilde{\mathbf{T}}(M, \widehat{y}_0) \triangleq \inf\{T \in (0, +\infty) \mid y(T; 0, \widehat{y}_0, u) = 0 \text{ and } u \in \mathscr{U}_M\}, \tag{5.3}$$

where

$$\mathscr{U}_M \triangleq \{u \in L^\infty(0, +\infty; U) \mid \|u(t)\|_U \leq M \text{ a.e. } t \in (0, +\infty)\}.$$

In this problem,

- a tetrad $(0, \widehat{y}_0, T, u)$ is called admissible, if $T \in (0, +\infty)$, $u \in \mathscr{U}_M$ and $y(T; 0, \widehat{y}_0, u) = 0$;
- a tetrad $(0, \widehat{y}_0, \widetilde{\mathbf{T}}(M, \widehat{y}_0), \widetilde{u}^*)$ is called optimal if

$$\widetilde{\mathbf{T}}(M, \widehat{y}_0) \in (0, +\infty), \, \widetilde{u}^* \in \mathscr{U}_M \text{ and } y(\widetilde{\mathbf{T}}(M, \widehat{y}_0); 0, \widehat{y}_0, \widetilde{u}^*) = 0;$$

- when $(0, \widehat{y}_0, \widetilde{\mathbf{T}}(M, \widehat{y}_0), \widetilde{u}^*)$ is an optimal tetrad, $\widetilde{\mathbf{T}}(M, \widehat{y}_0)$ and $\widetilde{u}^*$ are called the optimal time and an optimal control, respectively.

The following three facts can be checked easily:

- $\mathbf{T}(M, \widehat{y}_0) = \widetilde{\mathbf{T}}(M, \widehat{y}_0)$;
- The infimum in (5.2) can be reached if and only if the infimum in (5.3) can be reached;
- The zero extension of each optimal control to $(TP)_M^{\widehat{y}_0}$ over $(0, +\infty)$ is an optimal control to $(\widetilde{TP})_M^{\widehat{y}_0}$; the restriction of each optimal control to $(\widetilde{TP})_M^{\widehat{y}_0}$ over $(0, \mathbf{T}(M, \widehat{y}_0))$ is an optimal control to $(TP)_M^{\widehat{y}_0}$.

Because of the above three facts, $(TP)_M^{\widehat{y}_0}$ and $(\widetilde{TP})_M^{\widehat{y}_0}$ can be treated as the same problem.

The corresponding minimal norm problem $(NP)_T^{y_0}$ (with $T > 0$ and $y_0 \in Y$ arbitrarily fixed) reads:

$$(NP)_T^{y_0} \qquad N(T, y_0) \triangleq \inf\{\|v\|_{L^\infty(0,T;U)} \mid y(T; 0, y_0, v) = 0\}. \qquad (5.4)$$

In this problem,

- $N(T, y_0)$ is called the minimal norm;
- a control $v \in L^\infty(0, T; U)$ is called admissible, if $y(T; 0, y_0, v) = 0$;
- a control $v^* \in L^\infty(0, T; U)$ is called optimal, if it is admissible and satisfies that $\|v^*\|_{L^\infty(0,T;U)} = N(T, y_0)$.

Moreover, for each $y_0 \in Y$, the map $T \to N(T, y_0)$ is called a minimal norm function, which is denoted by $N(\cdot, y_0)$.

In this section, we aim to introduce connections between $(TP)_M^{\widehat{y}_0}$ and $(NP)_T^{\widehat{y}_0}$. From (5.4), we see that for each $y_0 \in Y$, the minimal norm function $N(\cdot, y_0)$ is decreasing over $(0, +\infty)$. Hence, for each $y_0 \in Y$, it is well defined that

$$\widehat{N}(y_0) \triangleq \lim_{T \to +\infty} N(T, y_0). \qquad (5.5)$$

We now give the definition about the equivalence between $(TP)_M^{\widehat{y}_0}$ and $(NP)_T^{\widehat{y}_0}$.

Definition 5.1 Let $M \geq 0$ and $T \in (0, +\infty)$. Problems $(TP)_M^{\widehat{y}_0}$ and $(NP)_T^{\widehat{y}_0}$ are said to be equivalent if the following three conditions hold:

(i) Both $(\widetilde{TP})_M^{\widehat{y}_0}$ and $(NP)_T^{\widehat{y}_0}$ have optimal controls;
(ii) The restriction of each optimal control to $(\widetilde{TP})_M^{\widehat{y}_0}$ over $(0, T)$ is an optimal control to $(NP)_T^{\widehat{y}_0}$;
(iii) The zero extension of each optimal control to $(NP)_T^{\widehat{y}_0}$ over $(0, +\infty)$ is an optimal control to $(\widetilde{TP})_M^{\widehat{y}_0}$.

The main result of this section is as follows.

Theorem 5.1 *The function* $T \to N(T, \widehat{y}_0)$ *is strictly decreasing and continuous from* $(0, +\infty)$ *onto* $(\widehat{N}(\widehat{y}_0), +\infty)$. *Moreover, if write*

$$\mathscr{E}_{\widehat{y}_0} \triangleq \{(N(T, \widehat{y}_0), T) \in \mathbb{R}^2 \mid T \in (0, +\infty)\}, \tag{5.6}$$

then the following conclusions are true:

(i) *When* $(M, T) \in \mathscr{E}_{\widehat{y}_0}$, *problems* $(TP)_M^{\widehat{y}_0}$ *and* $(NP)_T^{\widehat{y}_0}$ *are equivalent;*

(ii) *When* $(M, T) \in [0, +\infty) \times (0, +\infty) \setminus \mathscr{E}_{\widehat{y}_0}$, *problems* $(TP)_M^{\widehat{y}_0}$ *and* $(NP)_T^{\widehat{y}_0}$ *are not equivalent.*

Before proving Theorem 5.1, some preliminaries are given in order.

Lemma 5.1 *Let* $y_0 \in Y$. *Then for each* $T \in (0, +\infty)$, *the problem* $(NP)_T^{y_0}$ *has at least one optimal control.*

Proof Arbitrarily fix $T \in (0, +\infty)$ and $y_0 \in Y$. Without loss of generality, we can assume that $y_0 \neq 0$, for otherwise the null control is an optimal control.

By (v) of $(\mathscr{A}_1)$ (given at the beginning of Section 5.1), there exists a control $\widehat{v} \in L^\infty(0, T; U)$ so that

$$y(T; 0, y_0, \widehat{v}) = 0,$$

i.e., $(NP)_T^{y_0}$ has admissible controls. So we can let $\{v_k\}_{k \geq 1} \subseteq L^\infty(0, T; U)$ satisfy that

$$y(T; 0, y_0, v_k) = 0 \quad \text{and} \quad \|v_k\|_{L^\infty(0,T;U)} \leq N(T, y_0) + k^{-1} \quad \text{for all } k \in \mathbb{N}^+. \tag{5.7}$$

By the inequality in (5.7), there exists a subsequence of $\{v_k\}_{k \geq 1}$, denoted in the same way, and $\widehat{v} \in L^\infty(0, T; U)$ so that

$$v_k \to \widehat{v} \quad \text{weakly star in } L^\infty(0, T; U) \quad \text{as } k \to +\infty. \tag{5.8}$$

From this and the inequality in (5.7), it follows that

$$\|\widehat{v}\|_{L^\infty(0,T;U)} \leq \liminf_{k \to +\infty} \|v_k\|_{L^\infty(0,T;U)} \leq N(T, y_0). \tag{5.9}$$

Meanwhile, by (5.8), we see that

$$y(T; 0, y_0, v_k) \to y(T; 0, y_0, \widehat{v}) \quad \text{weakly in } Y.$$

This, together with the equality in (5.7), implies that $y(T; 0, y_0, \widehat{v}) = 0$, i.e., $\widehat{v}$ is an admissible control to $(NP)_T^{y_0}$. Then by the optimality of $N(T, y_0)$ (see (5.4)) and (5.9), we find that $\widehat{v}$ is an optimal control to $(NP)_T^{y_0}$.

Hence, we end the proof of Lemma 5.1. $\square$

The following lemma is concerned with the continuity and monotonicity of minimal norm function $N(\cdot, y_0)$.

Lemma 5.2 *The function* $N(\cdot, \widehat{y}_0)$ *is continuous and strictly decreasing from* $(0, +\infty)$ *onto* $(\widehat{N}(\widehat{y}_0), +\infty)$.

Proof The proof is organized by the following four steps:
Step 1. We show that

$$N(T, \widehat{y}_0) \in (0, +\infty) \text{ for each } T \in (0, +\infty). \tag{5.10}$$

By contradiction, suppose that (5.10) was not true. Then we would have that

$$N(\widehat{T}, \widehat{y}_0) = 0 \text{ for some } \widehat{T} \in (0, +\infty).$$

This, together with Lemma 5.1, implies that the null control is the unique optimal control to $(NP)_{\widehat{T}}^{\widehat{y}_0}$. So we have that

$$e^{A\widehat{T}} \widehat{y}_0 = y(\widehat{T}; 0, \widehat{y}_0, 0) = 0.$$

This, along with (vi) of $(\mathscr{A}_1)$ (given at the beginning of Section 5.1), yields that $\widehat{y}_0 = 0$, which leads to a contradiction, since we assumed that $\widehat{y}_0 \neq 0$. Thus, (5.10) is true.
Step 2. We prove that when $0 < T_1 < T_2 < +\infty$,

$$N(T_1, \widehat{y}_0) > N(T_2, \widehat{y}_0). \tag{5.11}$$

To this end, we arbitrarily fix $T_1, T_2 \in (0, +\infty)$ so that $T_2 > T_1$. By Lemma 5.1, the problem $(NP)_{T_1}^{\widehat{y}_0}$ has an optimal control v_1^*. Then we have that

$$y(T_1; 0, \widehat{y}_0, v_1^*) = 0 \text{ and } \|v_1^*\|_{L^\infty(0,T_1;U)} = N(T_1, \widehat{y}_0). \tag{5.12}$$

Meanwhile, by (v) of $(\mathscr{A}_1)$ (given at the beginning of Section 5.1), there is $v_2 \in L^\infty(0, T_2 - T_1; U)$ so that

$$y(T_2 - T_1; 0, e^{AT_1}\widehat{y}_0, v_2) = 0. \tag{5.13}$$

Since $N(T_1, \widehat{y}_0) \in (0, +\infty)$ (see (5.10)), we can choose a constant $\lambda \in (0, 1)$ so that

$$\lambda \|v_2\|_{L^\infty(0,T_2-T_1;U)} < N(T_1, \widehat{y}_0)/2. \tag{5.14}$$

We now define a new control in the following manner:

$$v_\lambda(t) \triangleq \begin{cases} (1 - \lambda)v_1^*(t), \ t \in (0, T_1), \\ \lambda v_2(t - T_1), \ t \in (T_1, T_2). \end{cases} \tag{5.15}$$

By (5.15), we find that

$$y(T_2; 0, \widehat{y}_0, v_\lambda) = e^{AT_2}\widehat{y}_0 + (1 - \lambda) \int_0^{T_1} e^{A(T_2-t)} B v_1^*(t) dt$$

$$+ \lambda \int_{T_1}^{T_2} e^{A(T_2-t)} B v_2(t - T_1) dt$$

$$= (1 - \lambda) e^{A(T_2-T_1)} \left(e^{AT_1}\widehat{y}_0 + \int_0^{T_1} e^{A(T_1-t)} B v_1^*(t) dt \right)$$

$$+ \lambda \left[e^{A(T_2-T_1)} (e^{AT_1}\widehat{y}_0) + \int_0^{T_2-T_1} e^{A(T_2-T_1-t)} B v_2(t) dt \right].$$

This, together with the first equality in (5.12) and (5.13), implies that

$$y(T_2; 0, \widehat{y}_0, v_\lambda) = (1-\lambda) e^{A(T_2-T_1)} y(T_1; 0, \widehat{y}_0, v_1^*) + \lambda y(T_2-T_1; 0, e^{AT_1}\widehat{y}_0, v_2) = 0,$$

which indicates that v_λ is an admissible control to $(NP)_{T_2}^{\widehat{y}_0}$. Then by the optimality of $N(T_2, \widehat{y}_0)$ (see (5.4)) and (5.15), we get that

$$N(T_2, \widehat{y}_0) \leq \|v_\lambda\|_{L^\infty(0,T_2;U)} \leq \max \left\{ (1 - \lambda) \|v_1^*\|_{L^\infty(0,T_1;U)}, \lambda \|v_2\|_{L^\infty(0,T_2-T_1;U)} \right\}.$$

From this, the second equality in (5.12) and (5.14), we obtain (5.11) at once.

Step 3. We show that given $\eta \in (0, +\infty)$ and $T_1, T_2 \in (0, +\infty)$ with $\eta < T_1 < T_2 < \eta^{-1}$, there exists a constant $C(\eta) \in (0, +\infty)$ so that

$$N(T_1, \widehat{y}_0) \leq N(T_2, \widehat{y}_0) + C(\eta) \left[\|e^{A(T_2-T_1)}\widehat{y}_0 - \widehat{y}_0\|_Y + N(T_2, \widehat{y}_0)(T_2 - T_1) \right]. \tag{5.16}$$

For this purpose, we arbitrarily fix $\eta \in (0, +\infty)$ and $T_1, T_2 \in (0, +\infty)$ with $\eta < T_1 < T_2 < \eta^{-1}$. By Lemma 5.1, the problem $(NP)_{T_2}^{\widehat{y}_0}$ has an optimal control v_2^*. Then we have that

$$y(T_2; 0, \widehat{y}_0, v_2^*) = 0 \quad \text{and} \quad \|v_2^*\|_{L^\infty(0,T_2;U)} = N(T_2, \widehat{y}_0). \tag{5.17}$$

Write

$$\widetilde{y} \triangleq y(T_2 - T_1; 0, \widehat{y}_0, v_2^*). \tag{5.18}$$

By (v) of $(\mathscr{A}_1)$ (given at the beginning of Section 5.1), we can apply Theorem 1.20 to find $C(\eta) \in (0, +\infty)$ and $v_3 \in L^\infty(0, T_1; U)$ so that

$$y(T_1; 0, \widehat{y}_0 - \widetilde{y}, v_3) = 0 \quad \text{and} \quad \|v_3\|_{L^\infty(0,T_1;U)} \leq C(\eta) \|\widehat{y}_0 - \widetilde{y}\|_Y. \tag{5.19}$$

We now define a new control in the following manner:

$$v_4(t) \triangleq v_2^*(t + T_2 - T_1) + v_3(t), \quad t \in (0, T_1). \tag{5.20}$$

It can be easily verified that

$$
y(T_1; 0, \widehat{y}_0, v_4) = e^{AT_1}\widehat{y}_0 + \int_0^{T_1} e^{A(T_1-t)} B v_2^*(t + T_2 - T_1) dt
$$

$$
+ \int_0^{T_1} e^{A(T_1-t)} B v_3(t) dt
$$

$$
= \left[e^{AT_1} \widetilde{y} + \int_{T_2-T_1}^{T_2} e^{A(T_2-t)} B v_2^*(t) dt \right]
$$

$$
+ \left[e^{AT_1}(\widehat{y}_0 - \widetilde{y}) + \int_0^{T_1} e^{A(T_1-t)} B v_3(t) dt \right].
$$

This, together with (5.18) and the first equalities in both (5.17) and (5.19), implies that

$$
y(T_1; 0, \widehat{y}_0, v_4) = y(T_2; 0, \widehat{y}_0, v_2^*) + y(T_1; 0, \widehat{y}_0 - \widetilde{y}, v_3) = 0,
$$

which indicates that v_4 is an admissible control to $(NP)_{T_1}^{\widehat{y}_0}$. Then by the optimality of $N(T_1, \widehat{y}_0)$, we get that

$$
N(T_1, \widehat{y}_0) \le \|v_4\|_{L^\infty(0,T_1;U)}.
$$

This, together with (5.20), the second conclusions in both (5.17) and (5.19), yields that

$$
N(T_1, \widehat{y}_0) \le \|v_2^*\|_{L^\infty(0,T_2;U)} + \|v_3\|_{L^\infty(0,T_1;U)}
$$
$$
\le N(T_2, \widehat{y}_0) + C(\eta)\|\widehat{y}_0 - \widetilde{y}\|_Y.
$$

From the above inequality, (5.18) and the second equality in (5.17), it follows that

$$
N(T_1, \widehat{y}_0) \le N(T_2, \widehat{y}_0) + C(\eta)\left(\|\widehat{y}_0 - e^{A(T_2-T_1)}\widehat{y}_0\|_Y \right.
$$

$$
\left. + \left\| \int_0^{T_2-T_1} e^{A(T_2-T_1-t)} B v_2^*(t) dt \right\|_Y \right)
$$

$$
\le N(T_2, \widehat{y}_0) + C(\eta)\left[\|\widehat{y}_0 - e^{A(T_2-T_1)}\widehat{y}_0\|_Y + N(T_2, \widehat{y}_0)(T_2 - T_1) \right],
$$

which leads to (5.16).

Step 4. We prove that

$$
\lim_{T \to 0+} N(T, \widehat{y}_0) = +\infty. \tag{5.21}
$$

By contradiction, suppose that (5.21) was not true. Then there would exist a sequence $\{T_k\}_{k\geq 1} \subseteq (0, +\infty)$ with $T_k \to 0$ and a positive constant C so that

$$N(T_k, \widehat{y}_0) \leq C \quad \text{for all } k \in \mathbb{N}^+.$$

Thus, according to Lemma 5.1, each $(NP)_{T_k}^{\widehat{y}_0}$, with $k \in \mathbb{N}^+$, has an optimal control v_k^*. Then we have that

$$\|v_k^*\|_{L^\infty(0,T_k;U)} \leq C \quad \text{and} \quad y(T_k; 0, \widehat{y}_0, v_k^*) = 0.$$

These yield that

$$
\begin{aligned}
\widehat{y}_0 &= \lim_{k\to+\infty} y(T_k; 0, \widehat{y}_0, 0) = \lim_{k\to+\infty} \left[y(T_k; 0, \widehat{y}_0, v_k^*) - y(T_k; 0, 0, v_k^*) \right] \\
&= \lim_{k\to+\infty} y(T_k; 0, \widehat{y}_0, v_k^*) = 0,
\end{aligned}
$$

which contradicts the assumption that $\widehat{y}_0 \neq 0$. So (5.21) is true.

Finally, the desired results of this lemma follow from results in the above four steps, as well as the definition of $\widehat{N}(\widehat{y}_0)$ (see (5.5)). This ends the proof of Lemma 5.2.   □

Lemma 5.3 *The problem* $(\widetilde{TP})_M^{\widehat{y}_0}$ *has optimal controls if and only if* $M > \widehat{N}(\widehat{y}_0)$.

Proof Arbitrarily fix $\widehat{y}_0 \in Y \setminus \{0\}$. Then by Lemma 5.2 and the definition of $\widehat{N}(\widehat{y}_0)$ (see (5.5)), we have that

$$\widehat{N}(\widehat{y}_0) < +\infty.$$

The proof will be finished by the following two steps.

Step 1. We show that if $M > \widehat{N}(\widehat{y}_0)$, then $(\widetilde{TP})_M^{\widehat{y}_0}$ has an optimal control.

To this end, we arbitrarily fix $M > \widehat{N}(\widehat{y}_0)$. Then by the definition of $\widehat{N}(\widehat{y}_0)$, we can find $\widehat{T} \in (0, +\infty)$ so that

$$N(\widehat{T}, \widehat{y}_0) < \widehat{N}(\widehat{y}_0) + (M - \widehat{N}(\widehat{y}_0)) = M. \tag{5.22}$$

Meanwhile, according to Lemma 5.1, there exists $v^* \in L^\infty(0, \widehat{T}; U)$ so that

$$y(\widehat{T}; 0, \widehat{y}_0, v^*) = 0 \quad \text{and} \quad \|v^*\|_{L^\infty(0,\widehat{T};U)} = N(\widehat{T}, \widehat{y}_0). \tag{5.23}$$

Write $\widetilde{v}^*$ for the zero extension of v^* over $(0, +\infty)$. From (5.23) and (5.22), we see that $(0, \widehat{y}_0, \widehat{T}, \widetilde{v}^*)$ is an admissible tetrad to $(\widetilde{TP})_M^{\widehat{y}_0}$. So we can let $\{T_k\}_{k\geq 1} \subseteq (0, +\infty)$ and $\{v_k\}_{k\geq 1} \subseteq L^\infty(0, +\infty; U)$ satisfy that

$$T_k \to \widetilde{T}(M, \widehat{y}_0), \tag{5.24}$$

$$y(T_k; 0, \widehat{y}_0, v_k) = 0 \text{ and } \|v_k\|_{L^\infty(0,+\infty;U)} \leq M \text{ for all } k \in \mathbb{N}^+. \tag{5.25}$$

By the inequality in (5.25), there exists a subsequence of $\{v_k\}_{k\geq 1}$, denoted in the same manner, and $\widehat{v} \in L^\infty(0, +\infty; U)$ so that

$$v_k \to \widehat{v} \text{ weakly star in } L^\infty(0, +\infty; U) \text{ as } k \to +\infty, \tag{5.26}$$

and

$$\|\widehat{v}\|_{L^\infty(0,+\infty;U)} \leq \liminf_{k\to+\infty} \|v_k\|_{L^\infty(0,+\infty;U)} \leq M. \tag{5.27}$$

It follows from (5.24) and (5.26) that

$$y(T_k; 0, \widehat{y}_0, v_k) \to y(\widetilde{T}(M, \widehat{y}_0); 0, \widehat{y}_0, \widehat{v}) \text{ weakly in } Y,$$

which, combined with the equality in (5.25), indicates that $y(\widetilde{T}(M, \widehat{y}_0); 0, \widehat{y}_0, \widehat{v}) = 0$. This, together with (5.27), yields that $\widehat{v}$ is an optimal control to $(\widetilde{TP})_M^{\widehat{y}_0}$.

Step 2. We show that if $0 \leq M \leq \widehat{N}(\widehat{y}_0)$, then $(\widetilde{TP})_M^{\widehat{y}_0}$ has no optimal control.

By contradiction, suppose that it was not true. Then there would be M, with $0 \leq M \leq \widehat{N}(\widehat{y}_0)$, and a control u^* in $L^\infty(0, +\infty; U)$ so that

$$y(\widetilde{T}(M, \widehat{y}_0); 0, \widehat{y}_0, u^*) = 0 \text{ and } \|u^*\|_{L^\infty(0,+\infty;U)} \leq M. \tag{5.28}$$

Since $\widehat{y}_0 \neq 0$, the equality of (5.28) implies that $\widetilde{T}(M, \widehat{y}_0) \in (0, +\infty)$ and that $u^*|_{(0,\widetilde{T}(M,\widehat{y}_0))}$ is an admissible control to $(NP)_{\widetilde{T}(M,\widehat{y}_0)}^{\widehat{y}_0}$. Then, it follows from the optimality of $N(\widetilde{T}(M, \widehat{y}_0), \widehat{y}_0)$ that

$$N(\widetilde{T}(M, \widehat{y}_0), \widehat{y}_0) \leq \|u^*\|_{L^\infty(0,\widetilde{T}(M,\widehat{y}_0);U)}.$$

Since $M \leq \widehat{N}(\widehat{y}_0)$, the above inequality, combined with the inequality in (5.28), indicates that

$$N(\widetilde{T}(M, \widehat{y}_0), \widehat{y}_0) \leq M \leq \widehat{N}(\widehat{y}_0). \tag{5.29}$$

Now, it follows from Lemma 5.2 and (5.29) that

$$N(T, \widehat{y}_0) < N(\widetilde{T}(M, \widehat{y}_0), \widehat{y}_0) \leq \widehat{N}(\widehat{y}_0) \text{ for each } T > \widetilde{T}(M, \widehat{y}_0).$$

However, by Lemma 5.2 and the definition of $\widehat{N}(\widehat{y}_0)$ (see (5.5)), we find that $N(T, \widehat{y}_0) > \widehat{N}(\widehat{y}_0)$ for all $T > 0$. Thus, we get a contradiction. So the conclusion in Step 2 is true.

In summary, we end the proof of Lemma 5.3. □

Remark 5.1 Lemma 5.3 can also be proved by combining Lemma 5.2 with Theorem 3.6.

Corollary 5.1 *Let $(M, T) \in [0, +\infty) \times (0, +\infty)$. Then $M = N(T, \widehat{y_0})$ if and only if $T = \widetilde{\mathbf{T}}(M, \widehat{y_0})$.*

Proof Let $\widehat{y_0} \in Y \setminus \{0\}$ and $(M, T) \in [0, +\infty) \times (0, +\infty)$. We first show that

$$M = N(T, \widehat{y_0}) \quad \Rightarrow \quad T = \widetilde{\mathbf{T}}(M, \widehat{y_0}). \tag{5.30}$$

According to Lemma 5.1, there exists a control $u_1^* \in L^\infty(0, T; U)$ so that

$$y(T; 0, \widehat{y_0}, u_1^*) = 0 \quad \text{and} \quad \|u_1^*\|_{L^\infty(0,T;U)} = N(T, \widehat{y_0}). \tag{5.31}$$

Write $\widetilde{u}_1^*$ for the zero extension of u_1^* over $(0, +\infty)$. Since $M = N(T, \widehat{y_0})$, it follows from (5.31) that $(0, \widehat{y_0}, T, \widetilde{u}_1^*)$ is an admissible tetrad to $(\widetilde{TP})_M^{\widehat{y_0}}$. Then by the optimality of $\widetilde{\mathbf{T}}(M, \widehat{y_0})$ (see (5.3)), we find that

$$\widetilde{\mathbf{T}}(M, \widehat{y_0}) \leq T < +\infty. \tag{5.32}$$

Meanwhile, since $M = N(T, \widehat{y_0})$, it follows by Lemma 5.2 that $M > \widehat{N}(\widehat{y_0})$. This, together with Lemma 5.3, yields that $\widetilde{\mathbf{T}}(M, \widehat{y_0}) \in (0, +\infty)$ and there exists a control $u_2^* \in L^\infty(0, +\infty; U)$ satisfying that

$$y(\widetilde{\mathbf{T}}(M, \widehat{y_0}); 0, \widehat{y_0}, u_2^*) = 0 \quad \text{and} \quad \|u_2^*\|_{L^\infty(0,+\infty;U)} \leq M. \tag{5.33}$$

Then by making use of (5.33), we see that $u_2^*|_{(0,\widetilde{\mathbf{T}}(M,\widehat{y_0}))}$ is an admissible control to $(NP)_{\widetilde{\mathbf{T}}(M,\widehat{y_0})}^{\widehat{y_0}}$ and

$$N(\widetilde{\mathbf{T}}(M, \widehat{y_0}), \widehat{y_0}) \leq \|u_2^*\|_{L^\infty(0,\widetilde{\mathbf{T}}(M,\widehat{y_0});U)} \leq M. \tag{5.34}$$

Because $M = N(T, \widehat{y_0})$, it follows by (5.34) and Lemma 5.2 that

$$\widetilde{\mathbf{T}}(M, \widehat{y_0}) \geq T.$$

This, together with (5.32), yields that $T = \widetilde{\mathbf{T}}(M, \widehat{y_0})$.

We next show that

$$T = \widetilde{\mathbf{T}}(M, \widehat{y_0}) \quad \Rightarrow \quad M = N(T, \widehat{y_0}). \tag{5.35}$$

Since $\widetilde{\mathbf{T}}(M, \widehat{y_0}) = T \in (0, +\infty)$, we have that $(\widetilde{TP})_M^{\widehat{y_0}}$ has an admissible tetrad. Then by similar arguments as those used in the proof of Lemma 5.3 (see Step 1 in Lemma 5.3), we can verify that $(\widetilde{TP})_M^{\widehat{y_0}}$ has an optimal control. This, together with Lemma 5.3, implies that

$$M > \widehat{N}(\widehat{y}_0),$$

which, combined with Lemma 5.2, indicates that

$$M = N(\widehat{T}, \widehat{y}_0) \text{ for some } \widehat{T} \in (0, +\infty). \tag{5.36}$$

From this and (5.30), it follows that

$$\widehat{T} = \widetilde{\mathbf{T}}(M, \widehat{y}_0). \tag{5.37}$$

Since we have assumed that $T = \widetilde{\mathbf{T}}(M, \widehat{y}_0)$, it follows from (5.37) and (5.36) that $M = N(T, \widehat{y}_0)$. Hence, (5.35) holds.

Finally, by (5.30) and (5.35), we finish the proof of Corollary 5.1.   □

We now are on the position to prove Theorem 5.1.

Proof (Proof of Theorem 5.1) Let $\mathscr{E}_{\widehat{y}_0}$ be given by (5.6). It follows from Lemma 5.2 that the function $N(\cdot, \widehat{y}_0)$ is strictly decreasing and continuous from $(0, +\infty)$ onto $(\widehat{N}(\widehat{y}_0), +\infty)$. The conclusions (i) and (ii) of Theorem 5.1 will be proved one by one.

(i) Arbitrarily fix $(M, T) \in \mathscr{E}_{\widehat{y}_0}$. Three facts are given in order.

First, by (5.6), Lemma 5.2, and Corollary 5.1, we have that

$$M = N(T, \widehat{y}_0) \in (\widehat{N}(\widehat{y}_0), +\infty) \text{ and } T = \widetilde{\mathbf{T}}(M, \widehat{y}_0) > 0. \tag{5.38}$$

The above two equalities, together with Lemmas 5.3 and 5.1, imply that both $(\widetilde{TP})_M^{\widehat{y}_0}$ and $(NP)_T^{\widehat{y}_0}$ have optimal controls.

Second, each optimal control u^* to $(\widetilde{TP})_M^{\widehat{y}_0}$ satisfies that

$$y(\widetilde{\mathbf{T}}(M, \widehat{y}_0); 0, \widehat{y}_0, u^*) = 0 \text{ and } \|u^*\|_{L^\infty(0,+\infty;U)} \le M. \tag{5.39}$$

From (5.39) and (5.38), one can easily see that $u^*|_{(0,T)}$ is an optimal control to $(NP)_T^{\widehat{y}_0}$.

Third, each optimal control v^* to $(NP)_T^{\widehat{y}_0}$ satisfies that

$$y(T; 0, \widehat{y}_0, v^*) = 0 \text{ and } \|v^*\|_{L^\infty(0,T;U)} - N(T, \widehat{y}_0). \tag{5.40}$$

Write $\widetilde{v}^*$ for the zero extension of v^* over $(0, +\infty)$. From (5.40) and (5.38), we see that $\widetilde{v}^*$ is an optimal control to $(\widetilde{TP})_M^{\widehat{y}_0}$.

Finally, from the above three facts and Definition 5.1, we see that $(TP)_M^{\widehat{y}_0}$ and $(NP)_T^{\widehat{y}_0}$ are equivalent. So the conclusion (i) in Theorem 5.1 is true.

(ii) By contradiction, suppose that the conclusion (ii) was not true. Then there would be

$$(\widehat{M}, \widehat{T}) \in [0, +\infty) \times (0, +\infty) \setminus \mathscr{E}_{\widehat{y}_0} \tag{5.41}$$

so that $(TP)_{\widehat{M}}^{\widehat{y_0}}$ and $(NP)_{\widehat{T}}^{\widehat{y_0}}$ are equivalent. We claim that

$$\widehat{T} = \widetilde{\mathbf{T}}(M, \widehat{y_0}). \tag{5.42}$$

When (5.42) is proved, it follows by Corollary 5.1 that $\widehat{M} = N(\widehat{T}, \widehat{y_0})$, which contradicts (5.41) and completes the proof of (ii).

We now turn to prove (5.42). Since $(TP)_{\widehat{M}}^{\widehat{y_0}}$ and $(NP)_{\widehat{T}}^{\widehat{y_0}}$ are equivalent, it follows by (i) of Definition 5.1 that $(NP)_{\widehat{T}}^{\widehat{y_0}}$ has an optimal control v^*. By the optimality of v^*, we find that

$$y(\widehat{T}; 0, \widehat{y_0}, v^*) = 0 \quad \text{and} \quad \|v^*\|_{L^\infty(0, \widehat{T}; U)} = N(\widehat{T}, \widehat{y_0}). \tag{5.43}$$

Moreover, from (iii) of Definition 5.1, we see that the zero extension of v^* over $(0, +\infty)$, denoted by $\widetilde{v}^*$, is an optimal control to $(\widetilde{TP})_{\widehat{M}}^{\widehat{y_0}}$. This, together with (5.43), yields that

$$\widehat{T} \geq \widetilde{\mathbf{T}}(M, \widehat{y_0}).$$

From this, we see that to show (5.42), it suffices to prove what follows is not true:

$$\widehat{T} > \widetilde{\mathbf{T}}(\widehat{M}, \widehat{y_0}). \tag{5.44}$$

By contradiction, suppose that (5.44) was true. From (i) of Definition 5.1, $(\widetilde{TP})_{\widehat{M}}^{\widehat{y_0}}$ has an optimal control $\widehat{u}^*$. Then, by the optimality of $\widehat{u}^*$, we see that

$$\widetilde{\mathbf{T}}(\widehat{M}, \widehat{y_0}) \in (0, +\infty), \ y(\widetilde{\mathbf{T}}(\widehat{M}, \widehat{y_0}); 0, \widehat{y_0}, \widehat{u}^*) = 0 \ \text{ and } \ \|\widehat{u}^*\|_{L^\infty(0, +\infty; U)} \leq \widehat{M}. \tag{5.45}$$

Arbitrarily take $u_0 \in U$ satisfying that

$$\|u_0\|_U = \widehat{M}, \tag{5.46}$$

and then define a new control

$$\widehat{u}(t) \triangleq \begin{cases} \widehat{u}^*(t), & \text{if } 0 < t \leq \widetilde{\mathbf{T}}(\widehat{M}, \widehat{y_0}), \\ u_0, & \text{if } t > \widetilde{\mathbf{T}}(\widehat{M}, \widehat{y_0}). \end{cases} \tag{5.47}$$

It follows from (5.45)–(5.47) that

$$y(\widetilde{\mathbf{T}}(\widehat{M}, \widehat{y_0}); 0, \widehat{y_0}, \widehat{u}) = 0 \quad \text{and} \quad \|\widehat{u}\|_{L^\infty(0, +\infty; U)} = \widehat{M}.$$

These indicate that $\widehat{u}$ is an optimal control to $(\widetilde{TP})_{\widehat{M}}^{\widehat{y_0}}$. This, along with the fact that $(TP)_{\widehat{M}}^{\widehat{y_0}}$ and $(NP)_{\widehat{T}}^{\widehat{y_0}}$ are equivalent, yields that $\widehat{u}|_{(0, \widehat{T})}$ is an optimal control to $(NP)_{\widehat{T}}^{\widehat{y_0}}$ (see (ii) of Definition 5.1). Thus we have that

$$\|\widehat{u}\|_{L^\infty(0,\widehat{T};U)} = N(\widehat{T}, \widehat{y}_0).$$

From this, (5.44) and (5.45)–(5.47), one can easily check that

$$\widehat{M} = N(\widehat{T}, \widehat{y}_0),$$

which, along with (5.6), indicates that

$$(\widehat{M}, \widehat{T}) \in \mathscr{E}_{\widehat{y}_0}.$$

This contradicts (5.41). Hence, (5.44) is not true. Thus we have proved (5.42). Hence, we complete the proof of Theorem 5.1.   □

At the end of this section, we give two examples which can be put into the framework $(\mathscr{A}_1)$ (given at the beginning of Section 5.1).

Example 5.1 Let $(A, B) \in \mathbb{R}^{n \times n} \times \mathbb{R}^{n \times m}$ with $n, m \in \mathbb{N}^+$, $Y \triangleq \mathbb{R}^n$ and $U \triangleq \mathbb{R}^m$. Assume that (A, B) satisfies Kalman controllability rank condition:

$$\text{rank } (B, AB, \ldots, A^{n-1}B) = n.$$

Consider the following controlled ordinary differential equation:

$$\dot{y}(t) = Ay(t) + Bu(t), \quad t \in (0, +\infty).$$

Then we can easily check that the corresponding time optimal control problem $(TP)_M^{\widehat{y}_0}$ can be put into the framework $(\mathscr{A}_1)$ (given at the beginning of Section 5.1). Here, we used Theorem 1.23. We end Example 5.1.

Example 5.2 Let Ω be a bounded domain in $\mathbb{R}^d$, $d \geq 1$, with a C^2 boundary $\partial\Omega$. Let $\omega \subseteq \Omega$ be a nonempty and open subset. Consider the following controlled heat equation:

$$\begin{cases} \partial_t y - \Delta y = \chi_\omega u & \text{in } \Omega \times (0, +\infty), \\ y = 0 & \text{on } \partial\Omega \times (0, +\infty), \\ y(0) \in L^2(\Omega), \end{cases}$$

where $u \in L^\infty(0, +\infty; L^2(\Omega))$. Let $Y \triangleq L^2(\Omega)$ and $U \triangleq L^2(\Omega)$; Let $A = \Delta$ with its domain $H^2(\Omega) \cap H_0^1(\Omega)$; Let $B = \chi_\omega$, where χ_ω (the characteristic function of ω) is treated as a linear and bounded operator on U. Then one can easily check that the corresponding time optimal control problem $(TP)_M^{\widehat{y}_0}$ can be put into the framework $(\mathscr{A}_1)$ (given at the beginning of Section 5.1). Here, we used Theorems 1.22, 1.21, and (i) of Remark 1.5 (after Theorem 1.22). Thus Theorem 5.1 holds for this example. With the aid of Theorem 5.1, one can obtain a necessary and sufficient condition on the optimal time and the optimal control for $(TP)_M^{\widehat{y}_0}$ in the current case (see [17]). We now end Example 5.2.

5.2 Minimal Time Controls, Minimal Norm Controls, and Time-Varying Systems: Part I

In this section, we will study the equivalence of the problem $(TP)_{min}^{Q_S,Q_E}$ and the corresponding minimal norm control problem under the following framework $(\mathscr{A}_2)$:

(i) The state space Y and the control space U are real separable Hilbert spaces.
(ii) The controlled system is as:

$$\dot{y}(t) = Ay(t) + D(t)y(t) + Bu(t), \quad t \in (0, +\infty), \tag{5.48}$$

where $A : \mathscr{D}(A) \subseteq Y \mapsto Y$ generates a C_0 semigroup $\{e^{At}\}_{t\geq 0}$ on Y, $D(\cdot) \in L_{loc}^1([0, +\infty); \mathscr{L}(Y))$ and $B \in \mathscr{L}(U, Y)$. Given $\tau \in [0, +\infty)$, $y_0 \in Y$ and $u \in L^\infty(\tau, +\infty; U)$, write $y(\cdot; \tau, y_0, u)$ for the solution of (5.48) over $[\tau, +\infty)$, with the initial condition that $y(\tau) = y_0$. Moreover, write $\{\Phi(t, s) : t \geq s \geq 0\}$ for the evolution operator generated by $A + D(\cdot)$ over Y.
(iii) Let $\mathbb{U} = B_M(0)$, where $B_M(0)$ is the closed ball in U, centered at 0 and of radius $M \geq 0$. (When $M = 0$, we agree that $B_M(0) \triangleq \{0\}$.)
(iv) Let $Q_S = \{0\} \times \{\widehat{y}_0\}$ with $\widehat{y}_0 \in Y \setminus Q$; $Q_E = (0, +\infty) \times Q$, where Q is arbitrarily taken from the following family of sets:

$$\mathscr{F} \triangleq \{\widehat{Q} \subseteq Y \mid \widehat{Q} \text{ is bounded, closed, convex}$$
$$\text{and has nonempty interior in } Y\}. \tag{5.49}$$

(v) The system (5.48) is L^∞-approximately controllable over each interval, i.e., for any $T_2 > T_1 \geq 0$, $y_0, y_1 \in Y$ and $\varepsilon \in (0, +\infty)$, there exists $u \in L^\infty(T_1, T_2; U)$ so that $\|y(T_2; T_1, y_0, u) - y_1\|_Y \leq \varepsilon$.

In this section, we simply write $(TP)_{min}^{Q_S,Q_E}$ for $(TP)_Q^{M,\widehat{y}_0}$ which reads:

$$(TP)_Q^{M,\widehat{y}_0} \qquad \mathbf{T}(M, \widehat{y}_0, Q) \triangleq \inf\{T \in (0, +\infty) \mid y(T; 0, \widehat{y}_0, u) \in Q,$$
$$u \in L(0, T; \mathbb{U})\}.$$

Similar to what we explained in the previous section (see (5.2) and (5.3)), we can treat $(TP)_Q^{M,\widehat{y}_0}$ as the following problem:

$$(\widetilde{TP})_Q^{M,\widehat{y}_0} \qquad \widetilde{\mathbf{T}}(M, \widehat{y}_0, Q) \triangleq \inf\{T \in (0, +\infty) \mid y(T; 0, \widehat{y}_0, u) \in Q$$
$$\text{and } u \in \mathscr{U}_M\}, \tag{5.50}$$

where

$$\mathscr{U}_M \triangleq \{u \in L^\infty(0, +\infty; U) \mid \|u(t)\|_U \leq M \text{ a.e. } t \in (0, +\infty)\}.$$

In this problem,

- a tetrad $(0, \widehat{y}_0, T, u)$ is called admissible, if $T \in (0, +\infty)$, $u \in \mathcal{U}_M$ and $y(T; 0, \widehat{y}_0, u) \in Q$;
- a tetrad $(0, \widehat{y}_0, \widetilde{T}(M, \widehat{y}_0, Q), u^*)$ is called optimal, if

$$\widetilde{T}(M, \widehat{y}_0, Q) \in (0, +\infty), \ u^* \in \mathcal{U}_M \ \text{and} \ y(\widetilde{T}(M, \widehat{y}_0, Q); 0, \widehat{y}_0, u^*) \in Q;$$

- when $(0, \widehat{y}_0, \widetilde{T}(M, \widehat{y}_0, Q), u^*)$ is an optimal tetrad, $\widetilde{T}(M, \widehat{y}_0, Q)$ and u^* are called the optimal time and an optimal control, respectively.

Moreover, given $\widehat{y}_0$ and Q, the map $M \to \widetilde{T}(M, \widehat{y}_0, Q)$ $(M \geq 0)$ is called a minimal time function.

We next introduce the corresponding minimal norm control problem in the following manner: Given $T \in (0, +\infty)$, $y_0 \in Y$ and $Q \in \mathcal{F}$, we set

$$(NP)_Q^{T, y_0} \qquad N(T, y_0, Q) \triangleq \inf\{\|v\|_{L^\infty(0,T;U)} \mid y(T; 0, y_0, v) \in Q\}. \qquad (5.51)$$

In this problem,

- $N(T, y_0, Q)$ is called the minimal norm;
- a control $v \in L^\infty(0, T; U)$ is called admissible, if $y(T; 0, y_0, v) \in Q$;
- a control $v^* \in L^\infty(0, T; U)$ is called optimal, if it is admissible and satisfies that $\|v^*\|_{L^\infty(0,T;U)} = N(T, y_0, Q)$.

Moreover, given y_0 and Q, the map $T \to N(T, y_0, Q)$ is called a minimal norm function, which is denoted by $N(\cdot, y_0, Q)$.

We now give the definition about the equivalence between $(TP)_Q^{M, \widehat{y}_0}$ and $(NP)_Q^{T, \widehat{y}_0}$.

Definition 5.2 Let $M \geq 0$, $T \in (0, +\infty)$, $Q \in \mathcal{F}$ and $\widehat{y}_0 \in Y \setminus Q$. Problems $(TP)_Q^{M, \widehat{y}_0}$ and $(NP)_Q^{T, \widehat{y}_0}$ are said to be equivalent if the following three conditions hold:

(i) Both $(\widetilde{TP})_Q^{M, \widehat{y}_0}$ and $(NP)_Q^{T, \widehat{y}_0}$ have optimal controls;

(ii) The restriction of each optimal control to $(\widetilde{TP})_Q^{M, \widehat{y}_0}$ over $(0, T)$ is an optimal control to $(NP)_Q^{T, \widehat{y}_0}$;

(iii) The zero extension of each optimal control to $(NP)_Q^{T, \widehat{y}_0}$ over $(0, +\infty)$ is an optimal control to $(\widetilde{TP})_Q^{M, \widehat{y}_0}$.

The main result of this section is as follows.

Theorem 5.2 Let $\widehat{y}_0 \in Y$ and $Q \in \mathcal{F}$ satisfy that $\widehat{y}_0 \notin Q$. Then the function $N(\cdot, \widehat{y}_0, Q)$ is continuous over $(0, +\infty)$. Moreover, if write

$$(\mathscr{GT})_{\widehat{y}_0, Q} \triangleq \{(M, T) \in [0, +\infty) \times (0, +\infty) \mid T = \widetilde{T}(M, \widehat{y}_0, Q)\} \qquad (5.52)$$

and

$$(\mathscr{K}N)_{\widehat{y}_0, Q} \triangleq \left\{ (M, T) \in [0, +\infty) \times (0, +\infty) \mid M = 0, N(T, \widehat{y}_0, Q) = 0 \right\},$$
(5.53)

then the following conclusions are true:

(i) *When* $(M, T) \in (\mathscr{G}T)_{\widehat{y}_0, Q} \setminus (\mathscr{K}N)_{\widehat{y}_0, Q}$, *problems* $(TP)_Q^{M, \widehat{y}_0}$ *and* $(NP)_Q^{T, \widehat{y}_0}$ *are equivalent and the null controls (over* $(0, \mathbf{T}(M, \widehat{y}_0, Q))$ *and* $(0, T)$, *respectively) are not optimal controls to these two problems, respectively.*

(ii) *When* $(M, T) \in (\mathscr{K}N)_{\widehat{y}_0, Q}$, *problems* $(TP)_Q^{M, \widehat{y}_0}$ *and* $(NP)_Q^{T, \widehat{y}_0}$ *are equivalent and the null controls (over* $(0, \mathbf{T}(M, \widehat{y}_0, Q))$ *and* $(0, T)$, *respectively) are the unique optimal controls to these two problems, respectively.*

(iii) *When* $(M, T) \in [0, +\infty) \times (0, +\infty) \setminus \left((\mathscr{G}T)_{\widehat{y}_0, Q} \cup (\mathscr{K}N)_{\widehat{y}_0, Q} \right)$, *problems* $(TP)_Q^{M, \widehat{y}_0}$ *and* $(NP)_Q^{T, \widehat{y}_0}$ *are not equivalent.*

We will give its proof later.

5.2.1 Properties of Minimal Time and Minimal Norm Functions

In this subsection, we mainly discuss some properties of minimal time functions and minimal norm functions (see Theorems 5.3 and 5.4). We start with the following three lemmas.

Lemma 5.4 *The following three conclusions are true:*

(i) *If there exists* $z \in Y$ *and* $T > \tau > 0$ *so that* $B^* \Phi(T, t)^* z = 0$ *for all* $t \in (\tau, T)$, *then* $z = 0$.

(ii) *Let* $T_2 > T_1 > 0$ *and* $y_d, y_0 \in Y$. *Then for each* $\varepsilon > 0$, *there exists a positive constant* $C_\varepsilon \triangleq C(T_1, T_2, y_d, y_0, \varepsilon)$ *so that for each* $T \in [T_1, T_2]$,

$$|\langle y_d - \Phi(T, 0)y_0, z \rangle_Y|$$
$$\leq C_\varepsilon \int_0^T \|B^* \Phi(T, t)^* z\|_U \, dt + \varepsilon \|z\|_Y \quad \text{for all } z \in Y.$$
(5.54)

(iii) *Let* $T_2 > T_1 > 0$, $y_0 \in Y$ *and* $Q \in \mathscr{F}$. *Then there exists a positive constant* $C \triangleq C(T_1, T_2, y_0, Q)$ *so that for each* $T \in [T_1, T_2]$, *there is* $u_T \in L^\infty(0, T; U)$ *so that*

$$y(T; 0, y_0, u_T) \in Q \quad \text{and} \quad \|u_T\|_{L^\infty(0,T;U)} \leq C.$$
(5.55)

Proof We prove the conclusions one by one.

(i) Suppose that there exists $z \in Y$ and $T > \tau > 0$ so that

$$B^* \Phi(T, t)^* z = 0 \quad \text{for all } t \in (\tau, T). \tag{5.56}$$

By (v) of $(\mathscr{A}_2)$ (given at the beginning of Section 5.2), for each $y_d \in Y$ and $\varepsilon \in (0, +\infty)$, there is $u \in L^\infty(0, T; U)$ so that

$$\|y(T; 0, 0, \chi_{(\tau,T)} u) - y_d\|_Y \leq \varepsilon.$$

This, along with (5.56), yields that for each $y_d \in Y$ and $\varepsilon \in (0, +\infty)$,

$$\begin{aligned}
\langle y_d, z \rangle_Y &= \langle y(T; 0, 0, \chi_{(\tau,T)} u), z \rangle_Y + \langle y_d - y(T; 0, 0, \chi_{(\tau,T)} u), z \rangle_Y \\
&\leq \int_\tau^T \langle u(t), B^* \Phi(T, t)^* z \rangle_U dt + \varepsilon \|z\|_Y \\
&\leq \varepsilon \|z\|_Y,
\end{aligned}$$

which leads to $z = 0$.

(ii) Let $T_2 > T_1 > 0$ and $y_d, y_0 \in Y$. By contradiction, suppose that (5.54) was not true. Then there would exist $\varepsilon_0 \in (0, +\infty)$, $\{t_k\}_{k \geq 1} \subseteq [T_1, T_2]$ and $\{z_k\}_{k \geq 1} \subseteq Y$ so that for each $k \in \mathbb{N}^+$,

$$|\langle y_d - \Phi(t_k, 0) y_0, z_k \rangle_Y| = 1 > k \int_0^{t_k} \|B^* \Phi(t_k, t)^* z_k\|_U dt + \varepsilon_0 \|z_k\|_Y, \tag{5.57}$$

which implies that $\{z_k\}_{k \geq 1}$ is bounded in Y. Since $\{t_k\}_{k \geq 1} \subseteq [T_1, T_2]$, there exists a subsequence of $\{k\}_{k \geq 1}$, still denoted in the same manner, $\widehat{t} \in [T_1, T_2]$ and $\widehat{z} \in Y$ so that

$$\lim_{k \to +\infty} t_k = \widehat{t} \quad \text{and} \quad z_k \to \widehat{z} \text{ weakly in } Y.$$

These, together with (5.57), imply that

$$1 = |\langle y_d - \Phi(\widehat{t}, 0) y_0, \widehat{z} \rangle_Y| \quad \text{and} \quad B^* \Phi(\widehat{t}, t)^* \widehat{z} = 0 \quad \text{for all } t \in (0, \widehat{t}).$$

The above conclusions, combined with the conclusion (i) of this lemma, lead to a contradiction. So the conclusion (ii) of this lemma is true.

(iii) Let $T_2 > T_1 > 0$, $y_0 \in Y$ and $Q \in \mathscr{F}$. By (5.49), there exists a closed ball $B_r(y_d)$ (centered at y_d and of radius $r \in (0, +\infty)$) so that

$$B_r(y_d) \subseteq Q. \tag{5.58}$$

Let C_r be given by (5.54) with $\varepsilon = r$. Arbitrarily fix $T \in [T_1, T_2]$. Define the following subspace of $L^1(0, T; U) \times Y$:

$$\mathcal{O} \triangleq \left\{ (C_r B^* \Phi(T, \cdot)^* z, rz) \in L^1(0, T; U) \times Y \mid z \in Y \right\}. \qquad (5.59)$$

And then define an operator $\mathcal{R} : \mathcal{O} \mapsto \mathbb{R}$ in the following manner:

$$\mathcal{R}\left(C_r B^* \Phi(T, \cdot)^* z, rz\right) \triangleq \langle \Phi(T, 0) y_0 - y_d, z \rangle_Y \quad \text{for each } z \in Y. \qquad (5.60)$$

From (5.60), we see that the map $\mathcal{R}$ is well defined and linear. By (5.54), we find that

$$\|\mathcal{R}\|_{\mathscr{L}(\mathcal{O}, \mathbb{R})} \leq 1. \qquad (5.61)$$

Here, the norm of $\mathcal{O}$ is inherited from that of the space $L^1(0, T; U) \times Y$.

Next, according to the Hahn-Banach Theorem (see Theorem 1.5), there exists a bounded linear functional

$$\widetilde{\mathcal{R}} : L^1(0, T; U) \times Y \mapsto \mathbb{R}$$

so that

$$\widetilde{\mathcal{R}} = \mathcal{R} \text{ on } \mathcal{O} \text{ and } \|\widetilde{\mathcal{R}}\|_{\mathscr{L}(L^1(0,T;U) \times Y, \mathbb{R})} = \|\mathcal{R}\|_{\mathscr{L}(\mathcal{O}, \mathbb{R})}. \qquad (5.62)$$

Then according to the Riesz Representation Theorem (see Theorem 1.4), there is a pair $(\widetilde{u}, \psi) \in L^\infty(0, T; U) \times Y$ so that

$$\widetilde{\mathcal{R}}(f, g) = \langle \widetilde{u}, f \rangle_{L^\infty(0,T;U), L^1(0,T;U)} + \langle \psi, g \rangle_Y$$

$$\text{for all } (f, g) \in L^1(0, T; U) \times Y. \qquad (5.63)$$

Now, it follows from (5.60), (5.59), (5.62), and (5.63) that for each $z \in Y$,

$$\langle \Phi(T, 0) y_0 - y_d, z \rangle_Y = \widetilde{\mathcal{R}}\left(C_r B^* \Phi(T, \cdot)^* z, rz\right)$$

$$= \langle \widetilde{u}, C_r B^* \Phi(T, \cdot)^* z \rangle_{L^\infty(0,T;U), L^1(0,T;U)} + \langle \psi, rz \rangle_Y.$$

This yields that for each $z \in Y$,

$$\langle y(T; 0, y_0, -C_r \widetilde{u}) - y_d, z \rangle_Y$$

$$= \langle \Phi(T, 0) y_0 - y_d, z \rangle_Y - \int_0^T \langle \widetilde{u}(t), C_r B^* \Phi(T, t)^* z \rangle_U \, dt$$

$$= \langle r\psi, z \rangle_Y,$$

which leads to

$$y(T; 0, y_0, -C_r \tilde{u}) = y_d + r\psi. \tag{5.64}$$

Finally, from (5.64) and (5.58), we see that in order to show (5.55), it suffices to prove the inequality:

$$\max\{\|\tilde{u}\|_{L^\infty(0,T;U)}, \|\psi\|_Y\} \le 1. \tag{5.65}$$

By (5.63), (5.62), and (5.61), we see that for each $f \in L^1(0, T; U)$ and $g \in Y$,

$$\int_0^T \langle \tilde{u}(t), f(t) \rangle_U \, dt + \langle \psi, g \rangle_Y = \widetilde{\mathscr{R}}(f, g) \le \|(f, g)\|_{L^1(0,T;U) \times Y},$$

which leads to (5.65).

Hence, we end the proof of Lemma 5.4. □

Lemma 5.5 *For any $T \in (0, +\infty)$, $y_0 \in Y$ and $Q \in \mathscr{F}$, the problem $(NP)_Q^{T,y_0}$ has at least one optimal control.*

Proof Since $Q \in \mathscr{F}$, it has a nonempty interior. Then it follows from (v) of $(\mathscr{A}_2)$ that $(NP)_Q^{T,y_0}$ has at least one admissible control. Next, by the same way as that used in the proof of Lemma 5.1, one can easily prove the existence of optimal controls to this problem. This completes the proof. □

Lemma 5.6 *Let $T \in (0, +\infty)$, $y_0 \in Y$ and $Q \in \mathscr{F}$. Then the following two conclusions are true:*

(i) *If $\Phi(T, 0)y_0 \notin Q$, then there exists $z \in Y \setminus \{0\}$ so that*

$$\sup_{\|u\|_{L^\infty(0,T;U)} \le N(T, y_0, Q)} \langle y(T; 0, y_0, u), z \rangle_Y \le \inf_{q \in Q} \langle q, z \rangle_Y. \tag{5.66}$$

(ii) *Each optimal control v^* to $(NP)_Q^{T,y_0}$ satisfies that*

$$\|v^*\|_{L^\infty(\tau,T;U)} = N(T, y_0, Q) \quad \text{for any } \tau \in [0, T). \tag{5.67}$$

Proof We prove the conclusions one by one.

(i) Define a subset of Y by

$$\mathscr{A} \triangleq \{y(T; 0, y_0, u) \mid \|u\|_{L^\infty(0,T;U)} \le N(T, y_0, Q)\}. \tag{5.68}$$

We claim that

$$\mathscr{A} \cap Q \neq \emptyset \quad \text{and} \quad \mathscr{A} \cap \overset{\circ}{Q} = \emptyset, \tag{5.69}$$

where $\overset{\circ}{Q}$ denotes the interior of Q in Y.

To show the first conclusion in (5.69), we use Lemma 5.5 to take an optimal control u_1 for the problem $(NP)_Q^{T,y_0}$. By the optimality of u_1, we have that

$$y(T; 0, y_0, u_1) \in Q \quad \text{and} \quad \|u_1\|_{L^\infty(0,T;U)} = N(T, y_0, Q).$$

These, combined with (5.68), lead to the first conclusion in (5.69).

We next prove the second conclusion in (5.69). By contradiction, suppose that it was not true. Then by (5.68), there would exist $u_2 \in L^\infty(0, T; U)$ so that

$$y(T; 0, y_0, u_2) \in \overset{\circ}{Q} \quad \text{and} \quad \|u_2\|_{L^\infty(0,T;U)} \le N(T, y_0, Q). \qquad (5.70)$$

Since $y(T; 0, y_0, u_2) \in \overset{\circ}{Q}$, we can find a constant $\lambda \in (0, 1)$ so that

$$y(T; 0, y_0, \lambda u_2) \in Q,$$

which implies that λu_2 is an admissible control to $(NP)_Q^{T,y_0}$. This, together with the optimality of $N(T, y_0, Q)$ and the inequality in (5.70), yields that

$$N(T, y_0, Q) \le \|\lambda u_2\|_{L^\infty(0,T;U)} \le \lambda N(T, y_0, Q).$$

Since $\lambda \in (0, 1)$, the above leads to

$$N(T, y_0, Q) = 0,$$

which indicates that $\Phi(T, 0)y_0 \in Q$. This contradicts the assumption that $\Phi(T, 0)y_0 \notin Q$. Hence, the second conclusion in (5.69) is true.

Since $\mathscr{A}$ and $\overset{\circ}{Q}$ are nonempty convex subsets in Y (see (5.68), the first conclusion in (5.69) and (5.49)), by the second conclusion in (5.69), we can apply the Hahn-Banach Theorem (see Theorem 1.11) to find $z \in Y \setminus \{0\}$ so that

$$\sup_{y \in \mathscr{A}} \langle y, z \rangle_Y \le \inf_{q \in Q} \langle q, z \rangle_Y.$$

This, along with (5.68), leads to (5.66).

(ii) Let v^* be an optimal control to $(NP)_Q^{T,y_0}$. Then $0 \le N(T, y_0, Q) < +\infty$. When $N(T, y_0, Q) = 0$, it is clear that v^* is the null control and (5.67) is true.

Next we show (5.67) when $N(T, y_0, Q) \in (0, +\infty)$. In this case, we have that

$$\Phi(T, 0)y_0 \notin Q. \qquad (5.71)$$

By contradiction, suppose that (5.67) was not true. Then there would be $\hat{\tau} \in [0, T)$ and an optimal control $\hat{v}^*$ to $(NP)_Q^{T,y_0}$ so that

$$\|\widehat{v}^*\|_{L^\infty(\widehat{\tau},T;U)} < N(T, y_0, Q). \tag{5.72}$$

Since $\widehat{v}^*$ is an optimal control to $(NP)_Q^{T,y_0}$, we have that

$$y(T; 0, y_0, \widehat{v}^*) \in Q \text{ and } \|\widehat{v}^*\|_{L^\infty(0,T;U)} = N(T, y_0, Q).$$

These, along with (5.71) and (i) of this lemma, yield that for some $z \in Y \setminus \{0\}$,

$$\max_{\|u\|_{L^\infty(0,T;U)} \le N(T,y_0,Q)} \langle y(T; 0, y_0, u), z \rangle_Y = \langle y(T; 0, y_0, \widehat{v}^*), z \rangle_Y,$$

which implies that

$$\max_{\|u\|_{L^\infty(0,T;U)} \le N(T,y_0,Q)} \int_0^T \langle u(t), B^*\Phi(T, t)^* z \rangle_U \, dt$$
$$= \int_0^T \langle \widehat{v}^*(t), B^*\Phi(T, t)^* z \rangle_U \, dt. \tag{5.73}$$

By (5.73), we can use the similar way as that used in the proof of (4.15) to get that

$$\langle \widehat{v}^*(t), B^*\Phi(T, t)^* z \rangle_U = \max_{\|w\|_U \le N(T,y_0,Q)} \langle w, B^*\Phi(T, t)^* z \rangle_U \text{ for a.e. } t \in (0, T).$$

The above, along with (5.72), yields that

$$B^*\Phi(T, t)^* z = 0 \text{ for all } t \in (\widehat{\tau}, T).$$

This, together with (i) of Lemma 5.4, implies that $z = 0$, which leads to a contradiction.

Hence, we complete the proof of Lemma 5.6. □

Let $y_0 \in Y$ and $Q \in \mathscr{F}$. For each $M \ge 0$, we define

$$\mathscr{I}_M^{y_0} \triangleq \{t \in (0, +\infty) \mid N(t, y_0, Q) \le M\}. \tag{5.74}$$

We agree that

$$\inf \mathscr{I}_M^{y_0} \triangleq +\infty, \text{ when } \mathscr{I}_M^{y_0} = \emptyset. \tag{5.75}$$

The following theorem presents some connection between the minimal time and the minimal norm functions. Such connection plays an important role in our studies.

Theorem 5.3 *Let $\widehat{y}_0 \in Y$ and $Q \in \mathscr{F}$ satisfy that $\widehat{y}_0 \notin Q$. Let $\mathscr{I}_M^{\widehat{y}_0}$, with $M \ge 0$, be defined by (5.74). Then*

$$\widetilde{\mathbf{T}}(M, \widehat{y}_0, Q) = \inf \mathscr{J}_M^{\widehat{y}_0} \ \text{ for all } \ M \geq 0. \tag{5.76}$$

Proof Arbitrarily fix $\widehat{y}_0 \in Y$ and $Q \in \mathscr{F}$ satisfying $\widehat{y}_0 \notin Q$. Let $M \geq 0$. Then either $\mathscr{J}_M^{\widehat{y}_0} = \emptyset$ or $\mathscr{J}_M^{\widehat{y}_0} \neq \emptyset$.

In the case that $\mathscr{J}_M^{\widehat{y}_0} = \emptyset$, we first claim that

$$y(t; 0, \widehat{y}_0, u) \notin Q \ \text{ for all } \ t \in (0, +\infty) \ \text{ and } \ u \in \mathscr{U}_M. \tag{5.77}$$

By contradiction, suppose that (5.77) was not true. Then there would exist $\widehat{t} \in (0, +\infty)$ and $\widehat{u} \in L^\infty(0, +\infty; U)$ so that

$$y(\widehat{t}; 0, \widehat{y}_0, \widehat{u}) \in Q \ \text{ and } \ \|\widehat{u}\|_{L^\infty(0,+\infty;U)} \leq M. \tag{5.78}$$

The first conclusion in (5.78) implies that $\widehat{u}|_{(0,\widehat{t})}$ is an admissible control to $(NP)_Q^{\widehat{t},\widehat{y}_0}$. This, along with the optimality of $N(\widehat{t}, \widehat{y}_0, Q)$ and the second conclusion in (5.78), yields that

$$N(\widehat{t}, \widehat{y}_0, Q) \leq \|\widehat{u}\|_{L^\infty(0,\widehat{t};U)} \leq M,$$

which, combined with (5.74), indicates that $\widehat{t} \in \mathscr{J}_M^{\widehat{y}_0}$. This leads to a contradiction since we are in the case that $\mathscr{J}_M^{\widehat{y}_0} = \emptyset$. So (5.77) is true.

Now, from (5.77), we see that $(\widetilde{TP})_Q^{M,\widehat{y}_0}$ has no admissible tetrad. Thus, we have that

$$\widetilde{\mathbf{T}}(M, \widehat{y}_0, Q) = +\infty.$$

This, together with (5.75), leads to (5.76).

In the case where $\mathscr{J}_M^{\widehat{y}_0} \neq \emptyset$, we first show that

$$\widetilde{\mathbf{T}}(M, \widehat{y}_0, Q) \leq \inf \mathscr{J}_M^{\widehat{y}_0}. \tag{5.79}$$

To this end, we arbitrarily fix $\widehat{t} \in \mathscr{J}_M^{\widehat{y}_0}$. Then it follows by (5.74) that

$$N(\widehat{t}, \widehat{y}_0, Q) \leq M.$$

Thus $(NP)_Q^{\widehat{t},\widehat{y}_0}$ has an optimal control v. Write $\widetilde{v}$ for the zero extension of v over $(0, +\infty)$. One can easily check that

$$y(\widehat{t}; 0, \widehat{y}_0, \widetilde{v}) \in Q \ \text{ and } \ \|\widetilde{v}\|_{L^\infty(0,+\infty;U)} = N(\widehat{t}, \widehat{y}_0, Q) \leq M. \tag{5.80}$$

From (5.80), we see that $(0, \widehat{y}_0, \widehat{t}, \widetilde{v})$ is an admissible tetrad to $(\widetilde{TP})_Q^{M,\widehat{y}_0}$. This, along with the optimality of $\widetilde{\mathbf{T}}(M, \widehat{y}_0, Q)$, yields that

$$\widetilde{\mathbf{T}}(M, \widehat{y}_0, Q) \leq \widehat{t}.$$

Since $\widehat{t}$ is arbitrarily taken from the set $\mathscr{I}_M^{\widehat{y}_0}$, the above inequality leads to (5.79). We next show the reverse of (5.79). For this purpose, we define

$$\mathscr{T}_{(M, \widehat{y}_0, Q)} \triangleq \{t \in (0, +\infty) \mid y(t; 0, \widehat{y}_0, u) \in Q \text{ and } u \in \mathscr{U}_M\}. \quad (5.81)$$

From (5.80), it follows that

$$\mathscr{T}_{(M, \widehat{y}_0, Q)} \neq \emptyset.$$

Then by (5.81), we see that given $\widetilde{t} \in \mathscr{T}_{(M, \widehat{y}_0, Q)}$, there is a control $\widetilde{u} \in L^\infty(0, +\infty; U)$ so that

$$y(\widetilde{t}; 0, \widehat{y}_0, \widetilde{u}) \in Q \text{ and } \|\widetilde{u}\|_{L^\infty(0, +\infty; U)} \leq M. \quad (5.82)$$

The first conclusion in (5.82) implies that $\widetilde{u}|_{(0, \widetilde{t})}$ is an admissible control to $(NP)_Q^{\widetilde{t}, \widehat{y}_0}$. This, along with the optimality of $N(\widetilde{t}, \widehat{y}_0, Q)$ and the second conclusion in (5.82), yields that

$$N(\widetilde{t}, \widehat{y}_0, Q) \leq \|\widetilde{u}\|_{L^\infty(0, \widetilde{t}; U)} \leq M,$$

which, combined with (5.74), indicates that $\widetilde{t} \in \mathscr{I}_M^{\widehat{y}_0}$. Hence,

$$\inf \mathscr{I}_M^{\widehat{y}_0} \leq \widetilde{t}.$$

Since $\widetilde{t}$ was arbitrarily taken from $\mathscr{T}_{(M, \widehat{y}_0, Q)}$, the above implies that

$$\inf \mathscr{I}_M^{\widehat{y}_0} \leq \inf \mathscr{T}_{(M, \widehat{y}_0, Q)}.$$

This, along with (5.81) and (5.50), leads to the reverse of (5.79).

Finally, (5.76) follows from (5.79) and its reverse at once. This ends the proof of Theorem 5.3. $\qquad\qquad \square$

Remark 5.2 For better understanding of Theorem 5.3, we explain it with the aid of Figure 5.1 (see below), where the curve denotes the graph of the minimal norm function. Suppose that the minimal norm function is continuous over $(0, +\infty)$ (which will be proved in Theorem 5.4). A beam (which is parallel to the t-axis and has the distance M with the t-axis) moves from the left to the right. The first time point at which this beam reaches the curve is the optimal time to $(\widetilde{TP})_Q^{M, \widehat{y}_0}$. Thus, we can treat Theorem 5.3 as a "falling sun theorem" (see, for instance, raising sun lemma-Lemma 3.5 and Figure 5 on Pages 121–122 in [12]): If one thinks of the sun falling down the west (at the left) with the rays of light parallel to the t-axis,

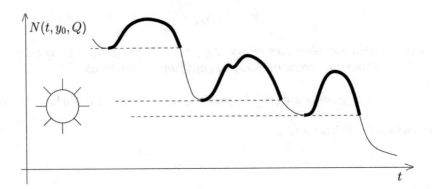

Fig. 5.1 Falling sum theorem

then the points $(\widetilde{\mathbf{T}}(M, \widehat{y}_0, Q), M)$, with $M \geq 0$, are precisely the points which are in the bright part on the curve. (These points constitute the part outside bold on the curve in Figure 5.1.)

To study some properties about minimal norm function $N(\cdot, y_0, Q)$ (with $y_0 \in Y$ and $Q \in \mathscr{F}$), we consider the following minimization problem:

$$(JP)_Q^{T,y_0} \qquad V(T, y_0, Q) \triangleq \inf_{z \in Y} J_Q^{T,y_0}(z), \qquad (5.83)$$

where $T \in (0, +\infty)$ and $J_Q^{T,y_0} : Y \mapsto \mathbb{R}$ is defined by

$$J_Q^{T,y_0}(z) \triangleq \frac{1}{2}\left(\int_0^T \|B^*\Phi(T, t)^*z\|_U \,dt\right)^2 \\ + \langle y_0, \Phi(T, 0)^*z\rangle_Y + \sup_{q \in Q}\langle q, -z\rangle_Y, \quad z \in Y. \qquad (5.84)$$

For the problem $(JP)_Q^{T,y_0}$, we have the following proposition:

Proposition 5.1 *Let $T_2 > T_1 > 0$, $T \in [T_1, T_2]$, $y_0 \in Y$, and $Q \in \mathscr{F}$. Let $B_r(y_d)$ be a closed ball in Y, centered at y_d and of radius $r > 0$, so that $B_r(y_d) \subseteq Q$. Then the following conclusions are true:*

(i) There exists a constant $C \triangleq C(T_1, T_2, y_0, Q) \in (0, +\infty)$ so that

$$J_Q^{T,y_0}(z) \geq \frac{r}{2}\|z\|_Y - C \text{ for all } z \in Y. \qquad (5.85)$$

(ii) There exists $z^* \in Y$ so that

$$V(T, y_0, Q) = J_Q^{T,y_0}(z^*). \qquad (5.86)$$

Besides, the above z^* can be chosen as 0 if and only if $\Phi(T, 0)y_0 \in Q$.

(iii) It holds that

$$V(T, y_0, Q) = -\frac{1}{2} N(T, y_0, Q)^2. \tag{5.87}$$

Proof We will prove the conclusions (i)–(iii) one by one.

(i) Arbitrarily fix $z \in Y$. From (5.84), we see that

$$J_Q^{T,y_0}(z) \geq \frac{1}{2} \left(\int_0^T \|B^*\Phi(T,t)^*z\|_U \, dt \right)^2$$

$$+ \langle y_0, \Phi(T,0)^*z \rangle_Y + \sup_{q \in B_r(y_d)} \langle q, -z \rangle_Y$$

$$= \frac{1}{2} \left(\int_0^T \|B^*\Phi(T,t)^*z\|_U \, dt \right)^2$$

$$+ \langle \Phi(T,0)y_0 - y_d, z \rangle_Y + r\|z\|_Y. \tag{5.88}$$

Meanwhile, by (ii) of Lemma 5.4, there is $C \triangleq C(T_1, T_2, y_0, y_d, r) \in (0, +\infty)$ so that

$$\langle \Phi(T,0)y_0 - y_d, z \rangle_Y \geq -C \int_0^T \|B^*\Phi(T,t)^*z\|_U \, dt - \frac{r}{2}\|z\|_Y.$$

This, together with (5.88), implies that

$$J_Q^{T,y_0}(z) \geq \frac{1}{2} \left(\int_0^T \|B^*\Phi(T,t)^*z\|_U \, dt \right)^2$$

$$- C \int_0^T \|B^*\Phi(T,t)^*z\|_U \, dt + \frac{r}{2}\|z\|_Y \tag{5.89}$$

$$\geq \frac{r}{2}\|z\|_Y - \frac{1}{2}C^2.$$

Because $B_r(y_d)$ was arbitrarily taken from Y, the above C depends only on $T_1, T_2, y_0,$ and Q, i.e., $C = C(T_1, T_2, y_0, Q)$. This, together with (5.89), gives (5.85) immediately.

(ii) By (5.83), (5.84), (5.85), and by (i) and (ii) of ($\mathscr{A}_2$) (given at the beginning of Section 5.2), we can use Theorem 1.13 to find $z^* \in Y$ holding (5.86).

We now prove that $z^* = 0 \Rightarrow \Phi(T,0)y_0 \in Q$. Indeed, if $z^* = 0$, then by (5.83), (5.84), and (5.86), we have that for each $z \in Y$,

$$0 \leq \lim_{\lambda \to 0+} \frac{1}{\lambda} \left[J_Q^{T,y_0}(\lambda z) - J_Q^{T,y_0}(0) \right] = \langle \Phi(T,0)y_0, z \rangle_Y + \sup_{q \in Q} \langle q, -z \rangle_Y.$$

This implies that

$$\sup_{q \in Q} \langle q, z \rangle_Y \geq \langle \Phi(T,0)y_0, z \rangle_Y \quad \text{for all } z \in Y.$$

From this and Theorem 1.11, we can easily check that $\Phi(T,0)y_0 \in Q$.

We next show that $\Phi(T,0)y_0 \in Q$ implies that z^* can be chosen as 0. In fact, if $\Phi(T,0)y_0 \in Q$, then

$$\langle y_0, \Phi(T,0)^*z\rangle_Y + \sup_{q\in Q}\langle q, -z\rangle_Y$$

$$\geq \langle y_0, \Phi(T,0)^*z\rangle_Y + \langle \Phi(T,0)y_0, -z\rangle_Y = 0 \quad \text{for all } z \in Y.$$

From this, (5.83) and (5.84), we see that

$$V(T, y_0, Q) = 0 = J_Q^{T,y_0}(0).$$

Hence, we can choose $z^* = 0$.

In summary, the conclusion (ii) has been proved.

(iii) We first show that

$$V(T, y_0, Q) \geq -\frac{1}{2}N(T, y_0, Q)^2. \tag{5.90}$$

By Lemma 5.5, $(NP)_Q^{T,y_0}$ has an optimal control v^*. By the optimality of v^*, we have that

$$y(T; 0, y_0, v^*) \in Q \quad \text{and} \quad \|v^*\|_{L^\infty(0,T;U)} = N(T, y_0, Q). \tag{5.91}$$

It follows from the first conclusion in (5.91) that for each $z \in Y$,

$$\sup_{q\in Q}\langle q, -z\rangle_Y \geq \langle y(T; 0, y_0, v^*), -z\rangle_Y$$

$$= -\langle y_0, \Phi(T,0)^*z\rangle_Y - \int_0^T \langle v^*(t), B^*\Phi(T,t)^*z\rangle_U dt.$$

This, together with the second conclusion in (5.91), implies that for each $z \in Y$,

$$\sup_{q\in Q}\langle q, -z\rangle_Y \geq -\langle y_0, \Phi(T,0)^*z\rangle_Y - N(T, y_0, Q)\int_0^T \|B^*\Phi(T,t)^*z\|_U dt$$

$$\geq -\langle y_0, \Phi(T,0)^*z\rangle_Y - \frac{1}{2}\Big(\int_0^T \|B^*\Phi(T,t)^*z\|_U dt\Big)^2$$

$$-\frac{1}{2}N(T, y_0, Q)^2.$$

The above, along with (5.83) and (5.84), yields (5.90).

We next show that

$$V(T, y_0, Q) \leq -\frac{1}{2}N(T, y_0, Q)^2. \tag{5.92}$$

Notice that either $\Phi(T,0)y_0 \in Q$ or $\Phi(T,0)y_0 \notin Q$.

In the case that $\Phi(T, 0)y_0 \in Q$, we have that $y(T; 0, y_0, 0) \in Q$. This implies that the null control is an optimal control to $(NP)_Q^{T,y_0}$. So we have that

$$N(T, y_0, Q) = 0. \tag{5.93}$$

Meanwhile, by (5.83) and (5.84), we see that

$$V(T, y_0, Q) \leq J_Q^{T,y_0}(0) = 0,$$

which, combined with (5.93), indicates (5.92) in this case.

In the case where $\Phi(T, 0)y_0 \notin Q$, we can use (i) of Lemma 5.6 to find $z \in Y \setminus \{0\}$ so that

$$\sup_{\|u\|_{L^\infty(0,T;U)} \leq N(T,y_0,Q)} \langle y(T; 0, y_0, u), z \rangle_Y \leq \inf_{q \in Q} \langle q, z \rangle_Y. \tag{5.94}$$

This yields that

$$\langle y_0, \Phi(T, 0)^* z \rangle_Y + N(T, y_0, Q) \int_0^T \|B^* \Phi(T, t)^* z\|_U \, dt$$

$$\leq \inf_{q \in Q} \langle q, z \rangle_Y = -\sup_{q \in Q} \langle q, -z \rangle_Y.$$

The above, along with (5.84), shows that for each $\lambda \geq 0$,

$$J_Q^{T,y_0}(\lambda z) \leq \frac{\lambda^2}{2} \left(\int_0^T \|B^* \Phi(T, t)^* z\|_U \, dt \right)^2$$

$$-\lambda N(T, y_0, Q) \int_0^T \|B^* \Phi(T, t)^* z\|_U \, dt \tag{5.95}$$

$$= \frac{1}{2} \left[\lambda \int_0^T \|B^* \Phi(T, t)^* z\|_U \, dt - N(T, y_0, Q) \right]^2$$

$$-\frac{1}{2} N(T, y_0, Q)^2.$$

Meanwhile, since $z \neq 0$, we can use (i) of Lemma 5.4 to find that

$$B^* \Phi(T, \cdot)^* z \neq 0 \quad \text{in} \quad L^1(0, T; U).$$

This, along with (5.83) and (5.95), yields (5.92).

Finally, (5.87) follows from (5.90) and (5.92) immediately.

Thus, we complete the proof of Proposition 5.1. □

The following result is mainly concerned with the continuity of the minimal norm function $N(\cdot, y_0, Q)$.

Theorem 5.4 *Let* $y_0 \in Y$ *and* $Q \in \mathscr{F}$. *Then the minimal norm function* $t \to N(t, y_0, Q)$ *is continuous over* $(0, +\infty)$. *Furthermore, if* $y_0 \notin Q$, *then* $\lim\limits_{t \to 0+} N(t, y_0, Q) = +\infty$.

Proof Arbitrarily fix $y_0 \in Y$ and $Q \in \mathscr{F}$. The proof is organized by three steps.

Step 1. We show that for each $T \in (0, +\infty)$,

$$N(T, y_0, Q) \leq \liminf_{t \to T} N(t, y_0, Q). \tag{5.96}$$

Arbitrarily fix $T \in (0, +\infty)$, and then arbitrarily take a sequence $\{t_k\}_{k \geq 1} \subseteq (T/2, 2T)$ so that $t_k \to T$. It follows from (iii) of Lemma 5.4 (where $T_1 = T/2$ and $T_2 = 2T$) that

$$\sup_{k \in \mathbb{N}^+} N(t_k, y_0, Q) < +\infty. \tag{5.97}$$

By Lemma 5.5, each $(NP)_Q^{t_k, y_0}$, with $k \geq 1$, has an optimal control u_k. By the optimality of u_k, we have that

$$y(t_k; 0, y_0, u_k) \in Q \text{ and } \|u_k\|_{L^\infty(0, t_k; U)} = N(t_k, y_0, Q). \tag{5.98}$$

For each $k \geq 1$, let $\tilde{u}_k$ be the zero extension of u_k over $(0, +\infty)$. Then the second conclusion in (5.98), together with (5.97), implies that $\{\tilde{u}_k\}_{k \geq 1}$ is bounded in $L^\infty(0, +\infty; U)$. Thus there exists a subsequence of $\{k_\ell\}_{\ell \geq 1}$ of $\{k\}_{k \geq 1}$ and $\tilde{u} \in L^\infty(0, +\infty; U)$ so that

$$\tilde{u}_{k_\ell} \to \tilde{u} \text{ weakly star in } L^\infty(0, +\infty; U) \text{ as } \ell \to +\infty.$$

From this, we find that

$$y(t_{k_\ell}; 0, y_0, \tilde{u}_{k_\ell}) \to y(T; 0, y_0, \tilde{u}) \text{ weakly in } Y \text{ as } \ell \to +\infty \tag{5.99}$$

and

$$\|\tilde{u}\|_{L^\infty(0, +\infty; U)} \leq \liminf_{\ell \to +\infty} \|\tilde{u}_{k_\ell}\|_{L^\infty(0, +\infty; U)}. \tag{5.100}$$

Since Q is convex and closed, it follows by (5.98), (5.99), and (5.100) that

$$y(T; 0, y_0, \tilde{u}) \in Q \text{ and } \|\tilde{u}\|_{L^\infty(0, +\infty; U)} \leq \liminf_{\ell \to +\infty} N(t_{k_\ell}, y_0, Q).$$

From these and the optimality of $N(T, y_0, Q)$, we see that

$$N(T, y_0, Q) \leq \|\tilde{u}\|_{L^\infty(0, +\infty; U)} \leq \liminf_{\ell \to +\infty} N(t_{k_\ell}, y_0, Q),$$

which leads to (5.96).

Step 2. We show that for each $T \in (0, +\infty)$,

$$N(T, y_0, Q) \geq \limsup_{t \to T} N(t, y_0, Q). \tag{5.101}$$

According to (iii) of Proposition 5.1, (5.101) is equivalent to the following inequality:

$$V(T, y_0, Q) \leq \liminf_{t \to T} V(t, y_0, Q) \quad \text{for each } T \in (0, +\infty). \tag{5.102}$$

To show (5.102), we arbitrarily fix $T \in (0, +\infty)$, and then arbitrarily take a sequence $\{t_k\}_{k \geq 1} \subseteq (T/2, 2T)$ so that $t_k \to T$. It follows from (iii) of Proposition 5.1 that

$$V(t_k, y_0, Q) \leq 0 \quad \text{for all } k \geq 1. \tag{5.103}$$

Meanwhile, according to (ii) of Proposition 5.1, each $(JP)_Q^{t_k, y_0}$ (with $k \geq 1$) has a minimizer z_k^*. Then, by (5.86), (5.85), and (5.103), we find that

$$\sup_{k \in \mathbb{N}^+} \|z_k^*\|_Y < +\infty.$$

Thus, there exists a subsequence $\{k_\ell\}_{\ell \geq 1}$ of $\{k\}_{k \geq 1}$ and $\widehat{z} \in Y$ so that

$$z_{k_\ell}^* \to \widehat{z} \quad \text{weakly in } Y \text{ as } \ell \to +\infty,$$

which, combined with (5.84), indicates that

$$J_Q^{T, y_0}(\widehat{z}) \leq \liminf_{\ell \to +\infty} J_Q^{t_{k_\ell}, y_0}(z_{k_\ell}^*).$$

From this, (5.83), and (5.86), it follows that

$$V(T, y_0, Q) \leq \liminf_{\ell \to +\infty} V(t_{k_\ell}, y_0, Q),$$

which leads to (5.102).

Step 3. We prove that $\lim_{T \to 0+} N(T, y_0, Q) = +\infty$ when $y_0 \notin Q$.

By contradiction, suppose that it was not true. Then there would exist $\{T_k\}_{k \geq 1} \subseteq (0, +\infty)$ so that

$$\lim_{k \to +\infty} T_k = 0 \quad \text{and} \quad \sup_{k \in \mathbb{N}^+} N(T_k, y_0, Q) < +\infty.$$

These yield that

$$y_0 = \lim_{k \to +\infty} \Phi(T_k, 0) y_0 = \lim_{k \to +\infty} y(T_k; 0, y_0, u_k) \in Q,$$

where u_k is an optimal control to $(NP)_Q^{T_k, y_0}$. (The existence of u_k is ensured by Lemma 5.5.) This contradicts the fact that $y_0 \notin Q$.

Hence, we end the proof of Theorem 5.4. $\square$

Based on Theorems 5.4 and 5.3, we can prove the following Proposition 5.2, which will be used in the proof of Theorem 5.2.

Proposition 5.2 *Let $\widehat{y}_0 \in Y$ and $Q \in \mathscr{F}$ satisfy that $\widehat{y}_0 \notin Q$. Then the following two conclusions are true:*

(i) For each $M \in [0, +\infty)$, $\widetilde{T}(M, \widehat{y}_0, Q) \in (0, +\infty]$.
(ii) For each $M \in [0, +\infty)$ with $\widetilde{T}(M, \widehat{y}_0, Q) < +\infty$,

$$N(\widetilde{T}(M, \widehat{y}_0, Q), \widehat{y}_0, Q) = M. \tag{5.104}$$

Proof We prove the conclusions one by one.

(i) By contradiction, we suppose that the conclusion (i) was not true. Then there would be $M_0 \in [0, +\infty)$ so that

$$\widetilde{T}(M_0, \widehat{y}_0, Q) = 0.$$

This, along with (5.50), yields that there exist two sequences $\{t_k\}_{k \geq 1} \subseteq (0, +\infty)$ and $\{u_k\}_{k \geq 1} \subseteq \mathscr{U}_{M_0}$ so that

$$\lim_{k \to +\infty} t_k = 0, \quad y(t_k; 0, \widehat{y}_0, u_k) \in Q$$

$$\text{and } \|u_k\|_{L^\infty(0, +\infty; U)} \leq M_0 \text{ for all } k \in \mathbb{N}^+. \tag{5.105}$$

By (5.105), we obtain that

$$\widehat{y}_0 = \lim_{k \to +\infty} y(t_k; 0, \widehat{y}_0, u_k) \in Q,$$

which contradicts the assumption that $\widehat{y}_0 \notin Q$. So the conclusion (i) is true.

(ii) Arbitrarily fix $M \in [0, +\infty)$ so that $\widetilde{T}(M, \widehat{y}_0, Q) < +\infty$. By the conclusion (i) in this proposition, we have that $\widetilde{T}(M, \widehat{y}_0, Q) \in (0, +\infty)$. This, along with (5.76) (see Theorem 5.3), yields that $\mathscr{I}_M^{y_0} \neq \emptyset$. Thus, by (5.76) and (5.74), there is a sequence $\{t_k\}_{k \geq 1} \subseteq (0, +\infty)$ so that

$$\lim_{k \to +\infty} t_k = \widetilde{T}(M, \widehat{y}_0, Q) \text{ and } N(t_k, \widehat{y}_0, Q) \leq M \text{ for all } k \in \mathbb{N}^+.$$

Since $\widetilde{\mathbf{T}}(M, \widehat{y}_0, Q) \in (0, +\infty)$, the above conclusions, together with the continuity of the minimal norm function $N(\cdot, \widehat{y}_0, Q)$ at $\widetilde{\mathbf{T}}(M, \widehat{y}_0, Q)$ (see Theorem 5.4), yield that

$$N(\widetilde{\mathbf{T}}(M, \widehat{y}_0, Q), \widehat{y}_0, Q) \leq M. \tag{5.106}$$

We next prove (5.104). By contradiction, suppose that it was not true. Then by (5.106), we would have that

$$N(\widetilde{\mathbf{T}}(M, \widehat{y}_0, Q), \widehat{y}_0, Q) < M.$$

This, along with the continuity of the minimal norm function $N(\cdot, \widehat{y}_0, Q)$ at $\widetilde{\mathbf{T}}(M, \widehat{y}_0, Q)$, yields that there is $\delta_0 \in (0, \widetilde{\mathbf{T}}(M, \widehat{y}_0, Q))$ so that

$$N(\widetilde{\mathbf{T}}(M, \widehat{y}_0, Q) - \delta_0, \widehat{y}_0, Q) < M.$$

Then it follows from (5.74) that

$$\widetilde{\mathbf{T}}(M, \widehat{y}_0, Q) - \delta_0 \in \mathscr{I}_M^{\widehat{y}_0},$$

which contradicts (5.76). Hence, (5.104) is true.

Thus, we complete the proof of Proposition 5.2. $\square$

5.2.2 Proof of the Main Result

We now on the position to prove Theorem 5.2.

Proof (Proof of Theorem 5.2) Let $\widehat{y}_0 \in Y$ and $Q \in \mathscr{F}$ satisfy that $\widehat{y}_0 \notin Q$. First of all, by Theorem 5.4, the function $N(\cdot, \widehat{y}_0, Q)$ is continuous over $(0, +\infty)$. We now show the conclusions (i)–(iii) one by one.

(i) Arbitrarily fix (M, T) so that

$$(M, T) \in (\mathscr{G}T)_{\widehat{y}_0, Q} \setminus (\mathscr{K}N)_{\widehat{y}_0, Q}. \tag{5.107}$$

Then it follows from (5.52) that

$$0 < T = \widetilde{\mathbf{T}}(M, \widehat{y}_0, Q) < +\infty. \tag{5.108}$$

Claim One: Both $(NP)_Q^{T, \widehat{y}_0}$ and $(\widetilde{TP})_Q^{M, \widehat{y}_0}$ have optimal controls.

Indeed, since $T \in (0, +\infty)$ (see (5.108)), it follows from Lemma 5.5 that $(NP)_Q^{T, \widehat{y}_0}$ has at least one optimal control. Meanwhile, since $\widetilde{\mathbf{T}}(M, \widehat{y}_0, Q) < +\infty$ (see (5.108)), it follows that $(\widetilde{TP})_Q^{M, \widehat{y}_0}$ has at least one admissible control.

Then by a very similar way to that used in the proof of Lemma 5.3 (see Step 1 in Lemma 5.3), we can prove that $(\widetilde{TP})_Q^{M,\widehat{y}_0}$ has at least one optimal control.

Claim Two: For an arbitrarily fixed minimal optimal control u_1^* to $(\widetilde{TP})_Q^{M,\widehat{y}_0}$, $u_1^*|_{(0,T)}$ is an optimal control to $(NP)_Q^{T,\widehat{y}_0}$.

Indeed, by the optimality of u_1^* and (5.108), we have that

$$y(T; 0, \widehat{y}_0, u_1^*) = y(\widetilde{\mathbf{T}}(M, \widehat{y}_0, Q); 0, \widehat{y}_0, u_1^*) \in Q; \quad \|u_1^*\|_{L^\infty(0,+\infty;U)} \le M. \tag{5.109}$$

By the first conclusion in (5.109), we see that $u_1^*|_{(0,T)}$ is an admissible control to $(NP)_Q^{T,\widehat{y}_0}$. This, along with the optimality of $N(T, \widehat{y}_0, Q)$ and the second conclusion in (5.109), yields that

$$N(T, \widehat{y}_0, Q) \le \|u_1^*|_{(0,T)}\|_{L^\infty(0,T;U)} \le M. \tag{5.110}$$

Meanwhile, by (5.108), we can apply (ii) of Proposition 5.2 to find that

$$N(T, \widehat{y}_0, Q) = N(\widetilde{\mathbf{T}}(M, \widehat{y}_0, Q), \widehat{y}_0, Q) = M. \tag{5.111}$$

From (5.110) and (5.111), we see that

$$\|u_1^*|_{(0,T)}\|_{L^\infty(0,T;U)} = N(T, \widehat{y}_0, Q).$$

Since $u_1^*|_{(0,T)}$ is an admissible control to $(NP)_Q^{T,\widehat{y}_0}$, the above shows that $u_1^*|_{(0,T)}$ is an optimal control to $(NP)_Q^{T,\widehat{y}_0}$.

Claim Three: For an arbitrarily fixed optimal control v_1^* to $(NP)_Q^{T,\widehat{y}_0}$, the zero extension of v_1^* over $(0, +\infty)$, denoted by $\widetilde{v}_1^*$, is an optimal control to $(\widetilde{TP})_Q^{M,\widehat{y}_0}$.

Indeed, by the optimality of v_1^*, one can easily check that

$$y(T; 0, \widehat{y}_0, \widetilde{v}_1^*) \in Q \text{ and } \|\widetilde{v}_1^*\|_{L^\infty(0,+\infty;U)} = N(T, \widehat{y}_0, Q).$$

From these, (5.108) and (5.111), we find that $\widetilde{v}_1^*$ is an optimal control to $(\widetilde{TP})_Q^{M,\widehat{y}_0}$.

Now, by the above three claims and Definition 5.2, we see that the problems $(TP)_Q^{M,\widehat{y}_0}$ and $(NP)_Q^{T,\widehat{y}_0}$ are equivalent.

Finally, we claim that

$$N(T, \widehat{y}_0, Q) \ne 0. \tag{5.112}$$

When (5.112) is proved, we can easily check that the null control (defined on $(0, T)$) is not an optimal control to $(NP)_Q^{T,\widehat{y}_0}$. Then by the equivalence of

$(TP)_Q^{M,\widehat{y}_0}$ and $(NP)_Q^{T,\widehat{y}_0}$, we can easily prove that the null control (defined on $(0, +\infty)$) is not an optimal control to $(\widetilde{TP})_Q^{M,\widehat{y}_0}$. This implies that the null control (defined on $(0, \mathbf{T}(M, \widehat{y}_0, Q)))$ is not an optimal control to $(TP)_Q^{M,\widehat{y}_0}$.

The remainder is to show (5.112). By contradiction, suppose that (5.112) was not true. Then we would have that

$$N(T, \widehat{y}_0, Q) = 0.$$

This, together with (5.111), yields that $M = 0$. From this and (5.53), we get that

$$(M, T) \in (\mathscr{K}N)_{\widehat{y}_0, Q},$$

which contradicts (5.107). Therefore, (5.112) is true. This ends the proof of the conclusion (i) in Theorem 5.2.

(ii) Without loss of generality, we can assume that $(\mathscr{K}N)_{\widehat{y}_0, Q} \neq \emptyset$. Arbitrarily fix

$$(M, T) \in (\mathscr{K}N)_{\widehat{y}_0, Q}. \tag{5.113}$$

By Definition 5.2, we see that in order to prove the conclusion (ii), it suffices to show that the null controls (defined on $(0, T)$ and $(0, +\infty)$, respectively) are the unique optimal controls to $(NP)_Q^{T,\widehat{y}_0}$ and $(\widetilde{TP})_Q^{M,\widehat{y}_0}$, respectively. To this end, we observe from (5.113) and (5.53) that

$$T \in (0, +\infty), \quad M = 0 \quad \text{and} \quad N(T, \widehat{y}_0, Q) = 0. \tag{5.114}$$

By the first conclusion of (5.114) and Lemma 5.5, we see that $(NP)_Q^{T,\widehat{y}_0}$ has an optimal control. This, along with the last conclusion in (5.114), implies that the null control (defined on $(0, T)$) is the unique optimal control to $(NP)_Q^{T,\widehat{y}_0}$. From this, it follows that

$$y(T; 0, \widehat{y}_0, 0) \in Q,$$

which implies that $(0, \widehat{y}_0, T, 0)$ is an admissible tetrad for $(\widetilde{TP})_Q^{M,\widehat{y}_0}$. Then by a very similar way as that used in the proof of Lemma 5.3 (see Step 1 in Lemma 5.3), we can prove that $(\widetilde{TP})_Q^{M,\widehat{y}_0}$ has an optimal control. This, along with the second conclusion in (5.114), yields that the null control is the unique optimal control to $(\widetilde{TP})_Q^{M,\widehat{y}_0}$. Hence, we end the proof of the conclusion (ii) in Theorem 5.2.

(iii) By contradiction, suppose that the conclusion (iii) was not true. Then there would be a pair

$$(M, T) \in [0, +\infty) \times (0, +\infty) \setminus \big((\mathscr{G}T)_{\widehat{y}_0, Q} \cup (\mathscr{K}N)_{\widehat{y}_0, Q}\big) \tag{5.115}$$

so that $(TP)_Q^{M,\widehat{y}_0}$ and $(NP)_Q^{T,\widehat{y}_0}$ are equivalent. The key to get a contradiction is to prove that

$$\widetilde{\mathbf{T}}(M, \widehat{y}_0, Q) = T. \tag{5.116}$$

When this is proved, we see from (5.116) and (5.52) that

$$(M, T) \in (\mathscr{G}T)_{\widehat{y}_0, Q}.$$

(Notice that $T \in (0, +\infty)$ and $0 \le M < +\infty$.) This contradicts (5.115). Hence, the conclusion (iii) is true.

The remainder is to prove (5.116). First of all, we show that

$$0 < \widetilde{\mathbf{T}}(M, \widehat{y}_0, Q) \le T. \tag{5.117}$$

Indeed, since $(TP)_Q^{M,\widehat{y}_0}$ and $(NP)_Q^{T,\widehat{y}_0}$ are equivalent, two facts are derived from Definition 5.2: First, $(NP)_Q^{T,\widehat{y}_0}$ has an optimal control v_2^*; Second, the zero extension of v_2^* over $(0, +\infty)$, denoted by $\widetilde{v}_2^*$, is an optimal control to $(\widetilde{TP})_Q^{M,\widehat{y}_0}$. From these two facts, we can easily check that

$$\widetilde{\mathbf{T}}(M, \widehat{y}_0, Q) \le T.$$

This, along with (i) of Proposition 5.2, leads to (5.117).

Next, by contradiction, we suppose that (5.116) was not true. Then by (5.117) and (5.115), we would have that

$$0 < \widetilde{\mathbf{T}}(M, \widehat{y}_0, Q) < T < +\infty. \tag{5.118}$$

It follows from (5.118) that $(\widetilde{TP})_Q^{M,\widehat{y}_0}$ has at least one admissible control. Then by a very similar way to that used in the proof of Lemma 5.3 (see Step 1 in Lemma 5.3), we can prove that $(\widetilde{TP})_Q^{M,\widehat{y}_0}$ has an optimal control u^*. Thus we have that

$$y(\widetilde{\mathbf{T}}(M, \widehat{y}_0, Q); 0, \widehat{y}_0, u^*) \in Q \quad \text{and} \quad \|u^*\|_{L^\infty(0,+\infty;U)} \le M. \tag{5.119}$$

Arbitrarily take $\widehat{u} \in U \setminus \{0\}$ with $\|\widehat{u}\|_U = 1$. Define

$$\widetilde{u}^*(t) \triangleq \begin{cases} u^*(t), & t \in (0, \widetilde{\mathbf{T}}(M, \widehat{y}_0, Q)), \\ 0, & t \in [\widetilde{\mathbf{T}}(M, \widehat{y}_0, Q), +\infty); \end{cases}$$

$$\widehat{u}_M^*(t) \triangleq \begin{cases} u^*(t), & t \in (0, \widetilde{\mathbf{T}}(M, \widehat{y}_0, Q)), \\ M\widehat{u}, & t \in [\widetilde{\mathbf{T}}(M, \widehat{y}_0, Q), +\infty). \end{cases}$$

From these and (5.119), we see that $\widetilde{u}^*$ and $\widehat{u}_M^*$ are optimal controls to $(\widetilde{TP})_Q^{M,\widehat{y}_0}$. By the equivalence of $(TP)_Q^{M,\widehat{y}_0}$ and $(NP)_Q^{T,\widehat{y}_0}$, we find that $\widetilde{u}^*|_{(0,T)}$ and $\widehat{u}_M^*|_{(0,T)}$ are optimal controls to $(NP)_Q^{T,\widehat{y}_0}$. This, along with (5.118) and (ii) of Lemma 5.6, implies that

$$N(T, \widehat{y}_0, Q) = \|\widetilde{u}^*\|_{L^\infty((\widetilde{T}(M,\widehat{y}_0,Q)+T)/2,T;U)} = 0$$

and

$$N(T, \widehat{y}_0, Q) = \|\widehat{u}_M^*\|_{L^\infty((\widetilde{T}(M,\widehat{y}_0,Q)+T)/2,T;U)} = M.$$

From these, it follows that

$$N(T, \widehat{y}_0, Q) = M = 0.$$

This, along with (5.53), yields that

$$(M, T) \in (\mathscr{K}N)_{\widehat{y}_0,Q},$$

which contradicts (5.115). Thus, (5.116) is true. This ends the proof of the conclusion (iii) in Theorem 5.2.

In summary, we conclude that the conclusions (i), (ii), and (iii) are true. This completes the proof of Theorem 5.2. □

At the end of this subsection, we introduce an example, where the minimal norm function $T \to N(T, \widehat{y}_0, Q)$ is not decreasing, the set $(\mathscr{K}N)_{\widehat{y}_0,Q}$ is not empty, and the set $(\mathscr{G}T)_{\widehat{y}_0,Q}$ is not connected. This example may help us to understand Theorem 5.2 better.

Example 5.3 Let Ω be a bounded domain in $\mathbb{R}^d, d \geq 1$, with a C^2 boundary $\partial\Omega$. Let $\omega \subseteq \Omega$ be a nonempty and open subset. Consider the following controlled heat equation:

$$\begin{cases} \partial_t y - \Delta y = \chi_\omega u & \text{in } \Omega \times (0, +\infty), \\ y = 0 & \text{on } \partial\Omega \times (0, +\infty), \\ y(0) = \widehat{y}_0 \in L^2(\Omega), \end{cases}$$

where $u \in L^\infty(0, +\infty; L^2(\Omega))$. Let $Y = L^2(\Omega)$ and $U = L^2(\Omega)$; Let $A = \Delta$ with its domain $H^2(\Omega) \cap H_0^1(\Omega)$; Let $B = \chi_\omega$, where χ_ω is treated as a linear and bounded operator on U.

One can easily check that the corresponding minimal time control problem $(TP)_Q^{M,\widehat{y}_0}$, as well as the minimal norm control problem (see (5.51)), can be put into the framework $(\mathscr{A}_2)$ (given at the beginning of Section 5.2). Here, we used Theorems 1.22 and 1.19 and the backward uniqueness of parabolic equations. The authors in [11] constructed $Q \in \mathscr{F}$ and $\widehat{y}_0 \in L^2(\Omega) \setminus Q$ so that the following propositions are true (see [11]):

(i) The function $N(\cdot, \widehat{y}_0, Q)$ is not decreasing;
(ii) The set $(\mathscr{K}N)_{\widehat{y}_0, Q} \neq \emptyset$ and the set $(\mathscr{G}T)_{\widehat{y}_0, Q}$ is not connected.

The way to construct such $Q \in \mathscr{F}$ and $\widehat{y}_0 \in L^2(\Omega) \setminus Q$ is quite technical. We omit the details here. Now we end Example 5.3.

5.3 Minimal Time Controls, Minimal Norm Controls, and Time-Varying Systems: Part II

In this section, we will study the equivalence of the problem $(TP)_{min}^{Q_S, Q_E}$ and the corresponding minimal norm control problem under the following framework $(\mathscr{A}_3)$:

(i) The state space Y and the control space U are real separable Hilbert spaces.
(ii) The controlled system is as:

$$\dot{y}(t) = Ay(t) + D(t)y(t) + Bu(t), \quad t \in (0, +\infty), \tag{5.120}$$

where $A : \mathscr{D}(A) \subseteq Y \mapsto Y$ generates a C_0 semigroup $\{e^{At}\}_{t \geq 0}$ on Y, $D(\cdot) \in L_{loc}^1([0, +\infty); \mathscr{L}(Y))$ and $B \in \mathscr{L}(U, Y)$. Given $\tau \in [0, +\infty)$, $y_0 \in Y$ and $u \in L^\infty(\tau, +\infty; U)$, write $y(\cdot; \tau, y_0, u)$ for the solution of (5.120) over $[\tau, +\infty)$, with the initial condition that $y(\tau) = y_0$. Moreover, write $\{\Phi(t, s) : t \geq s \geq 0\}$ for the evolution operator generated by $A + D(\cdot)$ over Y.

(iii) Let $\mathbb{U} = B_M(0)$, where $B_M(0)$ is the closed ball in U, centered at 0 and of radius $M \geq 0$. (When $M = 0$, we agree that $B_M(0) \triangleq \{0\}$.)

(iv) Let $Q_S = \{0\} \times \{\widehat{y}_0\}$ and $Q_E = (0, +\infty) \times \{0\}$, where $\widehat{y}_0 \in Y \setminus \{0\}$ is fixed.

(v) The system (5.120) is L^∞-null controllable over each interval, i.e., for any $T_2 > T_1 \geq 0$ and $y_0 \in Y$, there exists $u \in L^\infty(T_1, T_2; U)$ so that $y(T_2; T_1, y_0, u) = 0$.

(vi) It holds that

$$Y_T = Z_T \quad \text{for each } T \in (0, +\infty), \tag{5.121}$$

where Y_T and Z_T are defined in the following manner:

$$Y_T \triangleq \overline{X_T}^{\|\cdot\|_{L^1(0, T; U)}}, \quad \text{with } X_T \triangleq \{B^*\Phi(T, \cdot)^*z \mid z \in Y\};$$
$$Z_T \triangleq \{B^*\varphi \in L^1(0, T; U) \mid \text{for each } s \in (0, T), \text{ there exists }$$
$$z_s \in Y \text{ so that } \varphi(\cdot) = \Phi(s, \cdot)^*z_s \text{ over } [0, s]\}. \tag{5.122}$$

(vii) The system (5.120) has the backward uniqueness property, i.e., if $\Phi(T, 0)y_0 = 0$ for some $y_0 \in Y$ and $T \geq 0$, then $y_0 = 0$.

In this section, we simply write $(TP)_{min}^{Q_S, Q_E}$ for $(TP)_M^{\widehat{y}_0}$ which reads:

$$(TP)_M^{\widehat{y}_0}\ \mathbf{T}(M, \widehat{y}_0) \triangleq \inf\left\{T \in (0, +\infty) \mid y(T; 0, \widehat{y}_0, u) = 0\ \text{ and }\ u \in L(0, T; \mathbb{U})\right\}.$$

Similar to what we explained at the beginning of Section 5.1, $(TP)_M^{\widehat{y}_0}$ can be treated as the following problem:

$$(\widetilde{TP})_M^{\widehat{y}_0}\ \ \widetilde{\mathbf{T}}(M, \widehat{y}_0) \triangleq \inf\left\{T \in (0, +\infty) \mid y(T; 0, \widehat{y}_0, u) = 0,\ u \in \mathscr{U}_M\right\},\quad (5.123)$$

where

$$\mathscr{U}_M \triangleq \left\{u \in L^\infty(0, +\infty; U) \mid \|u(t)\|_U \leq M\ \text{ a.e. }\ t > 0\right\}.$$

In this problem,

- a tetrad $(0, \widehat{y}_0, T, u)$ is called admissible, if

$$T \in (0, +\infty),\ u \in \mathscr{U}_M\ \text{ and }\ y(T; 0, \widehat{y}_0, u) = 0;$$

- a tetrad $(0, \widehat{y}_0, \widetilde{\mathbf{T}}(M, \widehat{y}_0), u^*)$ is called optimal, if

$$\widetilde{\mathbf{T}}(M, \widehat{y}_0) \in (0, +\infty),\ u^* \in \mathscr{U}_M\ \text{ and }\ y(\widetilde{\mathbf{T}}(M, \widehat{y}_0), 0, \widehat{y}_0, u^*) = 0;$$

- when $(0, \widehat{y}_0, \widetilde{\mathbf{T}}(M, \widehat{y}_0), u^*)$ is an optimal tetrad, $\widetilde{\mathbf{T}}(M, \widehat{y}_0)$ and u^* are called the optimal time and an optimal control, respectively.

The corresponding minimal norm problem $(NP)_T^{y_0}$, with $T \in (0, +\infty)$ and $y_0 \in Y$, is defined in the following manner:

$$(NP)_T^{y_0}\qquad N(T, y_0) \triangleq \inf\{\|v\|_{L^\infty(0,T;U)} \mid y(T; 0, y_0, v) = 0\}.\qquad (5.124)$$

In this problem,

- $N(T, y_0)$ is called the minimal norm;
- a control $v \in L^\infty(0, T; U)$ is called admissible, if $y(T; 0, y_0, v) = 0$;
- a control $v^* \in L^\infty(0, T; U)$ is called optimal, if it is admissible and $\|v^*\|_{L^\infty(0,T;U)} = N(T, y_0)$.

Moreover, for each $y_0 \in Y$, the map $T \to N(T, y_0)$ is called a minimal norm function, which is denoted by $N(\cdot, y_0)$. From (5.124), we see that for each $y_0 \in Y$, the function $N(\cdot, y_0)$ is decreasing over $(0, +\infty)$. Hence, for each $y_0 \in Y$, it is well defined that

$$\widehat{N}(y_0) \triangleq \lim_{T \to +\infty} N(T, y_0). \tag{5.125}$$

We now give the definition about the equivalence between $(TP)_M^{\widehat{y}_0}$ and $(NP)_T^{\widehat{y}_0}$.

Definition 5.3 Let $M \geq 0$ and $T \in (0, +\infty)$. Problems $(TP)_M^{\widehat{y}_0}$ and $(NP)_T^{\widehat{y}_0}$ are said to be equivalent if the following three conditions hold:

(i) Both $(\widetilde{TP})_M^{\widehat{y}_0}$ and $(NP)_T^{\widehat{y}_0}$ have optimal controls;
(ii) The restriction of each optimal control to $(\widetilde{TP})_M^{\widehat{y}_0}$ over $(0, T)$ is an optimal control to $(NP)_T^{\widehat{y}_0}$;
(iii) The zero extension of each optimal control to $(NP)_T^{\widehat{y}_0}$ over $(0, +\infty)$ is an optimal control to $(\widetilde{TP})_M^{\widehat{y}_0}$.

The main result of this section is as follows.

Theorem 5.5 *The function* $N(\cdot, \widehat{y}_0)$ *is strictly decreasing and continuous from* $(0, +\infty)$ *onto* $(\widehat{N}(\widehat{y}_0), +\infty)$. *Moreover, if let*

$$\mathscr{E}_{\widehat{y}_0} \triangleq \{(N(T, \widehat{y}_0), T) \in \mathbb{R}^2 \mid T \in (0, +\infty)\}, \tag{5.126}$$

then the following conclusions are true:

(i) *When* $(M, T) \in \mathscr{E}_{\widehat{y}_0}$, *problems* $(TP)_M^{\widehat{y}_0}$ *and* $(NP)_T^{\widehat{y}_0}$ *are equivalent;*
(ii) *When* $(M, T) \in [0, +\infty) \times (0, +\infty) \setminus \mathscr{E}_{\widehat{y}_0}$, *problems* $(TP)_M^{\widehat{y}_0}$ *and* $(NP)_T^{\widehat{y}_0}$ *are not equivalent.*

Before proving Theorem 5.5, some preliminaries (which are similar to the preliminaries of Theorem 5.1 in Section 5.1) will be given in order:

Lemma 5.7 *For each* $T \in (0, +\infty)$ *and* $y_0 \in Y$, *the problem* $(NP)_T^{y_0}$ *has an optimal control.*

Proof Its proof is the same as that of Lemma 5.1. We omit it here. $\square$

Based on Lemma 5.7, we can give an explicit expression of $N(T, y_0)$.

Lemma 5.8 *For each* $T \in (0, +\infty)$ *and* $y_0 \in Y$, *it holds that*

$$N(T, y_0) = \sup_{z \in Y \setminus \{0\}} \frac{\langle y_0, \Phi(T, 0)^* z \rangle_Y}{\|B^* \Phi(T, \cdot)^* z\|_{L^1(0,T;U)}}.$$

Proof Arbitrarily fix $T \in (0, +\infty)$ and $y_0 \in Y$. Because of (v), as well as (i) and (ii) of $(\mathscr{A}_3)$ (given at the beginning of Section 5.3), we can apply Theorem 1.20 to see that for any $T_2 > T_1 \geq 0$, there exists a constant $C(T_1, T_2) \in (0, +\infty)$ so that

$$\|\Phi(T_2, T_1)^* z\|_Y \leq C(T_1, T_2)\|B^* \Phi(T_2, \cdot)^* z\|_{L^1(T_1, T_2; U)} \quad \text{for all } z \in Y. \tag{5.127}$$

This implies that

$$\|B^*\Phi(T,\cdot)^*z\|_{L^1(0,T;U)} \neq 0 \quad \text{for all } z \in Y \setminus \{0\}, \tag{5.128}$$

and

$$M \triangleq \sup_{z \in Y \setminus \{0\}} \frac{\langle y_0, \Phi(T,0)^*z \rangle_Y}{\|B^*\Phi(T,\cdot)^*z\|_{L^1(0,T;U)}} < +\infty. \tag{5.129}$$

We now claim that

$$M \leq N(T, y_0). \tag{5.130}$$

By Lemma 5.7, $(NP)_T^{y_0}$ has an optimal control u^*. By the optimality of u^*, we have that

$$y(T; 0, y_0, u^*) = 0 \quad \text{and} \quad \|u^*\|_{L^\infty(0,T;U)} = N(T, y_0).$$

These imply that for each $z \in Y$,

$$\langle y_0, \Phi(T,0)^*z \rangle_Y = \langle y(T; 0, y_0, u^*), z \rangle_Y - \int_0^T \langle u^*(t), B^*\Phi(T,t)^*z \rangle_U \, dt$$

$$\leq N(T, y_0) \int_0^T \|B^*\Phi(T,t)^*z\|_U \, dt.$$

This, together with (5.128) and (5.129), gives (5.130).

We next show that

$$M \geq N(T, y_0). \tag{5.131}$$

For this purpose, we set

$$X_T \triangleq \{B^*\Phi(T,\cdot)^*z|_{(0,T)} \mid z \in Y\} \subseteq L^1(0, T; U).$$

It is clear that X_T is a linear subspace of $L^1(0, T; U)$. We define a map $\mathscr{F} : X_T \mapsto \mathbb{R}$ in the following manner:

$$\mathscr{F}\big(B^*\Phi(T,\cdot)^*z|_{(0,T)}\big) \triangleq \langle y_0, \Phi(T,0)^*z \rangle_Y \quad \text{for all } z \in Y. \tag{5.132}$$

From this and (5.127), one can check that $\mathscr{F}$ is well defined and linear. Furthermore, it follows from (5.129) and (5.132) that

$$\big|\mathscr{F}\big(B^*\Phi(T,\cdot)^*z|_{(0,T)}\big)\big| \leq M \int_0^T \|B^*\Phi(T,t)^*z\|_U \, dt \quad \text{for all } z \in Y.$$

From this, we can apply the Hahn-Banach Theorem and the Riesz Representation Theorem (see Theorems 1.5 and 1.4) to get $\widehat{u} \in L^\infty(0, T; U)$, with

$$\|\widehat{u}\|_{L^\infty(0,T;U)} \leq M \tag{5.133}$$

so that

$$\mathscr{F}\left(B^*\Phi(T, \cdot)^* z|_{(0,T)}\right) = \int_0^T \langle \widehat{u}(t), B^*\Phi(T, t)^* z\rangle_U dt \quad \text{for all } z \in Y. \tag{5.134}$$

From (5.134) and (5.132), we see that

$$\langle y(T; 0, y_0, -\widehat{u}), z\rangle_Y = \langle y_0, \Phi(T, 0)^* z\rangle_Y - \int_0^T \langle \widehat{u}(t), B^*\Phi(T, t)^* z\rangle_U dt$$

$$= 0 \quad \text{for all } z \in Y,$$

which indicates that

$$y(T; 0, y_0, -\widehat{u}) = 0.$$

This implies that $-\widehat{u}$ is an admissible control to $(NP)_T^{y_0}$, which, combined with (5.133), indicates that

$$N(T, y_0) \leq \|\widehat{u}\|_{L^\infty(0,T;U)} \leq M.$$

This leads to (5.131).

Finally, the desired equality in Lemma 5.8 follows from (5.131), (5.130), and (5.129) at once. This completes the proof of Lemma 5.8.   □

The following proposition is concerned with the continuity and monotonicity of the minimal norm function $N(\cdot, y_0)$.

Proposition 5.3 *Let* $y_0 \in Y \setminus \{0\}$. *Then the function* $N(\cdot, y_0)$ *is strictly decreasing and continuous from* $(0, +\infty)$ *onto* $(\widehat{N}(y_0), +\infty)$.

Proof Let $y_0 \in Y \setminus \{0\}$. The proof is organized by several steps.

Step 1. We show that $N(\cdot, y_0)$ is strictly decreasing from $(0, +\infty)$ to $(0, +\infty)$. First of all, we claim that

$$N(T, y_0) \in (0, +\infty) \quad \text{for each } T \in (0, +\infty). \tag{5.135}$$

By contradiction, suppose that (5.135) was not true. Then there would be $\widehat{T} > 0$ so that

$$N(\widehat{T}, y_0) = 0. \tag{5.136}$$

By Lemma 5.7, $(NP)_{\widehat{T}}^{y_0}$ has an optimal control $\widehat{u}^*$. From (5.136), we see that $\widehat{u}^* = 0$. Then by the optimality of $\widehat{u}^*$, we have that

$$\Phi(\widehat{T}, 0)y_0 = y(\widehat{T}; 0, y_0, 0) = 0,$$

which, combined with (vii) of $(\mathscr{A}_3)$ (given at the beginning of Section 5.3), indicates that $y_0 = 0$. This contradicts the assumption that $y_0 \neq 0$. Therefore, (5.135) is true.

Next, we arbitrarily fix $T_2 > T_1 > 0$. According to Lemma 5.7, $(NP)_{T_1}^{y_0}$ has an optimal control v_1. By the optimality of v_1, we see that

$$y(T_1; 0, y_0, v_1) = 0 \quad \text{and} \quad \|v_1\|_{L^\infty(0,T_1;U)} = N(T_1, y_0). \tag{5.137}$$

Meanwhile, by (v) of $(\mathscr{A}_3)$ (given at the beginning of Section 5.3), there exists $v_2 \in L^\infty(0, T_2; U)$ so that

$$y(T_2; 0, y_0, \chi_{(T_1,T_2)} v_2) = 0. \tag{5.138}$$

By (5.135), there exists a constant $\lambda \in (0, 1)$ so that

$$\lambda \|v_2\|_{L^\infty(0,T_2;U)} \leq (1 - \lambda)N(T_1, y_0). \tag{5.139}$$

Write $\widetilde{v}_1$ for the zero extension of v_1 over $(0, T_2)$. Define a new control as follows:

$$v_3(t) \triangleq (1 - \lambda)\widetilde{v}_1 + \lambda \chi_{(T_1,T_2)} v_2(t), \quad t \in (0, T_2). \tag{5.140}$$

Then, from (5.140), (5.138), and the first equality in (5.137), we see that

$$y(T_2; 0, y_0, v_3) = (1 - \lambda)y(T_2; 0, y_0, \widetilde{v}_1) + \lambda y(T_2; 0, y_0, \chi_{(T_1,T_2)} v_2)$$
$$= 0.$$

Thus, v_3 is an admissible control to $(NP)_{T_2}^{y_0}$. This, together with the optimality of $N(T_2, y_0)$, implies that

$$N(T_2, y_0) \leq \|v_3\|_{L^\infty(0,T_2;U)}.$$

The above inequality, along with (5.140), (5.139), and the second equality in (5.137), yields that

$$N(T_2, y_0) \leq (1 - \lambda)N(T_1, y_0).$$

Since $N(T_1, y_0) \in (0, +\infty)$ (see (5.135)), the above inequality implies that

$$N(T_2, y_0) < N(T_1, y_0).$$

This completes the proof of Step 1.

Step 2. We prove the right-continuity of $N(\cdot, y_0)$.

Arbitrarily fix $T \in (0, +\infty)$, and then arbitrarily choose a sequence $\{T_k\}_{k \geq 1} \subseteq (T, T+1)$ so that

$$\lim_{k \to +\infty} T_k = T.$$

By Lemma 5.7, each $(NP)_{T_k}^{y_0}$ (with $k \geq 1$) has an optimal control u_k^*. By the optimality of u_k^*, we see that

$$y(T_k; 0, y_0, u_k^*) = 0 \quad \text{and} \quad \|u_k^*\|_{L^\infty(0, T_k; U)} = N(T_k, y_0) \quad \text{for all } k \geq 1. \quad (5.141)$$

Meanwhile, it follows from the conclusion of Step 1 that

$$\sup_{k \in \mathbb{N}^+} N(T_k, y_0) < +\infty. \quad (5.142)$$

Write $\widetilde{u}_k^*$ for the zero extension of u_k^* over $(0, +\infty)$. By (5.142) and (5.141), the sequence $\{\widetilde{u}_k^*\}_{k \geq 1}$ is bounded in $L^\infty(0, +\infty; U)$. Then there exists a subsequence $\{\widetilde{u}_{k_\ell}^*\}_{\ell \geq 1}$ and $\widehat{u} \in L^\infty(0, +\infty; U)$ so that

$$\widetilde{u}_{k_\ell}^* \to \widehat{u} \quad \text{weakly star in } L^\infty(0, +\infty; U) \quad \text{as } \ell \to +\infty.$$

From this and (5.141), we can directly check that

$$y(T_{k_\ell}; 0, y_0, \widetilde{u}_{k_\ell}^*) \to y(T; 0, y_0, \widehat{u}) \quad \text{weakly in } Y, \quad y(T; 0, y_0, \widehat{u}) = 0$$

and

$$\|\widehat{u}\|_{L^\infty(0, +\infty; U)} \leq \liminf_{\ell \to +\infty} \|\widetilde{u}_{k_\ell}^*\|_{L^\infty(0, +\infty; U)} \leq \liminf_{\ell \to +\infty} N(T_{k_\ell}, y_0).$$

By the above two conclusions and the optimality of $N(T, y_0)$, we get that

$$N(T, y_0) \leq \|\widehat{u}\|_{L^\infty(0, T; U)} \leq \liminf_{\ell \to +\infty} N(T_{k_\ell}, y_0).$$

This, along with the monotonicity of $N(\cdot, y_0)$ (see (i) of this proposition), leads to the right-continuity of $N(\cdot, y_0)$.

Step 3. We show that $\lim_{T \to 0+} N(T, y_0) = +\infty$.

By contradiction, suppose that it was not true. Then there would be a sequence $\{t_k\}_{k \geq 1} \subseteq (0, +\infty)$, with $\lim_{k \to +\infty} t_k = 0$, so that

$$\sup_{k \in \mathbb{N}^+} N(t_k, y_0) < +\infty. \quad (5.143)$$

According to Lemma 5.7, each $(NP)_{t_k}^{y_0}$ (with $k \geq 1$) has an optimal control v_k^*. By the optimality of v_k^*, we see that

$$y(t_k; 0, y_0, v_k^*) = 0 \quad \text{and} \quad \|v_k^*\|_{L^\infty(0, t_k; U)} = N(t_k, y_0) \quad \text{for all } k \geq 1. \tag{5.144}$$

Now, it follows from (5.144) and (5.143) that

$$y_0 = \lim_{k \to +\infty} \left[y(t_k; 0, y_0, v_k^*) - y(t_k; 0, 0, v_k^*) \right]$$
$$= - \lim_{k \to +\infty} y(t_k; 0, 0, v_k^*) = 0,$$

which contradicts the assumption that $y_0 \neq 0$.

Step 4. We prove the left-continuity of $N(\cdot, y_0)$.

For this purpose, we arbitrarily fix a $T \in (0, +\infty)$, and then arbitrarily take $\{T_n\}_{n \geq 1} \subseteq (T/2, T)$ so that $T_n \nearrow T$. By Lemma 5.8, we can choose $\{z_n\}_{n \geq 1} \subseteq Y \setminus \{0\}$ so that for each $n \in \mathbb{N}^+$,

$$\int_0^{T_n} \|B^* \Phi(T_n, t)^* z_n\|_U dt = 1 \tag{5.145}$$

and

$$N(T_n, y_0) - 1/n \leq \langle y_0, \Phi(T_n, 0)^* z_n \rangle_Y. \tag{5.146}$$

We now claim that there exists a subsequence $\{n_k\}_{k \geq 1} \subseteq \{n\}_{n \geq 1}$ and $B^* \varphi \in Z_T$ (given by (5.122)) so that for each $\delta \in (0, T]$,

$$\Phi(T_{n_k}, T - \delta)^* z_{n_k} \to \varphi(T - \delta) \quad \text{weakly in } Y \text{ as } k \to +\infty. \tag{5.147}$$

Indeed, on one hand, if we write $\psi_n(\cdot)$ for the zero extension of $B^* \Phi(T_n, \cdot)^* z_n$ over $(0, T)$, then from (5.145), we find that for each $\ell \in \mathbb{N}^+$, $\{\psi_n(\cdot)|_{(T_\ell, T)}\}_{n \geq \ell}$ is bounded in $L^1(T_\ell, T; U)$. From this, we can use Theorem 1.20 to see that for each $\ell \in \mathbb{N}^+$, $\{\Phi(T_n, T_\ell)^* z_n\}_{n \geq \ell+1}$ is bounded in Y. (More precisely, in (1.85), we choose T_2, T_1 and z as $T_{\ell+1}$, T_ℓ and $\Phi(T_n, T_{\ell+1})^* z_n$, respectively.) This, together with the diagonal law, implies that there exists a subsequence $\{m_m\}_{m \geq 1} \subseteq \{n\}_{n \geq 1}$, so that for each $\ell \in (0, +\infty)$, there is $\widehat{z}_\ell \in Y$ so that

$$\Phi(T_{m_m}, T_\ell)^* z_{m_m} \to \widehat{z}_\ell \quad \text{weakly in } Y \text{ as } m \to +\infty. \tag{5.148}$$

On the other hand, we arbitrarily fix $\ell_1, \ell_2 \in \mathbb{N}^+$. Then we have that

$$\Phi(T_{m_m}, T_{\ell_1})^* z_{m_m} = \Phi(T_{\ell_1 + \ell_2}, T_{\ell_1})^* \Phi(T_{m_m}, T_{\ell_1 + \ell_2})^* z_{m_m} \quad \text{for all } m \geq \ell_1 + \ell_2.$$

This, together with (5.148), implies that

$$\widehat{z}_{\ell_1} = \Phi(T_{\ell_1+\ell_2}, T_{\ell_1})^* \widehat{z}_{\ell_1+\ell_2}. \tag{5.149}$$

Define a function $\varphi : [0, T) \to Y$ in the following manner:

$$\varphi(t) \triangleq \Phi(T_\ell, t)^* \widehat{z}_\ell \quad \text{for all} \quad t \in [0, T_\ell] \quad \text{and} \quad \ell \in \mathbb{N}^+.$$

From the above, (5.149), (5.148), (5.145), and (5.122), we can verify the following conclusions:

$$\varphi \text{ is well defined over } [0, T), \quad B^*\varphi \in Z_T$$

and

$$\Phi(T_{m_m}, T_\ell)^* z_{m_m} \to \varphi(T_\ell) \quad \text{weakly in } Y \text{ as } m \to +\infty.$$

These lead to (5.147).

We next show that for the subsequence $\{n_k\}_{k \geq 1}$ given in (5.147),

$$B^*\varphi \in Y_T, \quad \int_0^T \|B^*\varphi(t)\|_U \, dt \leq 1 \quad \text{and} \quad \langle y_0, \Phi(T_{n_k}, 0)^* z_{n_k} \rangle_Y \to \langle y_0, \varphi(0) \rangle_Y. \tag{5.150}$$

In fact, since $B^*\varphi \in Z_T$, the first conclusion in (5.150) follows from (vi) of $(\mathscr{A}_3)$ (given at the beginning of Section 5.3). The second conclusion in (5.150) follows from (5.147) and (5.145). The third conclusion in (5.150) follows from (5.147).

Finally, by (5.146) and (5.150), we get that

$$\limsup_{k \to +\infty} N(T_{n_k}, y_0) \leq \lim_{k \to +\infty} \langle y_0, \Phi(T_{n_k}, 0)^* z_{n_k} \rangle_Y = \langle y_0, \varphi(0) \rangle_Y. \tag{5.151}$$

From the above, we can show that

$$\limsup_{k \to +\infty} N(T_{n_k}, y_0) \leq N(T, y_0) \int_0^T \|B^*\varphi(t)\|_U \, dt \leq N(T, y_0). \tag{5.152}$$

Indeed, on one hand, since $B^*\varphi \in Y_T = Z_T$ (see (5.150) and (5.121)), and because of (5.122), there is $\widetilde{z}_{T/2} \in Y$ and a sequence $\{\widetilde{z}_n\}_{n \geq 1} \subseteq Y$ so that

$$\varphi(\cdot) = \Phi(T/2, \cdot)^* \widetilde{z}_{T/2} \quad \text{over } [0, T/2] \tag{5.153}$$

and

$$\|B^*\Phi(T, \cdot)^* \widetilde{z}_n - B^*\varphi(\cdot)\|_{L^1(0,T;U)} \to 0 \quad \text{as } n \to +\infty. \tag{5.154}$$

On the other hand, according to (i) of Theorem 1.20, there is a positive constant $C(T)$ (independent of n) so that

$$\|\Phi(T/2,0)^*(\Phi(T,T/2)^*\tilde{z}_n - \tilde{z}_{T/2})\|_Y$$
$$\leq C(T)\|B^*\Phi(T/2,\cdot)^*(\Phi(T,T/2)^*\tilde{z}_n - \tilde{z}_{T/2})\|_{L^1(0,T/2;U)}$$
$$= C(T)\|B^*\Phi(T,\cdot)^*\tilde{z}_n - B^*\Phi(T/2,\cdot)^*\tilde{z}_{T/2}\|_{L^1(0,T/2;U)},$$

which, combined with (5.153) and (5.154), indicates that

$$\|\Phi(T,0)^*\tilde{z}_n - \varphi(0)\|_Y \to 0.$$

From the above, Lemma 5.8 and (5.154), it follows that

$$\langle y_0, \varphi(0)\rangle_Y = \lim_{n\to+\infty} \langle y_0, \Phi(T,0)^*\tilde{z}_n\rangle_Y$$
$$\leq N(T,y_0)\limsup_{n\to+\infty} \|B^*\Phi(T,\cdot)^*\tilde{z}_n\|_{L^1(0,T;U)}$$
$$= N(T,y_0)\|B^*\varphi\|_{L^1(0,T;U)}.$$

This, together with (5.151) and (5.150), yields (5.152). From (5.152) and the monotonicity of $N(\cdot, y_0)$ (see step 1), we conclude that $N(\cdot, y_0)$ is left-continuous.

Hence, we end the proof of Proposition 5.3. $\qquad\square$

With the help of Proposition 5.3, we can have the following existence result on optimal controls of $(\widetilde{TP})_M^{\widehat{y_0}}$:

Lemma 5.9 *The problem* $(\widetilde{TP})_M^{\widehat{y_0}}$ $(M \in (0, +\infty))$ *has optimal controls if and only if* $M > \widehat{N}(\widehat{y_0})$. *Here* $\widehat{N}(\widehat{y_0})$ *is given by* (5.125).

Proof The proof is quite similar to that of Lemma 5.3. We omit it here. $\qquad\square$

Based on Lemma 5.9 and Proposition 5.3, we have the following consequence:

Corollary 5.2 *Let* $(M, T) \in [0, +\infty) \times (0, +\infty)$. *Then* $M = N(T, \widehat{y_0})$ *if and only if* $T = \widetilde{\mathbf{T}}(M, \widehat{y_0})$.

Proof The proof is quite similar to that of Corollary 5.1. We omit it here. $\qquad\square$

We now are in the position to prove Theorem 5.5.

Proof (Proof of Theorem 5.5) By the above several lemmas, as well as Proposition 5.3 and Corollary 5.2, we can use the same way as that used in the proof of Theorem 5.1 to prove Theorem 5.5. We omit the details here. $\qquad\square$

We end this section with the next example, which can be put into the framework of this section.

Example 5.4 Let Ω be a bounded domain in $\mathbb{R}^d$, $d \geq 1$, with a C^2 boundary $\partial\Omega$. Let $\omega \subseteq \Omega$ be a nonempty and open subset with its characteristic function χ_ω. Let $a_1 \in L^\infty(\Omega)$ and $a_2 \in L^\infty(0, +\infty)$. Consider the controlled heat equation:

$$\begin{cases} \partial_t y - \Delta y + (a_1(x) + a_2(t))y = \chi_\omega u & \text{in } \Omega \times (0, +\infty), \\ y = 0 & \text{on } \partial\Omega \times (0, +\infty), \\ y(0) = y_0 \in L^2(\Omega), \end{cases} \tag{5.155}$$

where $u \in L^\infty(0, +\infty; L^2(\Omega))$. Let $Y = U = L^2(\Omega)$; Let $A = \Delta$ with its domain $H^2(\Omega) \cap H_0^1(\Omega)$; Let $D(\cdot) = -a_1 - a_2(\cdot)$; Let $B = \chi_\omega$, where χ_ω is treated as a linear and bounded operator on U. Thus, the Equation (5.155) can be rewritten as

$$\begin{cases} \dot{y}(t) = Ay(t) + D(t)y(t) + Bu(t), \ t \in (0, +\infty), \\ y(0) = y_0. \end{cases}$$

To verify that the above setting can be put into the framework ($\mathscr{A}_3$) (given at the beginning of Section 5.3), we only need to show the following two facts:

(a) The above $D(\cdot)$ satisfies that

$$D \in L_{loc}^1([0, +\infty); \mathscr{L}(L^2(\Omega))). \tag{5.156}$$

(b) (vi) of ($\mathscr{A}_3$) (given at the beginning of Section 5.3) is true.

(Here we used Theorems 1.22, 1.21, and Remark 1.5 after Theorem 1.22.) We will prove (a) and (b) one by one.

We start with proving (a). Choose a sequence of step functions $\{a_{2,k}\}_{k \geq 1} \subseteq L^1(0, +\infty)$ so that

$$\lim_{k \to +\infty} |a_{2,k}(t) - a_2(t)| = 0 \ \text{ for a.e. } t \in (0, +\infty). \tag{5.157}$$

It is clear that for each $k \in \mathbb{N}^+$,

$$D_k(\cdot) \triangleq -a_1 - a_{2,k}(\cdot) \in L_{loc}^1([0, +\infty); \mathscr{L}(L^2(\Omega))). \tag{5.158}$$

This, along with (5.157), yields that for a.e. $t \in (0, +\infty)$,

$$\lim_{k \to +\infty} \|D(t) - D_k(t)\|_{\mathscr{L}(L^2(\Omega))} = \lim_{k \to +\infty} |a_{2,k}(t) - a_2(t)| = 0,$$

which, combined with (5.158), indicates that $D(\cdot)$ is strongly measurable from $(0, +\infty)$ to $\mathscr{L}(L^2(\Omega))$. Now, (5.156) can be easily checked. Hence, the conclusion (a) is true.

To prove the conclusion (b), we arbitrarily fix $T > 0$. We first show that

$$Y_T \subseteq Z_T. \ \text{ (see (5.122) for the definitions of } Y_T, Z_T.) \tag{5.159}$$

For this purpose, we arbitrarily take $\psi \in Y_T$. By the definition of Y_T (see (5.122)), there exists a sequence $\{z_n\}_{n\geq1} \subseteq Y$ so that

$$\chi_\omega \Phi(T, \cdot)^* z_n \to \psi \quad \text{strongly in} \quad L^1(0, T; U). \tag{5.160}$$

Let $\{T_k\}_{k\geq1} \subseteq (0, T)$ be such that $T_k \nearrow T$. For each $k \in \mathbb{N}^+$, by Theorem 1.20, we see that there exists $\hat{z}_k \in Y$ so that

$$\Phi(T, T_k)^* z_n \to \hat{z}_k \quad \text{as} \quad n \to +\infty. \tag{5.161}$$

This yields that

$$\hat{z}_k = \Phi(T_{k+\ell}, T_k)^* \hat{z}_{k+\ell} \quad \text{for each} \quad k, \ell \in \mathbb{N}^+. \tag{5.162}$$

For each $k \geq 1$, we define the following function:

$$\varphi(t) \triangleq \Phi(T_k, t)^* \hat{z}_k, \quad t \in [0, T_k]. \tag{5.163}$$

By (5.162) and (5.163), we see that $\varphi \in C([0, T); Y)$ is well defined. Moreover, it follows from (5.161) and (5.163) that

$$\chi_\omega \Phi(T, \cdot)^* z_n \to \chi_\omega \varphi(\cdot) \quad \text{strongly in} \quad L^1(0, T_k; U).$$

This, together with (5.160), implies that

$$\psi(\cdot) = \chi_\omega \varphi(\cdot) \in L^1(0, T; U),$$

which, along with (5.163) and the definition of Z_T (see (5.122)), leads to (5.159). We next show that

$$Z_T \subseteq Y_T. \tag{5.164}$$

Observe that $\psi \in C([0, T); L^2(\Omega)) \cap L^1(0, T; L^2(\omega))$ solves the equation:

$$\begin{cases} \partial_t \psi + \Delta\psi - (a_1(x) + a_2(t))\psi = 0 & \text{in} \ \Omega \times (0, T), \\ \psi = 0 & \text{on} \ \partial\Omega \times (0, T) \end{cases} \tag{5.165}$$

if and only if $\varphi \in C([0, T); L^2(\Omega)) \cap L^1(0, T; L^2(\omega))$ solves

$$\begin{cases} \partial_t \varphi + \Delta\varphi - a_1(x)\varphi = 0 & \text{in} \ \Omega \times (0, T), \\ \varphi = 0 & \text{on} \ \partial\Omega \times (0, T), \end{cases} \tag{5.166}$$

where the function φ is defined by

$$\varphi(x, t) \triangleq \exp\left(\int_t^T a_2(s)ds\right)\psi(x, t), \quad (x, t) \in \Omega \times (0, T). \tag{5.167}$$

Given $\chi_\omega \widehat{\psi} \in Z_T$, let $\widehat{\varphi}$ be given by (5.167) where $\psi = \widehat{\psi}$. Let $\{T_k\}_{k \geq 1} \subseteq (0, T)$ be such that $T_k \nearrow T$. Write φ_k for the solution of the Equation (5.166) with the terminal condition $\varphi_k(T) = \widehat{\varphi}(T_k)$ (which belongs to $L^2(\Omega)$). Let ψ_k be given by (5.167) where $\varphi = \varphi_k$. Then, $\psi_k \in C([0, T]; L^2(\Omega))$ solves (5.165). We claim that

$$\chi_\omega \psi_k \to \chi_\omega \widehat{\psi} \quad \text{strongly in} \quad L^1(0, T; L^2(\omega)). \tag{5.168}$$

When (5.168) is proved, we get that $\chi_\omega \widehat{\psi} \in Y_T$, which leads to (5.164).

The remainder is to show (5.168). Clearly, (5.168) is equivalent to

$$\chi_\omega \varphi_k \to \chi_\omega \widehat{\varphi} \quad \text{strongly in} \quad L^1(0, T; L^2(\omega)). \tag{5.169}$$

Let $\widetilde{\varphi}$ satisfy that

$$\begin{cases} \partial_t \widetilde{\varphi} + \Delta \widetilde{\varphi} - a_1(x)\widetilde{\varphi} = 0 & \text{in } \Omega \times (-T, T), \\ \widetilde{\varphi} = 0 & \text{on } \partial\Omega \times (-T, T) \end{cases} \tag{5.170}$$

and

$$\widetilde{\varphi}(x, t) = \widehat{\varphi}(x, t), \quad (x, t) \in \Omega \times (0, T). \tag{5.171}$$

It is clear that

$$\widetilde{\varphi} \in C\big([-T, T); L^2(\Omega)\big) \cap L^1\big(-T, T; L^2(\omega)\big). \tag{5.172}$$

Because the equations satisfied by $\widetilde{\varphi}$ and φ_k are time invariant, one can easily check that

$$\varphi_k(t) = \widetilde{\varphi}(t - (T - T_k)), \quad \text{when } t \in (0, T). \tag{5.173}$$

By (5.172), we see that given $\varepsilon \in (0, +\infty)$, there are two positive constants $\delta(\varepsilon)$ and $\eta(\varepsilon) = \eta(\varepsilon, \delta(\varepsilon))$ so that

$$\|\chi_\omega \widetilde{\varphi}\|_{L^1(a,b; L^2(\omega))} \leq \varepsilon, \quad \text{when } (a, b) \subseteq (-T, T), \ |a - b| \leq \delta(\varepsilon) \tag{5.174}$$

and

$$\|\widetilde{\varphi}(a) - \widetilde{\varphi}(b)\|_{L^2(\Omega)} \leq \varepsilon, \quad \text{when } (a, b) \subseteq \big[-T, T - \delta(\varepsilon)\big], \ |a - b| \leq \eta(\varepsilon). \tag{5.175}$$

Let $k_0 = k_0(\varepsilon)$ verify that

$$0 < T - T_k \leq \eta(\varepsilon), \quad \text{when } k \geq k_0. \tag{5.176}$$

From (5.171) and (5.173) it follows that

$$\|\chi_\omega\varphi_k - \chi_\omega\widehat{\varphi}\|_{L^1(0,T;L^2(\omega))}$$
$$\leq \|\chi_\omega(\widetilde{\varphi}(\cdot - (T - T_k)) - \widetilde{\varphi}(\cdot))\|_{L^1(0,T-\delta(\varepsilon);L^2(\omega))}$$
$$+\|\chi_\omega\widetilde{\varphi}\|_{L^1(T-\delta(\varepsilon),T;L^2(\omega))} + \|\chi_\omega\widetilde{\varphi}(\cdot - (T - T_k))\|_{L^1(T-\delta(\varepsilon),T;L^2(\omega))}.$$

This, along with (5.175), (5.176), and (5.174), yields that

$$\|\chi_\omega\varphi_k - \chi_\omega\widehat{\varphi}\|_{L^1(0,T;L^2(\omega))} \leq (T - \delta(\varepsilon))\varepsilon + 2\varepsilon \leq (T + 2)\varepsilon, \quad \text{when } k \geq k_0,$$

which leads to (5.169), as well as (5.168). Hence, (5.164) is proved.

Finally, by (5.159) and (5.164), we obtain the conclusion (b). Hence, this example can be put into the framework ($\mathcal{A}_3$) (given at the beginning of Section 5.3). We now end Example 5.4.

5.4   Maximal Time, Minimal Norm, and Optimal Target Controls

In this section, we will study the equivalence among the maximal time control problem $(TP)_{max}^{Q_S,Q_E}$, an optimal target control problem and a minimal norm control problem under the following framework ($\mathcal{A}_4$):

(i) The state space Y and the control space U are real separable Hilbert spaces.

(ii) Let $T \in (0, +\infty)$. Let $A : \mathcal{D}(A) \subseteq Y \mapsto Y$ generate a C_0 semigroup $\{e^{At}\}_{t\geq 0}$ on Y. Let $D(\cdot) \in L^1(0, T; \mathcal{L}(Y))$ and $B \in \mathcal{L}(U, Y)$. The controlled system is as:

$$\dot{y}(t) = Ay(t) + D(t)y(t) + Bu(t), \quad t \in (0, T). \tag{5.177}$$

Given $\tau \in [0, T)$, $y_0 \in Y$ and $u \in L^\infty(\tau, T; U)$, write $y(\cdot; \tau, y_0, u)$ for the solution of (5.177) over $[\tau, T]$, with the initial condition that $y(\tau) = y_0$. Moreover, write $\{\Phi(t, s) : t \geq s \geq 0\}$ for the evolution system generated by $A + D(\cdot)$ over Y.

(iii) Let $\mathbb{U} = B_M(0)$, where $B_M(0)$ is the closed ball in U centered at 0 and of radius $M \geq 0$. (When $M = 0$, we agree that $B_M(0) \triangleq \{0\}$.)

(iv) Let $Q_S = \{(t, y(t; 0, y_0, 0)) : t \in [0, T)\}$ and $Q_E = \{(T, B_r(z_d))\}$, where $r \in [0, +\infty)$, and where $y_0 \in Y$ and $z_d \in Y$ are arbitrarily fixed so that

$$r_T \triangleq \|y(T; 0, y_0, 0) - z_d\|_Y > 0. \tag{5.178}$$

(v) The following unique continuation property holds: If there exist two constants $a, b \in (0, T)$ and $z \in Y$ so that $B^*\Phi(T, t)^*z = 0$ for each $t \in [a, b]$, then $z = 0$ in Y.

In this section, we simply write $(TP)_{max}^{Qs,QE}$ for $(TP)_M^r$ which reads as:

$(TP)_M^r$ $\quad \tau(M, r) \triangleq \sup\{\tau \in [0, T) \mid y(T; \tau, y(\tau; 0, y_0, 0), u) \in B_r(z_d)$
$$\text{and } u \in L(\tau, T; \mathbb{U})\}. \tag{5.179}$$

We next consider the following problem:

$(\widetilde{TP})_M^r$ $\quad \widetilde{\tau}(M, r) \triangleq \sup\{\tau \in [0, T) \mid y(T; 0, y_0, \chi_{(\tau,T)}u) \in B_r(z_d)$
$$\text{and } u \in \mathcal{U}_{M,\tau}\}, \tag{5.180}$$

where

$$\mathcal{U}_{M,\tau} \triangleq \{u \in L^\infty(0, T; U) \mid \|u(t)\|_U \leq M \text{ a.e. } t \in (\tau, T)\}.$$

In this problem,

- we call $(0, y_0, T, \chi_{(\tau,T)}u)$ an admissible tetrad, if

$$\tau \in [0, T), u \in \mathcal{U}_{M,\tau} \text{ and } y(T; 0, y_0, \chi_{(\tau,T)}u) \in B_r(z_d);$$

- we call $(0, y_0, T, \chi_{(\widetilde{\tau}(M,r),T)}\widetilde{u}^*)$ an optimal tetrad, if

$$\widetilde{\tau}(M, r) \in [0, T), \widetilde{u}^* \in \mathcal{U}_{M,\widetilde{\tau}(M,r)} \text{ and } y(T, 0, y_0, \chi_{(\widetilde{\tau}(M,r),T)}\widetilde{u}^*) \in B_r(z_d);$$

- when $(0, y_0, T, \chi_{(\widetilde{\tau}(M,r),T)}\widetilde{u}^*)$ is an optimal tetrad, $\widetilde{\tau}(M, r)$ and $\widetilde{u}^*$ are called the optimal time and an optimal control, respectively.

We can easily check the following three facts:

- $\tau(M, r) = \widetilde{\tau}(M, r)$;
- The supreme in (5.179) can be reached if and only if the supreme in (5.180) can be reached;
- The zero extension of each optimal control to $(TP)_M^r$ over $(0, T)$ is an optimal control to $(\widetilde{TP})_M^r$, while the restriction of each optimal control to $(\widetilde{TP})_M^r$ over $(\tau(M, r), T)$ is an optimal control to $(TP)_M^r$.

Based on the above-mentioned facts, we can treat $(TP)_M^r$ and $(\widetilde{TP})_M^r$ as the same problem.

We now introduce the corresponding optimal target control problems $(OP)_M^\tau$, with $M \geq 0$ and $\tau \in [0, T)$, in the following manner:

$(OP)_M^\tau$ $\quad \mathbf{r}(M, \tau) \triangleq \inf\{\|y(T; 0, y_0, \chi_{(\tau,T)}u) - z_d\|_Y \mid u \in \mathcal{U}_{M,\tau}\}.$
$$\tag{5.181}$$

In the problem $(OP)_M^\tau$, u^* is called an optimal control, if

$$u^* \in \mathcal{U}_{M,\tau} \text{ and } \|y(T; 0, y_0, \chi_{(\tau,T)}u^*) - z_d\|_Y = \mathbf{r}(M, \tau).$$

We next introduce the corresponding minimal norm control problem $(NP)_\tau^r$, with $\tau \in [0, T)$ and $r \in [0, +\infty)$, in the following manner:

$$(NP)_\tau^r \qquad \mathbf{M}(\tau, r) \triangleq \inf\{\|v\|_{L^\infty(\tau,T;U)} \mid y(T; 0, y_0, \chi_{(\tau,T)}v) \in B_r(z_d)$$
$$\text{and } v \in L^\infty(0, T; U)\}. \tag{5.182}$$

In the problem $(NP)_\tau^r$,

- we call $v \in L^\infty(0, T; U)$ an admissible control, if $y(T; 0, y_0, \chi_{(\tau,T)}v) \in B_r(z_d)$;
- we call v^* an optimal control, if

$$y(T; 0, y_0, \chi_{(\tau,T)}v^*) \in B_r(z_d) \quad \text{and} \quad \|v^*\|_{L^\infty(\tau,T;U)} = \mathbf{M}(\tau, r);$$

- when the set in the right-hand side of (5.182) is empty, we agree that $\mathbf{M}(\tau, r) = +\infty$.

The purpose of this section is to build up connections among $(TP)_M^r$, $(OP)_M^\tau$ and $(NP)_\tau^r$. We now give the definition about the equivalence for these problems.

Definition 5.4 Let $M \geq 0$, $\tau \in [0, T)$, and $r \in (0, +\infty)$.

(i) Problems $(TP)_M^r$ and $(OP)_M^\tau$ are said to be equivalent if $(\widetilde{TP})_M^r$ and $(OP)_M^\tau$ have the same optimal controls and

$$(\tau, r) = (\widetilde{\tau}(M, r), \mathbf{r}(M, \tau)).$$

(ii) Problems $(TP)_M^r$ and $(NP)_\tau^r$ are said to be equivalent if $(\widetilde{TP})_M^r$ and $(NP)_\tau^r$ have the same optimal controls and

$$(M, \tau) = (\mathbf{M}(\tau, r), \widetilde{\tau}(M, r)).$$

(iii) Problems $(OP)_M^\tau$ and $(NP)_\tau^r$ are said to be equivalent if they have the same optimal controls and

$$(M, r) = (\mathbf{M}(\tau, r), \mathbf{r}(M, \tau)).$$

The main result of this section is presented as follows:

Theorem 5.6 *The following conclusions are true:*

(i) *Given $\tau \in [0, T)$ and $M \in (0, \mathbf{M}(\tau, 0))$, problems $(OP)_M^\tau$, $(TP)_M^{\mathbf{r}(M,\tau)}$, and $(NP)_\tau^{\mathbf{r}(M,\tau)}$ are equivalent;*

(ii) *Given $\tau \in [0, T)$ and $r \in (0, r_T)$, problems $(NP)_\tau^r$, $(OP)_{\mathbf{M}(\tau,r)}^\tau$, and $(TP)_{\mathbf{M}(\tau,r)}^r$ are equivalent;*

(iii) *Given $M \in (0, +\infty)$ and $r \in [\mathbf{r}(M, 0), r_T) \cap (0, r_T)$, problems $(TP)_M^r$, $(NP)_{\widetilde{\tau}(M,r)}^r$ and $(OP)_M^{\widetilde{\tau}(M,r)}$ are equivalent.*

Remark 5.3 The proof will be given in Section 5.4.4. Throughout this section, we agree that the characteristic function of the empty set is identically zero, i.e., $\chi_\emptyset \triangleq 0$.

5.4.1 Existence of Optimal Controls for These Problems

In this subsection, we will discuss the existence of optimal controls to problems $(OP)_M^\tau$, $(NP)_\tau^r$ and $(\widetilde{TP})_M^r$. We start with introducing the following preliminary lemma:

Lemma 5.10 *The following two conclusions are true:*

(i) *Let τ_1 and τ_2 be so that $T > \tau_2 > \tau_1 \geq 0$. Let $y_d \in Y$. Then given $\varepsilon \in (0, +\infty)$, there is $C_\varepsilon \triangleq C(T, \tau_1, \tau_2, y_d, \varepsilon) \in (0, +\infty)$ so that for each $\tau \in [\tau_1, \tau_2]$,*

$$|\langle y_d, z\rangle_Y| \leq C_\varepsilon \int_\tau^T \|B^*\Phi(T, t)^* z\|_U \mathrm{d}t + \varepsilon\|z\|_Y \text{ for all } z \in Y. (5.183)$$

(ii) *For each $\tau \in [0, T)$, $y_d \in Y$ and $r \in (0, +\infty)$, there exists $u \in L^\infty(0, T; U)$ so that $y(T; 0, y_0, \chi_{(\tau,T)}u) \in B_r(y_d)$.*

Proof The proof of this lemma is similar to that of Lemma 5.4; we omit the details. ☐

The following theorem is concerned with the existence of optimal controls to $(OP)_M^\tau$, $(NP)_\tau^r$ and $(\widetilde{TP})_M^r$.

Theorem 5.7 *The following conclusions are true:*

(i) *For any $M \geq 0$ and $\tau \in [0, T)$, $(OP)_M^\tau$ has optimal controls.*
(ii) *(a) For any $\tau \in [0, T)$ and $r \in (0, +\infty)$, $(NP)_\tau^r$ has optimal controls and $\mathbf{M}(\tau, r) \leq \mathbf{M}(\tau, 0)$; (b) For any $\tau \in [0, T)$ and $r \in (0, r_T)$, it holds that $\mathbf{M}(\tau, r) \in (0, +\infty)$. (Here, r_T is given by (5.178).)*
(iii) *The problem $(\widetilde{TP})_M^r$, with $r \in (0, r_T)$ and $M \geq 0$, has optimal controls if and only if $M \geq \mathbf{M}(0, r)$.*

Proof We prove the conclusions one by one.

(i) Arbitrarily fix $M \geq 0$ and $\tau \in [0, T)$. By the definition of $\mathbf{r}(M, \tau)$ (see (5.181)), there is a sequence $\{u_k\}_{k\geq 1} \subseteq \mathscr{U}_{M,\tau}$ so that

$$\|y(T; 0, y_0, \chi_{(\tau,T)}u_k) - z_d\|_Y \to \mathbf{r}(M, \tau) \text{ as } k \to +\infty. \quad (5.184)$$

Meanwhile, since $\{\chi_{(\tau,T)}u_k\}_{k\geq 1}$ is bounded in $L^\infty(0, T; U)$, there exists a subsequence of $\{u_k\}_{k\geq 1}$, denoted in the same manner, and $\widetilde{u} \in L^\infty(0, T; U)$ so that

$$\chi_{(\tau,T)} u_k \to \chi_{(\tau,T)} \widetilde{u} \quad \text{weakly star in } L^\infty(0, T; U) \tag{5.185}$$

and

$$\|\widetilde{u}\|_{L^\infty(\tau,T;U)} \le \liminf_{k\to+\infty} \|u_k\|_{L^\infty(\tau,T;U)} \le M.$$

From the above, we see that $\widetilde{u} \in \mathscr{U}_{M,\tau}$. By (5.185), we obtain that

$$y(T; 0, y_0, \chi_{(\tau,T)} u_k) \to y(T; 0, y_0, \chi_{(\tau,T)} \widetilde{u}) \quad \text{weakly in } Y.$$

This, together with (5.184) and the definition of $\mathbf{r}(M, \tau)$, yields that

$$\mathbf{r}(M, \tau) \le \|y(T; 0, y_0, \chi_{(\tau,T)} \widetilde{u}) - z_d\|_Y$$
$$\le \liminf_{k\to+\infty} \|y(T; 0, y_0, \chi_{(\tau,T)} u_k) - z_d\|_Y = \mathbf{r}(M, \tau).$$

Since $\widetilde{u} \in \mathscr{U}_{M,\tau}$, it follows from the above that $\widetilde{u}$ is an optimal control to $(OP)_M^\tau$.

(ii) We first show (a). Arbitrarily fix $\tau \in [0, T)$ and $r \in (0, +\infty)$.

By (ii) of Lemma 5.10 and the definition of $\mathbf{M}(\tau, r)$ (see (5.182)), there is a sequence $\{u_k\}_{k\ge 1} \subseteq L^\infty(0, T; U)$ so that

$$\|y(T; 0, y_0, \chi_{(\tau,T)} u_k) - z_d\|_Y \le r \quad \text{and} \quad \lim_{k\to+\infty} \|u_k\|_{L^\infty(\tau,T;U)} = \mathbf{M}(\tau, r). \tag{5.186}$$

Then there exists a subsequence of $\{u_k\}_{k\ge 1}$, denoted in the same manner, and $\widetilde{u} \in L^\infty(0, T; U)$ so that

$$\chi_{(\tau,T)} u_k \to \chi_{(\tau,T)} \widetilde{u} \quad \text{weakly star in } L^\infty(0, T; U) \tag{5.187}$$

and

$$\|\widetilde{u}\|_{L^\infty(\tau,T;U)} \le \liminf_{k\to+\infty} \|u_k\|_{L^\infty(\tau,T;U)} = \mathbf{M}(\tau, r). \tag{5.188}$$

By (5.187), we have that

$$y(T; 0, y_0, \chi_{(\tau,T)} u_k) \to y(T; 0, y_0, \chi_{(\tau,T)} \widetilde{u}) \quad \text{weakly in } Y.$$

From this and the inequality in (5.186), we see that

$$\|y(T; 0, y_0, \chi_{(\tau,T)} \widetilde{u}) - z_d\|_Y \le \liminf_{k\to+\infty} \|y(T; 0, y_0, \chi_{(\tau,T)} u_k) - z_d\|_Y \le r.$$

This, together with (5.188), implies that $\widetilde{u}$ is an optimal control to $(NP)_\tau^r$. Besides, by (5.182), we can easily check that $\mathbf{M}(\tau, r) \le \mathbf{M}(\tau, 0)$.

We next show (b). Arbitrarily fix $\tau \in [0, T)$ and $r \in (0, r_T)$. By (a) of this theorem, we see that $\mathbf{M}(\tau, r) < +\infty$. Seeking for a contradiction, we suppose that

$$\mathbf{M}(\tau, r) = 0 \text{ for some } r \in (0, r_T).$$

Then we would have that

$$\|y(T; 0, y_0, 0) - z_d\|_Y \le r < r_T,$$

which, along with (5.178), leads to a contradiction.
(iii) Arbitrarily fix $r \in (0, r_T)$ and $M \ge 0$. We first show that

$$M \ge \mathbf{M}(0, r) \Rightarrow (\widetilde{TP})^r_M \text{ has optimal controls.}$$

By (ii) of this theorem, $(NP)^r_0$ has an optimal control v^*. By the optimality of v^*, we see that

$$\|v^*\|_{L^\infty(0,T;U)} = \mathbf{M}(0, r) \le M \text{ and } y(T; 0, y_0, v^*) \in B_r(z_d).$$

From these, it follows that $(\widetilde{TP})^r_M$ has admissible controls. Then by the definition of $\widetilde{\tau}(M, r)$, there are two sequences $\{t_k\}_{k \ge 1} \subseteq [0, T)$ and $\{u_k\}_{k \ge 1} \subseteq \mathcal{U}_{M,t_k}$ so that

$$\lim_{k \to +\infty} t_k = \widetilde{\tau}(M, r) \text{ and } y(T; 0, y_0, \chi_{(t_k,T)}u_k) \in B_r(z_d). \qquad (5.189)$$

Furthermore, there is a subsequence of $\{u_k\}_{k \ge 1}$, denoted in the same manner, and a control $\widetilde{u} \in L^\infty(0, T; U)$ so that

$$\chi_{(t_k,T)}u_k \to \chi_{(\widetilde{\tau}(M,r),T)}\widetilde{u} \text{ weakly star in } L^\infty(0, T; U).$$

Since $\{u_k\}_{k \ge 1} \subseteq \mathcal{U}_{M,t_k}$, the latter conclusion implies that

$$y(T; 0, y_0, \chi_{(t_k,T)}u_k) \to y(T; 0, y_0, \chi_{(\widetilde{\tau}(M,r),T)}\widetilde{u}) \text{ weakly in } Y \qquad (5.190)$$

and

$$\|\chi_{(\widetilde{\tau}(M,r),T)}\widetilde{u}\|_{L^\infty(0,T;U)} \le \liminf_{k \to +\infty} \|\chi_{(t_k,T)}u_k\|_{L^\infty(0,T;U)} \le M. \qquad (5.191)$$

From (5.190), (5.191), and the second conclusion of (5.189), we find that

$$y(T; 0, y_0, \chi_{(\widetilde{\tau}(M,r),T)}\widetilde{u}) \in B_r(z_d) \text{ and } \widetilde{u} \in \mathcal{U}_{M,\widetilde{\tau}(M,r)}. \qquad (5.192)$$

Since $r \in (0, r_T)$, by the first conclusion of (5.192) and (5.178), we see that $\widetilde{\tau}(M, r) \in [0, T)$. This, together with (5.192), yields that $\widetilde{u}$ is an optimal control to $(\widetilde{TP})_M^r$.

We next show that

$$(\widetilde{TP})_M^r \text{ has optimal controls} \Rightarrow M \geq \mathbf{M}(0, r).$$

For this purpose, we assume that u^* is an optimal control to $(\widetilde{TP})_M^r$. Then

$$\widetilde{\tau}(M, r) \in [0, T), \, y(T; 0, y_0, \chi_{(\widetilde{\tau}(M,r),T)}u^*) \in B_r(z_d) \text{ and } u^* \in \mathscr{U}_{M,\widetilde{\tau}(M,r)}. \tag{5.193}$$

By the second conclusion of (5.193), we have that $\chi_{(\widetilde{\tau}(M,r),T)}u^*$ is an admissible control to $(NP)_0^r$. Thus, it follows from the optimality of $\mathbf{M}(0, r)$ and the third conclusion of (5.193) that

$$\mathbf{M}(0, r) \leq M.$$

Hence, we complete the proof of Theorem 5.7. □

5.4.2 Connections Between Problems of Optimal Target Control and Minimal Norm Control

This subsection presents connections between $(OP)_M^\tau$ and $(NP)_\tau^r$. These will be used in the proof of Theorem 5.6. We begin with studying $(OP)_M^\tau$.

Lemma 5.11 *Let* $(\tau, M) \in [0, T) \times [0, \mathbf{M}(\tau, 0))$. *Then the following conclusions are true:*

(i) $\mathbf{r}(M, \tau) \in (0, +\infty)$.
(ii) u^* *is an optimal control for* $(OP)_M^\tau$ *if and only if* $u^* \in \mathscr{U}_{M,\tau}$ *and*

$$\int_\tau^T \langle B^*\varphi^*(t), u^*(t) \rangle_U \, dt = \max_{v(\cdot) \in \mathscr{U}_{M,\tau}} \int_\tau^T \langle B^*\varphi^*(t), v(t) \rangle_U \, dt, \tag{5.194}$$

where

$$\varphi^*(t) \triangleq \Phi(T, t)^*(z_d - y(T; 0, y_0, \chi_{(\tau,T)}u^*)), \quad t \in [0, T].$$

(iii) *Each optimal control* u^* *to* $(OP)_M^\tau$ *satisfies that*

$$\|u^*\|_{L^\infty(a,b;U)} = M \text{ for any interval } (a, b) \subseteq (\tau, T). \tag{5.195}$$

Proof Arbitrarily fix $(\tau, M) \in [0, T) \times [0, \mathbf{M}(\tau, 0))$. We will prove the conclusions (i), (ii), and (iii) one by one.

(i) According to (i) of Theorem 5.7, there is a control $u \in L^\infty(0, T; U)$ so that

$$\|y(T; 0, y_0, \chi_{(\tau,T)}u) - z_d\|_Y = \mathbf{r}(M, \tau) \text{ and } \|u\|_{L^\infty(\tau,T;U)} \leq M.$$

Hence, $\mathbf{r}(M, \tau) < +\infty$. Meanwhile, if $\mathbf{r}(M, \tau) = 0$, then by the definition of $\mathbf{M}(\tau, 0)$, we obtain that

$$M \geq \|u\|_{L^\infty(\tau,T;U)} \geq \mathbf{M}(\tau, 0),$$

which contradicts the assumption that $M \in [0, \mathbf{M}(\tau, 0))$. Thus, $\mathbf{r}(M, \tau) \in (0, +\infty)$.

(ii) Arbitrarily fix $u_1, u_2 \in \mathscr{U}_{M,\tau}$ and $\varepsilon \in [0, 1]$. Set

$$\begin{cases} u_\varepsilon \triangleq (1 - \varepsilon)u_1 + \varepsilon u_2; \\ y_i(t) \triangleq y(t; 0, y_0, \chi_{(\tau,T)}u_i), & t \in [0, T], \ i = 1, 2; \\ y_\varepsilon(t) \triangleq y(t; 0, y_0, \chi_{(\tau,T)}u_\varepsilon), & t \in [0, T]. \end{cases}$$

We can easily check that

$$\begin{aligned} &\|y_\varepsilon(T) - z_d\|_Y^2 - \|y_1(T) - z_d\|_Y^2 \\ &= \varepsilon^2 \|y(T; 0, 0, \chi_{(\tau,T)}(u_2 - u_1))\|_Y^2 \\ &\quad + 2\varepsilon \int_\tau^T \langle u_2 - u_1, B^*\Phi(T, t)^*(y_1(T) - z_d)\rangle_U dt. \end{aligned} \tag{5.196}$$

We now show the necessity. Let u^* be an optimal control to $(OP)_M^\tau$. Arbitrarily take $u \in \mathscr{U}_{M,\tau}$. By choosing $u_1 = u^*$ and $u_2 = u$ in (5.196), we find that

$$\int_\tau^T \langle u - u^*, B^*\Phi(T, t)^*(z_d - y(T; 0, y_0, \chi_{(\tau,T)}u^*)))\rangle_U dt \leq 0 \text{ for all } u \in \mathscr{U}_{M,\tau}.$$

This implies (5.194).

We next show the sufficiency. Suppose that $u^* \in \mathscr{U}_{M,\tau}$ satisfies (5.194). Arbitrarily fix $u \in \mathscr{U}_{M,\tau}$. Then by (5.194), and by taking $u_1 = u^*$ and $u_2 = u$ in (5.196), we see that

$$\begin{aligned} &\|y_\varepsilon(T) - z_d\|_Y^2 - \|y(T; 0, y_0, \chi_{(\tau,T)}u^*) - z_d\|_Y^2 \\ &= \varepsilon^2 \|y(T; 0, 0, \chi_{(\tau,T)}(u - u^*))\|_Y^2 \\ &\quad + 2\varepsilon \int_\tau^T \langle u - u^*, B^*\Phi(T, t)^*(y(T; 0, y_0, \chi_{(\tau,T)}u^*) - z_d)\rangle_U dt \\ &\geq 0. \end{aligned} \tag{5.197}$$

Choosing $\varepsilon = 1$ in (5.197), we obtain that

$$\|y(T; 0, y_0, \chi_{(\tau,T)}u) - z_d\|_Y^2 \geq \|y(T; 0, y_0, \chi_{(\tau,T)}u^*) - z_d\|_Y^2 \text{ for all } u \in \mathscr{U}_{M,\tau}.$$

This implies that u^* is an optimal control to $(OP)_M^\tau$.

(iii) By contradiction, suppose that the conclusion (iii) was not true. Then there would be an optimal control $\widehat{u}^*$ to $(OP)_M^\tau$ and $(\widehat{a}, \widehat{b}) \subseteq (\tau, T)$ so that

$$\|\widehat{u}^*\|_{L^\infty(\widehat{a},\widehat{b};U)} < M. \tag{5.198}$$

According to (ii) of this lemma, (5.194) holds for $u^* = \widehat{u}^*$. Then by the same way as that used in the proof of Theorem 4.1, we can verify that

$$\langle B^*\widehat{\varphi}^*(t), \widehat{u}^*(t)\rangle_U = \max_{v \in \mathbb{U}}\langle B^*\widehat{\varphi}^*(t), v\rangle_U \text{ for a.e. } t \in (\tau, T), \tag{5.199}$$

where

$$\widehat{\varphi}^*(t) \triangleq \Phi(T, t)^*(z_d - y(T; 0, y_0, \chi_{(\tau,T)}\widehat{u}^*)) \text{ for each } t \in [0, T].$$

This, along with (5.198), yields that

$$B^*\widehat{\varphi}^*(t) = 0 \text{ for each } t \in [\widehat{a}, \widehat{b}],$$

which, combined with (v) of $(\mathscr{A}_4)$ (given at the beginning of Section 5.4), indicates that

$$y(T; 0, y_0, \chi_{(\tau,T)}\widehat{u}^*) = z_d.$$

This contradicts the conclusion (i) of this lemma. Therefore, (5.195) is true. In summary, we end the proof of Lemma 5.11. $\qquad \square$

Remark 5.4 Several notes are given in order.

(i) We mention that $\mathbf{M}(\tau, 0) \in (0, +\infty]$ for each $\tau \in [0, T)$ (see (5.182) and (5.178)).

(ii) The conclusion (ii) in Lemma 5.11 still stands without the assumption that $M < \mathbf{M}(\tau, 0)$. That is, a control $u \in \mathscr{U}_{M,\tau}$ satisfying (5.194) must be an optimal control to $(OP)_M^\tau$. This can be observed from the proof of the conclusion (ii) in Lemma 5.11.

(iii) To ensure the conclusion (i) in Lemma 5.11, it is necessary to have that $0 \leq M < \mathbf{M}(\tau, 0)$.

Lemma 5.12 *Let $\tau \in [0, T)$. Then the following conclusions are true:*

(i) *The map $\mathbf{r}(\cdot, \tau)$ is strictly decreasing and homeomorphic from $[0, \mathbf{M}(\tau, 0))$ onto $(0, r_T]$.*

(ii) *For each $r \in (0, +\infty)$, it stands that $\mathbf{M}(\tau, r) < \mathbf{M}(\tau, 0)$.*

(iii) *The inverse of* $\mathbf{r}(\cdot, \tau)$ *is the map* $\mathbf{M}(\tau, \cdot)$, *i.e.,*

$$r = \mathbf{r}(\mathbf{M}(\tau, r), \tau) \quad \text{when} \ \ r \in (0, r_T], \tag{5.200}$$

and

$$M = \mathbf{M}(\tau, \mathbf{r}(M, \tau)) \quad \text{when} \ \ M \in [0, \mathbf{M}(\tau, 0)). \tag{5.201}$$

Proof We prove the conclusions one by one.

(i) The proof of the conclusion (i) is divided into the following three steps:

Step 1. We show that the map $\mathbf{r}(\cdot, \tau)$ is strictly decreasing.

Let $0 \le M_1 < M_2 < \mathbf{M}(\tau, 0)$. We claim that

$$\mathbf{r}(M_1, \tau) > \mathbf{r}(M_2, \tau).$$

Otherwise, we would have that

$$\mathbf{r}(M_1, \tau) \le \mathbf{r}(M_2, \tau).$$

Then by (i) of Theorem 5.7, there is an optimal control $u_1^* \in \mathscr{U}_{M_1, \tau}$ to $(OP)_{M_1}^\tau$ so that

$$\|y(T; 0, y_0, \chi_{(\tau,T)} u_1^*) - z_d\|_Y = \mathbf{r}(M_1, \tau) \le \mathbf{r}(M_2, \tau)$$

and

$$\|u_1^*\|_{L^\infty(\tau,T;U)} \le M_1 < M_2. \tag{5.202}$$

These imply that u_1^* is also an optimal control to $(OP)_{M_2}^\tau$. From this and (iii) of Lemma 5.11 it follows that

$$\|u_1^*(t)\|_{L^\infty(\tau,T;U)} = M_2.$$

This contradicts (5.202).

Step 2. We prove that the map $\mathbf{r}(\cdot, \tau)$ is Lipschitz continuous.

Let $0 \le M_1 < M_2 < \mathbf{M}(\tau, 0)$. According to (i) of Theorem 5.7, $(OP)_{M_2}^\tau$ has an optimal control u_2^*. Then it follows that

$$
\begin{aligned}
\mathbf{r}(M_2, \tau) &= \left\| \Phi(T, 0) y_0 + \int_\tau^T \Phi(T, t)^* B u_2^* dt - z_d \right\|_Y \\
&\ge \left\| \Phi(T, 0) y_0 + \int_\tau^T \Phi(T, t)^* B \left(\frac{M_1}{M_2} u_2^* \right) dt - z_d \right\|_Y \\
&\quad - \frac{M_2 - M_1}{M_2} \left\| \int_\tau^T \Phi(T, t)^* B u_2^* dt \right\|_Y.
\end{aligned}
\tag{5.203}
$$

Since $M_1 u_2^*/M_2 \in \mathcal{U}_{M_1,\tau}$, by (5.203) and the optimality of $\mathbf{r}(M_1, \tau)$, we get that

$$\mathbf{r}(M_2, \tau) \geq \mathbf{r}(M_1, \tau) - (M_2 - M_1)(T - \tau) \sup_{0 \leq t \leq T} \|\Phi(T, t)\|_{\mathscr{L}(H,H)} \|B\|_{\mathscr{L}(U,H)}.$$

The above inequality, together with the result of Step 1, yields that

$$|\mathbf{r}(M_1, \tau) - \mathbf{r}(M_2, \tau)| \leq (T - \tau)|M_1 - M_2| \sup_{0 \leq t \leq T} \|\Phi(T, t)\|_{\mathscr{L}(H,H)} \|B\|_{\mathscr{L}(U,H)}.$$

Step 3. We show that the map $\mathbf{r}(\cdot, \tau) : [0, \mathbf{M}(\tau, 0)) \to (0, r_T]$ is surjective and homeomorphic.

By Steps 1–2, we only need to show that

$$\mathbf{r}(0, \tau) = r_T \quad \text{and} \quad \lim_{M \to \mathbf{M}(\tau, 0)} \mathbf{r}(M, \tau) = 0. \tag{5.204}$$

The first equality in (5.204) follows from the definition of r_T (see (5.178)). We now show the second equality in (5.204). About $\mathbf{M}(\tau, 0)$, we have that either $\mathbf{M}(\tau, 0) < +\infty$ or $\mathbf{M}(\tau, 0) = +\infty$.

In the case that $\mathbf{M}(\tau, 0) < +\infty$, we arbitrarily fix $\varepsilon \in (0, +\infty)$. By the definition of $\mathbf{M}(\tau, 0)$, there exists $u_\varepsilon \in L^\infty(0, T; U)$ so that

$$y(T; 0, y_0, \chi_{(\tau,T)} u_\varepsilon) = z_d \quad \text{and} \quad \|u_\varepsilon\|_{L^\infty(\tau,T;U)} \leq \mathbf{M}(\tau, 0) + \varepsilon. \tag{5.205}$$

The inequality in (5.205) implies that $\frac{M}{\mathbf{M}(\tau,0)+\varepsilon} u_\varepsilon \in \mathcal{U}_{M,\tau}$. This, together with the optimality of $\mathbf{r}(M, \tau)$, yields that

$$\mathbf{r}(M, \tau) \leq \left\| y\left(T; 0, y_0, \chi_{(\tau,T)} \frac{M}{\mathbf{M}(\tau, 0) + \varepsilon} u_\varepsilon\right) - z_d \right\|_Y,$$

which, combined with (5.205), indicates that

$$\mathbf{r}(M, \tau) \leq \left\| \int_0^T \Phi(T, t) B \chi_{(\tau,T)} u_\varepsilon(t) dt \right\|_Y \frac{|\mathbf{M}(\tau, 0) + \varepsilon - M|}{\mathbf{M}(\tau, 0) + \varepsilon}$$
$$\leq (T - \tau)|\mathbf{M}(\tau, 0) + \varepsilon - M| \|B\|_{\mathscr{L}(U,H)} \sup_{0 \leq t \leq T} \|\Phi(T, t)\|_{\mathscr{L}(H)}.$$

Sending $\varepsilon \to 0+$ in the above inequality, we obtain that

$$\mathbf{r}(M, \tau) \leq (T - \tau)|\mathbf{M}(\tau, 0) - M| \|B\|_{\mathscr{L}(U,H)} \sup_{0 \leq t \leq T} \|\Phi(T, t)\|_{\mathscr{L}(H)}.$$

This implies the second equality in (5.204) in the first case.

In the case that $\mathbf{M}(\tau, 0) = +\infty$, we use (ii) of Lemma 5.10 to find that for each $\varepsilon \in (0, +\infty)$, there exists $u_\varepsilon \in L^\infty(0, T; U)$ so that

$$y(T; 0, y_0, \chi_{(\tau,T)} u_\varepsilon) \in B_\varepsilon(z_d). \tag{5.206}$$

Set $M_\varepsilon \triangleq \|u_\varepsilon\|_{L^\infty(\tau,T;U)}$. It follows from (5.206) and the optimality of $\mathbf{r}(M_\varepsilon, \tau)$ that

$$\mathbf{r}(M_\varepsilon, \tau) \leq \|y(T; 0, y_0, \chi_{(\tau,T)} u_\varepsilon) - z_d\|_Y \leq \varepsilon,$$

which, combined with the monotonicity of $\mathbf{r}(\cdot, \tau)$ over $[0, \mathbf{M}(\tau, 0))$, indicates the second equality in (5.204) for the second case. So we have proved (5.204). Thus, we finish the proof of the conclusion (i).

(ii) Let $r \in (0, +\infty)$ and $r_0 \triangleq \min(r, r_T)$. By (i) of this lemma and (i) of Theorem 5.7, we can find $M_0 \in [0, \mathbf{M}(\tau, 0))$ and an optimal control $u_0 \in \mathcal{U}_{M_0,\tau}$ to $(OP)^\tau_{M_0}$ so that

$$\|y(T; 0, y_0, \chi_{(\tau,T)} u_0) - z_d\|_Y = \mathbf{r}(M_0, \tau) = r_0 \leq r. \tag{5.207}$$

From (5.207) and the optimality of $\mathbf{M}(\tau, r)$, we see that

$$\mathbf{M}(\tau, r) \leq \|u_0\|_{L^\infty(\tau,T;U)} \leq M_0 < \mathbf{M}(\tau, 0) \quad \text{for all } r \in (0, +\infty).$$

Hence, the conclusion (ii) is true.

(iii) First of all, it follows by (5.182) and (5.178) that (5.200) holds for $r = r_T$.

We next show that (5.200) for an arbitrarily fixed $r \in (0, r_T)$. To this end, we let

$$\mathscr{A}_r \triangleq \{M \in [0, \mathbf{M}(\tau, 0)) \mid \mathbf{r}(M, \tau) > r\}$$

and

$$\mathscr{B}_r \triangleq \{M \in [0, \mathbf{M}(\tau, 0)) \mid \mathbf{r}(M, \tau) \leq r\}.$$

From (i) of this lemma, we see that $\mathscr{A}_r$ and $\mathscr{B}_r$ are nonempty and disjoint subsets of $[0, \mathbf{M}(\tau, 0))$, $\mathscr{A}_r \cup \mathscr{B}_r = [0, \mathbf{M}(\tau, 0))$ and

$$\sup \mathscr{A}_r = \inf \mathscr{B}_r \triangleq m. \tag{5.208}$$

We now claim that

$$m = \mathbf{M}(\tau, r). \tag{5.209}$$

Indeed, according to (ii) of Theorem 5.7 and (ii) of this lemma, $(NP)^r_\tau$ has an optimal control u_1. Then by the optimality of u_1, we have that

$$\|y(T; 0, y_0, \chi_{(\tau,T)} u_1) - z_d\|_Y \leq r \quad \text{and} \quad \|u_1\|_{L^\infty(\tau,T;U)} = \mathbf{M}(\tau, r) < \mathbf{M}(\tau, 0).$$

From these and the definition of $\mathbf{r}(\mathbf{M}(\tau, r), \tau)$, we see that $\mathbf{r}(\mathbf{M}(\tau, r), \tau) \leq r$, which indicates that

$$\mathbf{M}(\tau, r) \in \mathscr{B}_r \quad \text{and} \quad m \leq \mathbf{M}(\tau, r). \tag{5.210}$$

Meanwhile, for each $M \in (m, \mathbf{M}(\tau, 0))$, we have that $M \in \mathscr{B}_r$. Then according to (i) of Theorem 5.7, $(OP)_M^\tau$ has an optimal control u_2. By the optimality of u_2, we have that

$$\|y(T; 0, y_0, \chi_{(\tau,T)}u_2) - z_d\|_Y = \mathbf{r}(M, \tau) \leq r \quad \text{and} \quad \|u_2\|_{L^\infty(\tau,T;U)} \leq M.$$

These, together with the definition of $\mathbf{M}(\tau, r)$, imply that $\mathbf{M}(\tau, r) \leq M$, from which, it follows that

$$\mathbf{M}(\tau, r) \leq m. \tag{5.211}$$

Now, (5.209) follows from (5.210) and (5.211) immediately.

Then from (5.208), (5.209), and (i) of this lemma, we see that (5.200) holds for the case when $r \in (0, r_T)$.

To prove (5.201), we let $M \in [0, \mathbf{M}(\tau, 0))$. By (i) of this lemma, we see that $0 < \mathbf{r}(M, \tau) \leq r_T$. Thus, about $\mathbf{r}(M, \tau)$, we have that either $\mathbf{r}(M, \tau) \in (0, r_T)$ or $\mathbf{r}(M, \tau) = r_T$.

In the case where $\mathbf{r}(M, \tau) \in (0, r_T)$, it is clear that $M \in \mathscr{B}_{\mathbf{r}(M,\tau)}$, which, combined with (5.208) and (5.209), indicates that

$$\mathbf{M}(\tau, \mathbf{r}(M, \tau)) = \inf \mathscr{B}_{\mathbf{r}(M,\tau)} \leq M.$$

If $\mathbf{M}(\tau, \mathbf{r}(M, \tau)) < M$, then by the strictly decreasing property of $\mathbf{r}(\cdot, \tau)$, we have that

$$\mathbf{r}(\widehat{M}, \tau) > \mathbf{r}(M, \tau) \quad \text{for each} \ \widehat{M} \in (\mathbf{M}(\tau, \mathbf{r}(M, \tau)), M).$$

This implies that $\widehat{M} \in \mathscr{A}_{\mathbf{r}(M,\tau)}$, which, together with (5.208) and (5.209), yields that $\widehat{M} \leq \mathbf{M}(\tau, \mathbf{r}(M, \tau))$. This leads to a contradiction. Hence, (5.201) is true for the first case.

In the case that $\mathbf{r}(M, \tau) = r_T$, we can use (i) of this lemma to find that

$$M = 0. \tag{5.212}$$

Meanwhile, by (5.182) and (5.178), we obtain that $\mathbf{M}(\tau, r_T) = 0$. This, together with (5.212), yields that

$$M = \mathbf{M}(\tau, \mathbf{r}(M, \tau)).$$

Hence, (5.201) is true for the second case. Thus, we have proved (5.201).

In summary, we complete the proof of Lemma 5.12. $\qquad\qquad\square$

We end this subsection with the following theorem concerning the connections between the optimal target control problem and the minimal norm control problem.

Theorem 5.8 *The following conclusions are true:*

(i) *When $\tau \in [0, T)$ and $0 \leq M < \mathbf{M}(\tau, 0)$, each optimal control to $(OP)^\tau_M$ is also an optimal control to $(NP)^{\mathbf{r}(M,\tau)}_\tau$.*

(ii) *When $\tau \in [0, T)$ and $r \in (0, r_T]$, each optimal control to $(NP)^r_\tau$ is also an optimal control to $(OP)^\tau_{\mathbf{M}(\tau,r)}$.*

(iii) *When $\tau \in [0, T)$ and $r \in (0, r_T]$, each optimal control u^* to $(NP)^r_\tau$ satisfies that*

$$\|u^*\|_{L^\infty(a,b;U)} = \mathbf{M}(\tau, r) \ \ \text{for any interval } (a, b) \subseteq (\tau, T).$$

Proof We prove the conclusions one by one.

(i) Arbitrarily fix $\tau \in [0, T)$ and $0 \leq M < \mathbf{M}(\tau, 0)$. Then according to (i) of Theorem 5.7, $(OP)^\tau_M$ has an optimal control u_1. By the optimality of u_1, it follows that

$$\|y(T; 0, y_0, \chi_{(\tau,T)}u_1) - z_d\|_Y = \mathbf{r}(M, \tau) \ \ \text{and} \ \ \|u_1\|_{L^\infty(\tau,T;U)} \leq M < \mathbf{M}(\tau, 0).$$

From these and (5.201), we see that u_1 is an optimal control to $(NP)^{\mathbf{r}(M,\tau)}_\tau$.

(ii) Arbitrarily fix $\tau \in [0, T)$ and $r \in (0, r_T]$. Then according to (ii) of Theorem 5.7, $(NP)^r_\tau$ has an optimal control u_2. By the optimality of u_2, we have that

$$\|y(T; 0, y_0, \chi_{(\tau,T)}u_2) - z_d\|_Y \leq r \ \ \text{and} \ \ \|u_2\|_{L^\infty(\tau,T;U)} = \mathbf{M}(\tau, r).$$

These, along with (5.200), yield that u_2 is an optimal control to $(OP)^\tau_{\mathbf{M}(\tau,r)}$.

(iii) The conclusion (iii) follows from (ii) of this theorem and (iii) of Lemma 5.11. Hence, we end the proof of Theorem 5.8.   □

5.4.3 Connections Between Problems of Minimal Norm Control and Maximal Time Control

This subsection presents connections between $(NP)^r_\tau$ and $(\widetilde{TP})^r_M$. We start with the following Lemma 5.13. Since its proof is very similar to that of Proposition 5.1, we omit the details here.

Lemma 5.13 *Let $y_0 \in Y$ and $r \in (0, r_T)$. Let τ_1 and τ_2 satisfy that $0 \leq \tau_1 < \tau_2 < T$. Then the following two conclusions are true:*

(i) *For each $\tau \in [\tau_1, \tau_2]$, there exists $z^*_\tau \in Y \setminus \{0\}$ so that*

$$-\frac{1}{2}\mathbf{M}(\tau, r)^2 = \frac{1}{2}\left(\int_\tau^T \|B^*\Phi(T, t)^* z_\tau^*\|_U dt\right)^2$$
$$+\langle y_0, \Phi(T, 0)^* z_\tau^*\rangle_Y - \inf_{q\in B_r(z_d)} \langle q, z_\tau^*\rangle_Y. \tag{5.213}$$

(ii) *There exists a constant* $C(T, y_0, \tau_1, \tau_2, z_d, r) \in (0, +\infty)$ *so that* z_τ^* *(given in (5.213)) satisfies*

$$\|z_\tau^*\|_Y \leq C(T, y_0, \tau_1, \tau_2, z_d, r) \text{ for all } \tau \in [\tau_1, \tau_2].$$

The next lemma presents some properties of the function $\mathbf{M}(\cdot, r)$.

Lemma 5.14 *Let* $r \in (0, r_T)$. *Then the following conclusions are true:*

(i) *The function* $\mathbf{M}(\cdot, r)$ *is strictly increasing and homeomorphic from* $[0, T)$ *onto* $[\mathbf{M}(0, r), +\infty)$.
(ii) *The inverse of* $\mathbf{M}(\cdot, r)$ *is the function* $\tau(\cdot, r)$, *i.e.,*

$$M = \mathbf{M}(\widetilde{\tau}(M, r), r) \text{ for every } M \geq \mathbf{M}(0, r), \tag{5.214}$$

and

$$\tau = \widetilde{\tau}(\mathbf{M}(\tau, r), r) \text{ for every } \tau \in [0, T). \tag{5.215}$$

Proof (i) The proof is organized by four steps as follows:
Step 1. We show that the function $\mathbf{M}(\cdot, r)$ is strictly increasing.
Let $0 \leq \tau_1 < \tau_2 < T$. It suffices to show

$$\mathbf{M}(\tau_1, r) < \mathbf{M}(\tau_2, r). \tag{5.216}$$

Seeking for a contradiction, we suppose that

$$\mathbf{M}(\tau_1, r) \geq \mathbf{M}(\tau_2, r). \tag{5.217}$$

According to (ii) of Theorem 5.7, $(NP)_{\tau_2}^r$ has an optimal control u^*. Then by the optimality of u^* and (5.217), we have that

$$y(T; 0, y_0, \chi_{(\tau_2, T)}u^*) \in B_r(z_d) \text{ and } \|u^*\|_{L^\infty(\tau_2, T; U)} = \mathbf{M}(\tau_2, r) \leq \mathbf{M}(\tau_1, r).$$

From the above, we see that $\chi_{(\tau_2, T)}u^*$ is an optimal control to $(NP)_{\tau_1}^r$. Then it follows from (iii) of Theorem 5.8 and (ii) of Theorem 5.7 that

$$\|\chi_{(\tau_2, T)}u^*\|_{L^\infty(\tau_1, \tau_2; U)} = \mathbf{M}(\tau_1, r) \in (0, +\infty),$$

which leads to a contradiction, since $\chi_{(\tau_2, T)}u^*(t) = 0$ for a.e. $t \in (\tau_1, \tau_2)$. Hence, (5.216) is true and $\mathbf{M}(\cdot, r)$ is strictly increasing.

Step 2. We prove that the function $\mathbf{M}(\cdot, r)$ is left-continuous.

Let $0 \leq \tau_1 < \tau_2 < \cdots < \tau_k < \cdots < \tau < T$ and $\lim_{k \to +\infty} \tau_k = \tau$. It suffices to show that

$$\lim_{k \to +\infty} \mathbf{M}(\tau_k, r) = \mathbf{M}(\tau, r). \tag{5.218}$$

Suppose that (5.218) was not true. Then by Step 1, we would have that

$$\lim_{k \to +\infty} \mathbf{M}(\tau_k, r) = \mathbf{M}(\tau, r) - \delta \quad \text{for some} \ \ \delta \in (0, +\infty). \tag{5.219}$$

Meanwhile, according to (ii) of Theorem 5.7, each $(NP)^r_{\tau_k}$ (with $k \in \mathbb{N}^+$) has an optimal control u_k. Then we have that

$$\|u_k\|_{L^\infty(\tau_k, T; U)} = \mathbf{M}(\tau_k, r) < \mathbf{M}(\tau, r) \ \ \text{and} \ \ y(T; 0, y_0, \chi_{(\tau_k, T)} u_k) \in B_r(z_d). \tag{5.220}$$

Since $\lim_{k \to +\infty} \tau_k = \tau$, by the first conclusion in (5.220), there is a subsequence $\{u_{k_\ell}\}_{\ell \geq 1} \subseteq \{u_k\}_{k \geq 1}$ and a control $\widetilde{u} \in L^\infty(0, T; U)$ so that

$$\chi_{(\tau_{k_\ell}, T)} u_{k_\ell} \to \chi_{(\tau, T)} \widetilde{u} \quad \text{weakly star in} \ \ L^\infty(0, T; U). \tag{5.221}$$

This, along with the first conclusion in (5.220) and (5.219), implies that

$$\|\widetilde{u}\|_{L^\infty(\tau, T; U)} \leq \liminf_{\ell \to +\infty} \|u_{k_\ell}\|_{L^\infty(\tau_{k_\ell}, T; U)} = \liminf_{\ell \to +\infty} \mathbf{M}(\tau_{k_\ell}, r) = \mathbf{M}(\tau, r) - \delta. \tag{5.222}$$

Next, by (5.221), we can easily check that

$$y(T; 0, y_0, \chi_{(\tau_{k_\ell}, T)} u_{k_\ell}) \to y(T; 0, y_0, \chi_{(\tau, T)} \widetilde{u}) \quad \text{weakly in} \ \ Y.$$

This, together with the second conclusion in (5.220), yields that

$$y(T; 0, y_0, \chi_{(\tau, T)} \widetilde{u}) \in B_r(z_d).$$

From this and the optimality of $\mathbf{M}(\tau, r)$, we get that

$$\|\widetilde{u}\|_{L^\infty(\tau, T; U)} \geq \mathbf{M}(\tau, r),$$

which contradicts (5.222). Hence, (5.218) holds.

Step 3. We show that the function $\mathbf{M}(\cdot, r)$ is right-continuous.

Let $\tau \in [0, T)$. Take a sequence $\{\tau_k\}_{k \geq 1} \subseteq (\tau, T)$ so that $\lim_{k \to +\infty} \tau_k = \tau$. By Lemma 5.13, we find that for each $k \in \mathbb{N}^+$, there exists $z_k^* \in Y$ with $\|z_k^*\|_Y \leq C$ (where C is independent of k) so that

$$
-\frac{1}{2}\mathbf{M}(\tau_k, r)^2 = \frac{1}{2}\left(\int_{\tau_k}^{T} \|B^*\Phi(T, t)^* z_k^*\|_U \, dt\right)^2 \\
+ \langle y_0, \Phi(T, 0)^* z_k^*\rangle_Y - \inf_{q \in B_r(z_d)} \langle q, z_k^*\rangle_Y. \tag{5.223}
$$

Since $\{z_k^*\}_{k\geq 1}$ is bounded in Y, there exists a subsequence $\{z_{k_\ell}^*\}_{\ell\geq 1} \subseteq \{z_k^*\}_{k\geq 1}$ and $z \in Y$ so that

$$
z_{k_\ell}^* \to z \quad \text{weakly in } Y \text{ as } \ell \to +\infty.
$$

This, together with (5.223), yields that

$$
-\frac{1}{2}\limsup_{\ell\to+\infty}\mathbf{M}(\tau_{k_\ell}, r)^2 \geq \frac{1}{2}\left(\int_{\tau}^{T} \|B^*\Phi(T, t)^* z\|_U \, dt\right)^2 \\
+ \langle y_0, \Phi(T, 0)^* z\rangle_Y - \inf_{q \in B_r(z_d)} \langle q, z\rangle_Y. \tag{5.224}
$$

Meanwhile, by (ii) of Theorem 5.7, $(NP)_\tau^r$ has an optimal control v^*. Then by the optimality of v^*, we have that

$$
y(T; 0, y_0, \chi_{(\tau,T)}v^*) \in B_r(z_d) \quad \text{and} \quad \|v^*\|_{L^\infty(\tau,T;U)} = \mathbf{M}(\tau, r).
$$

From this and (5.224) it follows that

$$
-\frac{1}{2}\limsup_{\ell\to+\infty}\mathbf{M}(\tau_{k_\ell}, r)^2 \geq \int_{\tau}^{T} \langle v^*(t), B^*\Phi(T, t)^* z\rangle_U \, dt - \frac{1}{2}\mathbf{M}(\tau, r)^2 \\
+ \langle y_0, \Phi(T, 0)^* z\rangle_Y - \langle y(T; 0, y_0, \chi_{(\tau,T)}v^*), z\rangle_Y \\
= -\frac{1}{2}\mathbf{M}(\tau, r)^2,
$$

which indicates that

$$
\mathbf{M}(\tau, r) \geq \limsup_{\ell\to|\infty} \mathbf{M}(\tau_{k_\ell}, r).
$$

This, together with the conclusion in Step 1, yields that the function $\mathbf{M}(\cdot, r)$ is right-continuous over $[0, T)$.

Step 4. We prove that the function $\mathbf{M}(\cdot, r) : [0, T) \to [\mathbf{M}(0, r), +\infty)$ is surjective and homeomorphic.

By the results in Steps 1–3, we see that we only need to show

$$
\lim_{\tau\to T} \mathbf{M}(\tau, r) = +\infty. \tag{5.225}
$$

Seeking for a contradiction, suppose that (5.225) was not true. Then we would have that

$$\lim_{\tau \to T} \mathbf{M}(\tau, r) \leq N_0 \quad \text{for some} \quad N_0 \in (0, +\infty). \tag{5.226}$$

We take $\{\tau_k\}_{k \geq 1} \subseteq [0, T)$ with $\tau_k \to T$. According to (ii) of Theorem 5.7, each $(NP)^r_{\tau_k}$ (with $k \in \mathbb{N}^+$) has an optimal control u_k. By the optimality of u_k, we have that

$$y(T; 0, y_0, \chi_{(\tau_k, T)} u_k) \in B_r(z_d) \quad \text{and} \quad \|u_k\|_{L^\infty(\tau_k, T; U)} = \mathbf{M}(\tau_k, r). \tag{5.227}$$

Then it follows from (5.226) and (5.227) that $\{\chi_{(\tau_k, T)} u_k\}_{k \geq 1}$ is bounded in $L^\infty(0, T; U)$. Thus, there is a subsequence of $\{\chi_{(\tau_k, T)} u_k\}_{k \geq 1}$, still denoted in the same way, so that

$$\chi_{(\tau_k, T)} u_k \to 0 \quad \text{strongly in} \quad L^2(0, T; U),$$

from which, it follows that

$$y(T; 0, y_0, \chi_{(\tau_k, T)} u_k) \to y(T; 0, y_0, 0) \quad \text{strongly in} \quad Y.$$

This, together with (5.178) and the first conclusion in (5.227), yields that

$$r_T = \|y(T; 0, y_0, 0) - z_d\|_Y = \lim_{k \to +\infty} \|y(T; 0, y_0, \chi_{(\tau_k, T)} u_k) - z_d\|_Y \leq r,$$

which leads to a contradiction since $r < r_T$. Hence, (5.225) follows.

Therefore, we have proved the conclusion (i) in Lemma 5.14.

(ii) Let $M \geq \mathbf{M}(0, r)$. Set

$$\widehat{\mathscr{A}}_M \triangleq \{\tau \in [0, T) \mid \mathbf{M}(\tau, r) \leq M\} \quad \text{and} \quad \widehat{\mathscr{B}}_M \triangleq \{\tau \in [0, T) \mid \mathbf{M}(\tau, r) > M\}.$$

From (i) of this lemma, we see that $\widehat{\mathscr{A}}_M$ and $\widehat{\mathscr{B}}_M$ are nonempty and disjoint subsets of $[0, T)$, $\widehat{\mathscr{A}}_M \cup \widehat{\mathscr{B}}_M = [0, T)$ and

$$\sup \widehat{\mathscr{A}}_M = \inf \widehat{\mathscr{B}}_M \triangleq \widetilde{t}. \tag{5.228}$$

We now claim that

$$\widetilde{t} = \widetilde{\tau}(M, r). \tag{5.229}$$

Indeed, according to (iii) of Theorem 5.7, $(\widetilde{TP})^r_M$ has an optimal control u_1. Thus, we have that

$$y(T; 0, y_0, \chi_{(\widetilde{\tau}(M, r), T)} u_1) \in B_r(z_d) \quad \text{and} \quad \|u_1\|_{L^\infty(\widetilde{\tau}(M, r), T; U)} \leq M.$$

From these and the definition of $\mathbf{M}(\widetilde{\tau}(M, r), r)$, we obtain that $\mathbf{M}(\widetilde{\tau}(M, r), r) \leq M$. This, together with (5.228), yields that

$$\widetilde{\tau}(M, r) \in \widehat{\mathscr{A}}_M \quad \text{and} \quad \widetilde{\tau}(M, r) \leq \widetilde{t}. \tag{5.230}$$

About $\widetilde{t}$, we have that either $\widetilde{t} = 0$ or $\widetilde{t} > 0$.

In the case when $\widetilde{t} = 0$, we have that $\widetilde{t} \leq \widetilde{\tau}(M, r)$. This, along with the inequality in (5.230), yields (5.229) for this case.

In the case that $\widetilde{t} > 0$, one can easily check that

$$\tau \in \widehat{\mathscr{A}}_M, \quad \text{when} \quad \tau \in [0, \widetilde{t}). \tag{5.231}$$

Arbitrarily fix $\tau \in [0, \widetilde{t})$. According to (ii) of Theorem 5.7, $(NP)^r_\tau$ has an optimal control u_2. From the optimality of u_2, (5.231) and the definition of $\widehat{\mathscr{A}}_M$, we find that

$$y(T; 0, y_0, \chi_{(\tau, T)} u_2) \in B_r(z_d) \quad \text{and} \quad \|u_2\|_{L^\infty(\tau, T; U)} = \mathbf{M}(\tau, r) \leq M.$$

The above facts, combined with the definition of $\widetilde{\tau}(M, r)$, indicate that $\tau \leq \widetilde{\tau}(M, r)$, which leads to

$$\widetilde{t} \leq \widetilde{\tau}(M, r), \tag{5.232}$$

since τ was arbitrarily taken from $[0, \widetilde{t})$. Now, by (5.232) and the inequality in (5.230), we obtain (5.229) for the current case.

Thus, we have proved (5.229).

We next show that (5.214) and (5.215). It is clear that (5.214) follows from (5.228), (5.229), and (i) of this lemma. To prove (5.215), let $\tau \in [0, T)$. It is obvious that $\tau \in \widehat{\mathscr{A}}_{\mathbf{M}(\tau, r)}$, which, combined with (5.228) and (5.229), indicates that

$$\widetilde{\tau}(\mathbf{M}(\tau, r), r) = \sup \widehat{\mathscr{A}}_{\mathbf{M}(\tau, r)} \geq \tau.$$

By contradiction, suppose that $\widetilde{\tau}(\mathbf{M}(\tau, r), r) > \tau$. Then by the strictly increasing property of $\mathbf{M}(\cdot, r)$, we would have that

$$\mathbf{M}(\widehat{\tau}, r) > \mathbf{M}(\tau, r) \quad \text{for all} \quad \widehat{\tau} \in (\tau, \widetilde{\tau}(\mathbf{M}(\tau, r), r)).$$

This implies that

$$\widehat{\tau} \in \widehat{\mathscr{B}}_{\mathbf{M}(\tau, r)}.$$

From the latter, (5.228) and (5.229) it follows that

$$\widehat{\tau} \geq \widetilde{\tau}(\mathbf{M}(\tau, r), r),$$

which leads to a contradiction. Hence, (5.215) is true.

In summary, we end the proof of Lemma 5.14. $\square$

Lemma 5.15 *Let*

$$\mathscr{E}_1 \triangleq \{(M, r) \in (0, +\infty) \times (0, r_T) \mid r \geq \mathbf{r}(M, 0)\};$$

$$\mathscr{E}_2 \triangleq \{(M, r) \in (0, +\infty) \times (0, r_T) \mid M \geq \mathbf{M}(0, r)\}.$$

Then the following conclusions are true:

(i) It holds that $\mathscr{E}_1 = \mathscr{E}_2$.
(ii) When $(M, r) \in \mathscr{E}_1$,

$$M < \mathbf{M}(\widetilde{\tau}(M, r), 0). \tag{5.233}$$

Proof We prove the conclusions one by one.

(i) We first show that

$$\mathscr{E}_1 \subseteq \mathscr{E}_2. \tag{5.234}$$

To this end, we arbitrarily fix $(M, r) \in \mathscr{E}_1$. Since $r \geq \mathbf{r}(M, 0)$ and $\mathbf{M}(0, \cdot)$ is decreasing over $[0, +\infty)$, we have that

$$\mathbf{M}(0, r) \leq \mathbf{M}(0, \mathbf{r}(M, 0)). \tag{5.235}$$

Meanwhile, according to (i) of Theorem 5.7, $(OP)_M^0$ has an optimal control u^*. Then we have that

$$\|y(T; 0, y_0, u^*) - z_d\|_Y = \mathbf{r}(M, 0) \quad \text{and} \quad \|u^*\|_{L^\infty(0,T;U)} \leq M.$$

From these, we find that u^* is an admissible control to $(NP)_0^{\mathbf{r}(M,0)}$. This, along with the optimality of $\mathbf{M}(0, \mathbf{r}(M, 0))$, yields that

$$\mathbf{M}(0, \mathbf{r}(M, 0)) \leq M.$$

The above, together with (5.235), indicates that

$$M \geq \mathbf{M}(0, r),$$

which leads to (5.234).
    We next show that

$$\mathscr{E}_2 \subseteq \mathscr{E}_1. \tag{5.236}$$

For this purpose, we arbitrarily fix $(M, r) \in \mathscr{E}_2$. Since $r \in (0, r_T)$, it follows from (ii) of Theorem 5.7 and (5.200) that

$$\mathbf{M}(0, r) \in (0, +\infty) \quad \text{and} \quad r = \mathbf{r}(\mathbf{M}(0, r), 0). \tag{5.237}$$

Because $M \geq \mathbf{M}(0, r)$ and $\mathbf{r}(\cdot, \tau)$ is decreasing over $[0, +\infty)$, it follows from (5.237) that

$$M \geq \mathbf{M}(0, r) > 0 \quad \text{and} \quad r = \mathbf{r}(\mathbf{M}(0, r), 0) \geq \mathbf{r}(M, 0).$$

From these, we see that $(M, r) \in \mathscr{E}_1$. Hence, (5.236) is true. Finally, according to (5.234) and (5.236), the conclusion (i) is true.

(ii) To show (5.233), we arbitrarily fix $(M, r) \in \mathscr{E}_2$ (which implies that $r \in (0, r_T)$ and $M \geq \mathbf{M}(0, r)$). Then by (5.214) and (iii) of Theorem 5.7, we have that

$$M = \mathbf{M}(\widetilde{\tau}(M, r), r) \quad \text{and} \quad \widetilde{\tau}(M, r) \in [0, T). \tag{5.238}$$

Meanwhile, by Lemma 5.12, we see that $\mathbf{M}(\tau, \cdot)$ is strictly decreasing over $(0, r_T)$ for each $\tau \in [0, T)$. This, together with (5.238), yields that

$$M = \mathbf{M}(\widetilde{\tau}(M, r), r) < \mathbf{M}(\widetilde{\tau}(M, r), r/2),$$

which, combined with (ii) of Theorem 5.7, indicates that

$$M < \mathbf{M}(\widetilde{\tau}(M, r), r/2) \leq \mathbf{M}(\widetilde{\tau}(M, r), 0).$$

This yields (5.233).

Hence, we end the proof of Lemma 5.15. □

The connections between optimal norm and time optimal control problems are presented in the following theorem:

Theorem 5.9 *The following conclusions are true:*

(i) *When $\tau \in [0, T)$ and $r \in (0, r_T)$, each optimal control to $(NP)_\tau^r$ is an optimal control to $(\widetilde{TP})_{\mathbf{M}(\tau, r)}^r$.*

(ii) *Suppose that either*

$$(M, r) \in \{(M, r) \in (0, +\infty) \times (0, r_T) \mid r \geq \mathbf{r}(M, 0)\}$$

or

$$(M, r) \in \{(M, r) \in (0, +\infty) \times (0, r_T) \mid M \geq \mathbf{M}(0, r)\}.$$

Then each optimal control to $(\widetilde{TP})_M^r$ is an optimal control to $(NP)_{\widetilde{\tau}(M, r)}^r$.

Proof

(i) Arbitrarily fix $\tau \in [0, T)$ and $r \in (0, r_T)$. Then according to (ii) of Theorem 5.7, $(NP)^r_\tau$ has an optimal control u_1. By the optimality of u_1, we have that

$$y(T; 0, y_0, \chi_{(\tau,T)}u_1) \in B_r(z_d) \quad \text{and} \quad \|u_1\|_{L^\infty(\tau,T;U)} = \mathbf{M}(\tau, r).$$

From these and (5.215), we see that u_1 is an optimal control to $(\widetilde{TP})^r_{\mathbf{M}(\tau,r)}$.

(ii) Arbitrarily fix (M, r) as required. Then by Lemma 5.15, we have that $M \geq \mathbf{M}(0, r)$. Thus, according to (iii) of Theorem 5.7, $(\widetilde{TP})^r_M$ has an optimal control u_2. By the optimality of u_2, we have that

$$y(T; 0, y_0, \chi_{(\widetilde{\tau}(M,r),T)}u_2) \in B_r(z_d) \quad \text{and} \quad \|u_2\|_{L^\infty(\widetilde{\tau}(M,r),T;U)} \leq M.$$

From these and (5.214), we see that u_2 is an optimal control to $(NP)^r_{\widetilde{\tau}(M,r)}$.
Thus we complete the proof of Theorem 5.9. $\square$

5.4.4   Proof of the Main Result

This subsection presents the proof of Theorem 5.6.

Proof (Proof of Theorem 5.6) Throughout the proof, by $(P_1) = (P_2)$, we mean that problems (P_1) and (P_2) are equivalent in the sense of Definition 5.4; by $(P_1) \Rightarrow (P_2)$, we mean that each optimal control to (P_1) is also an optimal control to (P_2).

We organize the proof by three steps as follows:

Step 1. We show that

$$(OP)^\tau_M = (TP)^{\mathbf{r}(M,\tau)}_M = (NP)^{\mathbf{r}(M,\tau)}_\tau, \tag{5.239}$$

when $\tau \in [0, T)$ and $0 < M < \mathbf{M}(\tau, 0)$.

To this end, we arbitrarily fix (τ, M) as required. On one hand, it follows by (5.181) that $\mathbf{r}(M, \tau) \geq \mathbf{r}(M, 0)$. This, together with (i) of Lemma 5.11, Lemma 5.12, and Lemma 5.14, implies that

$$\tau \in [0, T), \ M \in (0, \mathbf{M}(\tau, 0)), \ \mathbf{r}(M, \tau) \in [\mathbf{r}(M, 0), r_T) \cap (0, r_T) \tag{5.240}$$

and

$$M = \mathbf{M}(\tau, \mathbf{r}(M, \tau)), \ \tau = \widetilde{\tau}(\mathbf{M}(\tau, \mathbf{r}(M, \tau)), \mathbf{r}(M, \tau)) = \widetilde{\tau}(M, \mathbf{r}(M, \tau)). \tag{5.241}$$

Using Theorem 5.8 and (5.240), we obtain that

$$(OP)^\tau_M \Rightarrow (NP)^{\mathbf{r}(M,\tau)}_\tau \quad \text{and} \quad (NP)^{\mathbf{r}(M,\tau)}_\tau \Rightarrow (OP)^\tau_{\mathbf{M}(\tau,\mathbf{r}(M,\tau))}.$$

From the above, the first conclusion in (5.241) and Definition 5.4, we see that

$$(OP)_M^\tau = (NP)_\tau^{\mathbf{r}(M,\tau)}. \tag{5.242}$$

On the other hand, by Theorem 5.9 and (5.240), we obtain that

$$(NP)_\tau^{\mathbf{r}(M,\tau)} \Rightarrow (\widetilde{TP})_{\mathbf{M}(\tau,\mathbf{r}(M,\tau))}^{\mathbf{r}(M,\tau)}$$

and

$$(\widetilde{TP})_{\mathbf{M}(\tau,\mathbf{r}(M,\tau))}^{\mathbf{r}(M,\tau)} \Rightarrow (NP)_{\widetilde{\tau}(\mathbf{M}(\tau,\mathbf{r}(M,\tau)),\mathbf{r}(M,\tau))}^{\mathbf{r}(M,\tau)}.$$

These, together with (5.241) and Definition 5.4, yield that

$$(NP)_\tau^{\mathbf{r}(M,\tau)} = (TP)_M^{\mathbf{r}(M,\tau)}. \tag{5.243}$$

Now, (5.239) follows from (5.242), (5.243), and Definition 5.4 at once.

Step 2. We prove that

$$(NP)_\tau^r = (OP)_{\mathbf{M}(\tau,r)}^\tau = (TP)_{\mathbf{M}(\tau,r)}^r, \tag{5.244}$$

when $\tau \in [0, T)$ and $r \in (0, r_T)$.

For this purpose, we arbitrarily fix (τ, r) as required. Then by (ii) of Theorem 5.7, Lemma 5.12, and Lemma 5.14, one can easily check that

$$\tau \in [0, T), \ r \in (0, r_T), \ \mathbf{M}(\tau, r) \in [\mathbf{M}(0, r), \mathbf{M}(\tau, 0)) \cap (0, \mathbf{M}(\tau, 0)) \tag{5.245}$$

and

$$r = \mathbf{r}(\mathbf{M}(\tau, r), \tau), \ \tau = \widetilde{\tau}(\mathbf{M}(\tau, r), r). \tag{5.246}$$

From Theorem 5.8 and (5.245), we have that

$$(NP)_\tau^r \Rightarrow (OP)_{\mathbf{M}(\tau,r)}^\tau \ \text{ and } \ (OP)_{\mathbf{M}(\tau,r)}^\tau \Rightarrow (NP)_\tau^{\mathbf{r}(\mathbf{M}(\tau,r),\tau)}.$$

These, along with (5.246) and Definition 5.4, yield that

$$(NP)_\tau^r = (OP)_{\mathbf{M}(\tau,r)}^\tau. \tag{5.247}$$

On the other hand, it follows by Theorem 5.9 and (5.245) that

$$(NP)_\tau^r \Rightarrow (\widetilde{TP})_{\mathbf{M}(\tau,r)}^r \ \text{ and } \ (\widetilde{TP})_{\mathbf{M}(\tau,r)}^r \Rightarrow (NP)_{\widetilde{\tau}(\mathbf{M}(\tau,r),r)}^r.$$

These, along with (5.246) and Definition 5.4, yield that

$$(NP)_\tau^r = (TP)_{\mathbf{M}(\tau,r)}^r. \tag{5.248}$$

Now, (5.244) follows from (5.247), (5.248), and Definition 5.4 at once.

Step 3. We show that

$$(TP)_M^r = (NP)_{\widetilde{\tau}(M,r)}^r = (OP)_M^{\widetilde{\tau}(M,r)}, \tag{5.249}$$

where $M \in (0, +\infty)$ and $r \in [\mathbf{r}(M, 0), r_T) \cap (0, r_T)$.

Arbitrarily fix (M, r) as required. Then by Lemma 5.15, (iii) of Theorem 5.7, Lemma 5.12, and Lemma 5.14, one can easily check that

$$\widetilde{\tau}(M, r) \in [0, T), \ r \in [\mathbf{r}(M, 0), r_T) \cap (0, r_T), \ 0 < M < \mathbf{M}(\widetilde{\tau}(M, r), 0) \tag{5.250}$$

and

$$M = \mathbf{M}(\widetilde{\tau}(M, r), r), \ r = \mathbf{r}(\mathbf{M}(\widetilde{\tau}(M, r), r), \widetilde{\tau}(M, r)) = \mathbf{r}(M, \widetilde{\tau}(M, r)). \tag{5.251}$$

By Theorem 5.9 and (5.250), we obtain that

$$(\widetilde{TP})_M^r \Rightarrow (NP)_{\widetilde{\tau}(M,r)}^r \ \text{ and } \ (NP)_{\widetilde{\tau}(M,r)}^r \Rightarrow (\widetilde{TP})_{\mathbf{M}(\widetilde{\tau}(M,r),r)}^r.$$

These, along with (5.251) and Definition 5.4, yield that

$$(TP)_M^r = (NP)_{\widetilde{\tau}(M,r)}^r. \tag{5.252}$$

On the other hand, by Theorem 5.8, (5.250), and (5.251), one can easily verify that

$$(NP)_{\widetilde{\tau}(M,r)}^r \Rightarrow (OP)_{\mathbf{M}(\widetilde{\tau}(M,r),r)}^{\widetilde{\tau}(M,r)}$$

and

$$(OP)_{\mathbf{M}(\widetilde{\tau}(M,r),r)}^{\widetilde{\tau}(M,r)} \Rightarrow (NP)_{\widetilde{\tau}(M,r)}^{\mathbf{r}(\mathbf{M}(\widetilde{\tau}(M,r),r),\widetilde{\tau}(M,r))}.$$

From these, (5.251) and Definition 5.4, we obtain that

$$(NP)_{\widetilde{\tau}(M,r)}^r = (OP)_M^{\widetilde{\tau}(M,r)}. \tag{5.253}$$

Now (5.249) follows from (5.252), (5.253), and Definition 5.4 at once.

Thus, we end the proof of Theorem 5.6.   □

Miscellaneous Notes

To our best knowledge, the terminology "equivalence of minimal time and minimal norm control problems" was first formally presented in [17]. (The definition of the equivalence was hidden in Theorem 1.1 of [17].) The equivalence theorems of different optimal control problems play important roles in the studies of these problems. With regard to their applications, we would like to mention the following facts:

- With the aid of the equivalence theorem in Section 5.1, one can derive the norm optimality from the time optimality (see [5]); one can build up a necessary and sufficient condition for both the optimal time and the optimal control of a minimal time control problem, through analyzing characteristics for the corresponding minimal norm control problem (see [17]); one can present a numerical algorithm to compute the optimal time of a minimal time control problem (see [9]); one can study a certain approximation property for a minimal time control problem (see [20]); one can obtain some regularity of the Bellman function (see [6]), which is heavily related to the controllability and has been studied in, for instance, [2] and [10].
- By the equivalence theorem in Section 5.2, one can get algorithms for both the optimal time and the optimal controls of some minimal time control problems (see [11]). The same can be said about the equivalence theorem in Section 5.3. By the equivalence theorem in Section 5.4, one can get algorithms for both the optimal time and the optimal controls of some maximal time control problems (see [15]).
- With the aid of the second equivalence theorem, one can design a feedback optimal norm control (see [15]).

Materials in Section 5.1 are generalized from the related materials in [17]. Materials in Section 5.2 are developed from the materials in [11]. Materials in Section 5.3 are developed from the materials in [18]. Materials in Section 5.4 are summarized from the materials in [15].

More notes on the equivalence theorems are given in order:

- The method to derive the equivalence theorem in Section 5.1 does not work for the equivalence theorem in Section 5.2, even when the system in Section 5.2 is time invariant. The main reason is that the minimal norm function in Section 5.2 is no longer decreasing.
- The method to derive the equivalence theorem in Section 5.1 does not work for the equivalence theorem in Section 5.3. The main reason is that the system in Section 5.3 is time-varying.
- The difference between problems in Section 5.2 and Section 5.3 is that they have different targets.
- Though it is assumed that $B \in \mathscr{L}(U, Y)$, many results of this chapter can be extended to the case when $B \in \mathscr{L}(U, Y_{-1})$ is admissible for the semigroup generated by A. (Here, $Y_{-1} \triangleq D(A^*)'$ is the dual space of $D(A^*)$.) The admissibility has been defined in [14]. For such extensions, we refer the readers to [16].
- We would like to mention several other related works. (a) We recommend H. O. Fattorini's book [5], where connections between minimal time and minimal norm optimal controls for abstract equations in Banach spaces were introduced.

(b) Some connections between minimal time and minimal norm controls were also addressed in [6] for abstract equations in Hilbert spaces. (c) Earlier, the relationship between minimal time and minimal norm optimal controls was studied for the heat equation with scalar controls, i.e., controls depending only on the time variable (see [8]). (d) When the target is the ball $B(0, \varepsilon)$ in $L^2(\Omega)$, the equivalence of minimal time and minimal norm control problems for some semi-linear heat equations was built up in [20]. (e) The equivalence between minimal time and minimal norm optimal controls was obtained for some ODEs, with rectangular control constraint, in [21]. This study improved the work [7] from the perspective of computation of optimal time for some ODEs with cube control constraint. (e) The equivalence theorems were derived for Schrödinger equations in [22].

We end this note with a few open problems which might interest readers:

- Can we develop some numerical algorithms for time optimal controls and the optimal time based on the equivalence built up in this chapter?
- Can we find more applications for the equivalence obtained in Section 5.2?
- How to extend equivalence in Section 5.3 to general time-varying systems?

References

1. O. Cârjă, On the minimal time function for distributed control systems in Banach spaces. J. Optim. Theory Appl. **44**, 397–406 (1984)
2. O. Cârjă, On continuity of the minimal time function for distributed control system. Boll. Un. Mat. Ital. A **4**, 293–302 (1985)
3. O. Cârjă, The minimal time function in infinite dimensions. SIAM J. Control Optim. **31**, 1103–1114 (1993)
4. H.O. Fattorini, Time-optimal control of solution of operational differential equations. J. SIAM Control **2**, 54–59 (1964)
5. H.O. Fattorini, *Infinite Dimensional Linear Control Systems, the Time Optimal and Norm Optimal Problems*. North-Holland Mathematics Studies, vol. 201 (Elsevier Science B.V., Amsterdam, 2005)
6. F. Gozzi, P. Loreti, Regularity of the minimum time function and minimum energy problems: the linear case. SIAM J. Control Optim. **37**, 1195–1221 (1999)
7. K. Ito, K. Kunisch, Semismooth newton methods for time-optimal control for a class of ODEs. SIAM J. Control Optim. **48**, 3997–4013 (2010)
8. W. Krabs, Optimal control of processes governed by partial differential equations, Part I: heating Processes. Z. Oper. Res. A-B **26**, 21–48 (1982)
9. X. Lü, L. Wang, Q. Yan, Computation of time optimal control problems governed by linear ordinary differential equations. J. Sci. Comput. **73**, 1–25 (2017)
10. N.N. Petrov, The Bellman problem for a time-optimality problem (Russian), Prikl. Mat. Meh. **34**, 820–826 (1970)
11. S. Qin, G. Wang, Equivalence between minimal time and minimal norm control problems for the heat equation. SIAM J. Control Optim. **56**, 981–1010 (2018)
12. E.M. Stein, R. Shakarchi, *Real Analysis, Measure Theory, Integration, and Hilbert Spaces*. Princeton Lectures in Analysis, vol. 3 (Princeton University Press, Princeton, NJ, 2005)

13. M. Tucsnak, G. Wang, C. Wu, Perturbations of time optimal control problems for a class of abstract parabolic systems. SIAM J. Control Optim. **54**, 2965–2991 (2016)
14. M. Tucsnak, G. Weiss, *Observation and Control for Operator Semigroups* (Birkhäuser Verlag, Basel, 2009)
15. G. Wang, Y. Xu, Equivalence of three different kinds of optimal control problems for heat equations and its applications. SIAM J. Control Optim. **51**, 848–880 (2013)
16. G. Wang, Y. Zhang, Decompositions and bang-bang problems. Math. Control Relat. Fields **7**, 73–170 (2017)
17. G. Wang, E. Zuazua, On the equivalence of minimal time and minimal norm controls for the internally controlled heat equations. SIAM J. Control Optim. **50**, 2938–2958 (2012)
18. G. Wang, Y. Xu, Y. Zhang, Attainable subspaces and the bang-bang property of time optimal controls for heat equations. SIAM J. Control Optim. **53**, 592–621 (2015)
19. H. Yu, Approximation of time optimal controls for heat equations with perturbations in the system potential. SIAM J. Control Optim. **52** 1663–1692 (2014)
20. H. Yu, Equivalence of minimal time and minimal norm control problems for semilinear heat equations. Syst. Control Lett. **73**, 17–24 (2014)
21. C. Zhang, The time optimal control with constraints of the rectangular type for linear time-varying ODEs. SIAM J. Control Optim. **51**, 1528–1542 (2013)
22. Y. Zhang, Two equivalence theorems of different kinds of optimal control problems for Schrödinger equations. SIAM J. Control Optim. **53**, 926–947 (2015)

Chapter 6
Bang-Bang Property of Optimal Controls

In this chapter, we will study the bang-bang property of some time optimal control problems. The bang-bang property of such a problem says, in plain language, that any optimal control reaches the boundary of the corresponding control constraint set at almost every time. This property not only is mathematically interesting, but also has important applications. For instance, the uniqueness of time optimal control is an immediate consequence of the bang-bang property in certain cases.

6.1 Bang-Bang Property in ODE Cases

In this section, we will study the bang-bang properties of problems $(TP)_{min}^{Q_S,Q_E}$ and $(TP)_{max}^{Q_S,Q_E}$ under the following framework $(\mathscr{B}_1)$:

(i) The state space Y and the control space U are $\mathbb{R}^n$ and $\mathbb{R}^m$ with $n, m \in \mathbb{N}^+$, respectively.

(ii) The controlled system is as:

$$\dot{y}(t) = Ay(t) + Bu(t), \quad t \in (0, +\infty), \tag{6.1}$$

where $A \in \mathbb{R}^{n \times n}$ and $B \in \mathbb{R}^{n \times m}$. Given $\tau \in [0, +\infty)$, $y_0 \in \mathbb{R}^n$ and $u \in L^\infty(\tau, +\infty; \mathbb{R}^m)$, we write $y(\cdot; \tau, y_0, u)$ for the solution of (6.1) over $[\tau, +\infty)$, with the initial condition that $y(\tau) = y_0$.

(iii) The control constraint set $\mathbb{U} = B_\rho(0)$ with $\rho > 0$ or $\mathbb{U} = \prod_{j=1}^m [-a_j, a_j]$ with $\{a_j\}_{j=1}^m \subseteq (0, +\infty)$. (Roughly speaking, $\mathbb{U}$ is either of the ball type or of the rectangular type.)

(iv) Sets Q_S and Q_E are disjoint nonempty subsets of $[0, +\infty) \times \mathbb{R}^n$ and $(0, +\infty) \times \mathbb{R}^n$, respectively.

© Springer International Publishing AG, part of Springer Nature 2018
G. Wang et al., *Time Optimal Control of Evolution Equations*, Progress in Nonlinear Differential Equations and Their Applications 92, https://doi.org/10.1007/978-3-319-95363-2_6

As what we did in Chapter 5, we always assume that $(TP)_{min}^{Q_S,Q_E}$ and $(TP)_{max}^{Q_S,Q_E}$ have optimal controls.

We now give the definition of the bang-bang property for problems $(TP)_{min}^{Q_S,Q_E}$ and $(TP)_{max}^{Q_S,Q_E}$ as follows.

Definition 6.1 Several definitions are given in order.

(i) An optimal control u^* (associated with an optimal tetrad $(t_S^*, y_0^*, t_E^*, u^*)$) to the problem $(TP)_{min}^{Q_S,Q_E}$ (or $(TP)_{max}^{Q_S,Q_E}$) is said to have the bang-bang property if

$$u^*(t) \in ex\mathbb{U} \quad \text{for a.e. } t \in (t_S^*, t_E^*), \tag{6.2}$$

where $ex\mathbb{U}$ denotes the extreme set of $\mathbb{U}$. (In this case, we also say that u^* is a bang-bang control.)

(ii) The problem $(TP)_{min}^{Q_S,Q_E}$ (or $(TP)_{max}^{Q_S,Q_E}$) is said to have the bang-bang property, if any optimal control satisfies the bang-bang property.

Remark 6.1 In the case when $\mathbb{U} = B_\rho(0)$ (i.e., the ball-type control constraint), the condition (6.2) is as:

$$\|u^*(t)\|_{\mathbb{R}^m} = \rho \quad \text{for a.e. } t \in (t_S^*, t_E^*).$$

In the case when $\mathbb{U} = \prod_{j=1}^{m}[-a_j, a_j]$ (i.e., the rectangle-type control constraint), the condition (6.2) is as:

$$|u_j^*(t)| = a_j \quad \text{for each } j \in \{1, \dots, m\} \text{ and for a.e. } t \in (t_S^*, t_E^*).$$

We next present a minimal time control problem, which does not hold the bang-bang property but has a bang-bang optimal control.

Example 6.1 Let

$$A = \begin{pmatrix} 0 & 0 \\ 0 & 1 \end{pmatrix}, \quad B = \begin{pmatrix} 1 \\ 0 \end{pmatrix}, \quad y_0 = \begin{pmatrix} 0 \\ 1 \end{pmatrix} \quad \text{and} \quad y_1 = \begin{pmatrix} 0 \\ e \end{pmatrix}.$$

Consider the problem $(TP)_{min}^{Q_S,Q_E}$, where the controlled equation is as:

$$\dot{y}(t) = Ay(t) + Bu(t), \quad t \in (0, +\infty) \quad \text{(with } y(t) \in \mathbb{R}^2 \text{ and } u(t) \in \mathbb{R})$$

and where

$$Q_S = \{0\} \times \{y_0\}, \quad Q_E = (0, +\infty) \times \{y_1\}, \quad \mathbb{U} = [-1, 1].$$

One can easily check the following facts:

(i) The problem $(TP)_{min}^{Q_S,Q_E}$ can be put into the framework $(\mathscr{B}_1)$ (given at the beginning of Section 6.1);

(ii) The tetrad $(0, y_0, 1, 0)$ is optimal for $(TP)_{min}^{Q_S,Q_E}$ and the corresponding optimal control does not have the bang-bang property. Consequently, $(TP)_{min}^{Q_S,Q_E}$ does not have the bang-bang property;

(iii) The tetrad $(0, y_0, 1, u^*)$ is optimal for $(TP)_{min}^{Q_S,Q_E}$ and u^* holds the bang-bang property, where

$$u^*(t) \triangleq \begin{cases} 1, & t \in (0, 1/2), \\ -1, & t \in [1/2, 1). \end{cases}$$

We end Example 6.1.

6.1.1 Bang-Bang Property for Minimal Time Controls

This subsection aims to give some conditions to ensure the bang-bang property for $(TP)_{min}^{Q_S,Q_E}$. We first study the bang-bang property in the case that $\mathbb{U} = B_\rho(0)$. To this end, we introduce the following Kalman controllability rank condition:

$$\text{rank } (B, AB, \ldots, A^{n-1}B) = n. \tag{6.3}$$

The first main theorem (of this subsection) concerns the bang-bang property of minimal time control problems where controls have the ball-type constraint.

Theorem 6.1 *Let $\mathbb{U} = B_\rho(0)$ with $\rho > 0$. Assume that (6.3) holds. Let*

$$Q_S = \{0\} \times Y_S \quad \text{and} \quad Q_E = (0, +\infty) \times Y_E,$$

where Y_S and Y_E are disjoint nonempty subsets of $\mathbb{R}^n$. Then the problem $(TP)_{min}^{Q_S,Q_E}$ has the bang-bang property.

Proof We organize the proof by two steps as follows:

Step 1. We prove that $(TP)_{min}^{Q_S,Q_E}$ has the bang-bang property when Y_S and Y_E are different singleton subsets of $\mathbb{R}^n$.

Arbitrarily fix $y_0, y_1 \in \mathbb{R}^n$ with $y_0 \neq y_1$. Let

$$Y_S = \{y_0\} \quad \text{and} \quad Y_E = \{y_1\}.$$

Let $(0, y_0, t_E^*, u^*)$ be an optimal tetrad of $(TP)_{min}^{Q_S,Q_E}$. It is clear that $t_E^* > 0$. Recall that (see (3.137) in Chapter 3)

$$Y_R(t_E^*) \triangleq \{y(t_E^*; 0, y_0, u) \mid u \in L(0, t_E^*; \mathbb{U})\}.$$

Then by Theorem 4.2, we find that $Y_R(t_E^*)$ and $\{y_1\}$ are separable in $\mathbb{R}^n$. This, along with Theorem 4.1, yields that $(TP)_{min}^{Q_S,Q_E}$ satisfies the classical Pontryagin Maximum Principle. Then according to Definition 4.1, there exists $z^* \in \mathbb{R}^n \setminus \{0\}$ so that

$$\langle u^*(t), B^* e^{A^*(t_E^*-t)} z^* \rangle_{\mathbb{R}^m} = \max_{\|v\|_{\mathbb{R}^m} \leq \rho} \langle v, B^* e^{A^*(t_E^*-t)} z^* \rangle_{\mathbb{R}^m} \quad \text{for a.e. } t \in (0, t_E^*).$$

$$(6.4)$$

Meanwhile, it follows from (6.3) and the analyticity of the function $t \to B^* e^{A^*t} z^*$ that

$$B^* e^{A^*(t_E^*-t)} z^* \neq 0 \text{ in } \mathbb{R}^m \text{ for a.e. } t \in (0, t_E^*).$$

This, together with (6.4), yields that

$$\|u^*(t)\|_{\mathbb{R}^m} = \rho \text{ a.e. } t \in (0, t_E^*).$$

Then by Definition 6.1, $(0, y_0, t_E^*, u^*)$ satisfies the bang-bang property, consequently, so does $(TP)_{min}^{Q_S,Q_E}$.

Step 2. We prove that $(TP)_{min}^{Q_S,Q_E}$ has the bang-bang property when Y_S and Y_E are disjoint nonempty subsets of $\mathbb{R}^n$.

Let $(0, y_0^*, t_E^*, u^*)$ be an optimal tetrad to $(TP)_{min}^{Q_S,Q_E}$. We can directly check that $(0, y_0^*, t_E^*, u^*)$ is also an optimal tetrad to the problem $(TP)_{min}^{\widehat{Q}_S,\widehat{Q}_E}$, where

$$\widehat{Q}_S = \{0\} \times \{y_0^*\} \text{ and } \widehat{Q}_E = (0, +\infty) \times \{y(t_E^*; 0, y_0^*, u^*)\}.$$

Since Y_S and Y_E are disjoint, we have that $\{y_0^*\} \cap \{y(t_E^*; 0, y_0^*, u^*)\} = \emptyset$. Thus, we can use the conclusion of Step 1, as well as Definition 6.1, to find that u^* has the bang-bang property, consequently, so does $(TP)_{min}^{Q_S,Q_E}$.

Hence, we end the proof of Theorem 6.1.   □

The next two examples give auxiliary instructions on Theorem 6.1.

Example 6.2 Consider the problem $(TP)_{min}^{Q_S,Q_E}$, where the controlled equation is as:

$$\dot{y}(t) = -y(t) + u(t), \quad t \in (0, +\infty) \text{ (with } y(t) \in \mathbb{R} \text{ and } u(t) \in \mathbb{R}),$$

and where

$$Q_S = \{0\} \times \{2\}, \quad Q_E = (0, +\infty) \times [-1, 1], \quad \mathbb{U} = [-1, 1].$$

One can easily check the following facts:

(i) The problem $(TP)_{min}^{Q_S,Q_E}$ can be put into the framework $(\mathcal{B}_1)$ (given at the beginning of Section 6.1);

(ii) The tetrad $(0, 2, ln2, 0)$ is admissible for $(TP)_{min}^{Q_S,Q_E}$;

(iii) The problem $(TP)_{min}^{Q_S, Q_E}$ has an optimal control;
(iv) The assumptions in Theorem 6.1 hold.

Then by Theorem 6.1, $(TP)_{min}^{Q_S, Q_E}$ has the bang-bang property. We now end Example 6.2.

Example 6.3 Let

$$A = \begin{pmatrix} 1 & 0 \\ 0 & 1 \end{pmatrix}, \ B = \begin{pmatrix} 1 \\ 0 \end{pmatrix}, \ y_0 = \begin{pmatrix} 0 \\ 1 \end{pmatrix} \ \text{and} \ y_1 = \begin{pmatrix} 0 \\ e \end{pmatrix}.$$

Consider the problem $(TP)_{min}^{Q_S, Q_E}$, where the controlled equation is as:

$$\dot{y}(t) = Ay(t) + Bu(t), \ t \in (0, +\infty) \ \text{(with} \ y(t) \in \mathbb{R}^2 \ \text{and} \ u(t) \in \mathbb{R}),$$

and where

$$Q_S = \{0\} \times \{y_0\}, \quad Q_E = (0, +\infty) \times \{y_1\}, \quad U = [-1, 1].$$

One can easily check the following facts:

(i) The problem $(TP)_{min}^{Q_S, Q_E}$ can be put into the framework $(\mathscr{B}_1)$ (given at the beginning of Section 6.1);
(ii) It holds that rank $(B, AB) = 1$ (i.e., (6.3) does not hold);
(iii) The tetrad $(0, y_0, 1, 0)$ is optimal for $(TP)_{min}^{Q_S, Q_E}$ and the corresponding optimal control does not satisfy the bang-bang property.

We now end Example 6.3.

The second main theorem (of this subsection) concerns the bang-bang property for minimal time control problems with the rectangle-type control constraint.

Theorem 6.2 *Let* $m \geq 2$ *and* $B \triangleq (b_1, \ldots, b_m)$ *with* $\{b_j\}_{j=1}^m \subseteq \mathbb{R}^n$. *Let* $U = \prod_{j=1}^m [-a_j, a_j]$ *with* $\{a_j\}_{j=1}^m \subseteq (0, +\infty)$. *The following two conclusions are true:*

(i) If

$$rank(b_j, Ab_j, \ldots, A^{n-1}b_j) = n \ \text{for each} \ j \in \{1, \ldots, m\}, \tag{6.5}$$

and if

$$Q_S = \{0\} \times Y_S \ \text{and} \ Q_E = (0, +\infty) \times Y_E,$$

where Y_S *and* Y_E *are disjoint nonempty subsets of* $\mathbb{R}^n$, *then* $(TP)_{min}^{Q_S, Q_E}$ *holds the bang-bang property.*

(ii) If

$$rank(b_1, Ab_1, \ldots, A^{n-1}b_1) < rank(B, AB, \ldots, A^{n-1}B), \qquad (6.6)$$

then there exists $\widehat{y}_0 \in \mathbb{R}^n \setminus \{0\}$ so that the problem $(TP)_{min}^{Q_S, Q_E}$, with

$$Q_S = \{0\} \times \{\widehat{y}_0\} \quad and \quad Q_E = (0, +\infty) \times \{0\},$$

has at least one optimal control but does not hold the bang-bang property.

It deserves to mention what follows: The condition (6.5) means that for each $j \in \{1, \ldots, m\}$, (A, b_j) satisfies the Kalman controllability rank condition (6.3) (where B is replaced by b_j).

To prove the conclusion (ii) of Theorem 6.2, we need two lemmas. The first lemma is concerned with some kind of norm optimal control problem. To state it, we define

$$\|u\|_{\mathbb{R}_{\mathbb{U}}^m} \triangleq \max_{1 \leq j \leq m} \{|u_j|/a_j\} \quad \text{for each } u = (u_1, \ldots, u_m)^\top \in \mathbb{R}^m, \qquad (6.7)$$

where $\mathbb{U} = \prod_{j=1}^m [-a_j, a_j]$ with $\{a_j\}_{j=1}^m \subseteq (0, +\infty)$. It is obvious that $(\mathbb{R}^m, \|\cdot\|_{\mathbb{R}_{\mathbb{U}}^m})$ is a Banach space. Then for each $y_0 \in \mathbb{R}^n$ and $T > 0$, we introduce the following norm optimal control problem:

$$(NP)_{\mathbb{U}}^{T, y_0} \qquad N_{\mathbb{U}}(T, y_0) \triangleq \inf \{\|u\|_{L^\infty(0, T; \mathbb{R}_{\mathbb{U}}^m)} \mid y(T; 0, y_0, u) = 0\}. \qquad (6.8)$$

Several notes on $(NP)_{\mathbb{U}}^{T, y_0}$ are given in order:

- When the right-hand side of (6.8) is empty, we agree that $N_{\mathbb{U}}(T, y_0) = +\infty$;
- A control $v \in L^\infty(0, T; \mathbb{R}_{\mathbb{U}}^m)$ is called admissible, if $y(T; 0, y_0, u) = 0$;
- A control $v^* \in L^\infty(0, T; \mathbb{R}_{\mathbb{U}}^m)$ is called optimal, if it is admissible and satisfies that $\|v^*\|_{L^\infty(0, T; \mathbb{R}_{\mathbb{U}}^m)} = N_{\mathbb{U}}(T, y_0)$.

The first lemma needed is as follows:

Lemma 6.1 *Let $m \geq 2$ and $B = (b_1, \ldots, b_m)$ with $\{b_j\}_{j=1}^m \subseteq \mathbb{R}^n$. Let $\mathbb{U} = \prod_{j=1}^m [-a_j, a_j]$ with $\{a_j\}_{j=1}^m \subseteq (0, +\infty)$. Assume that*

$$rank(b_1, Ab_1, \ldots, A^{n-1}b_1) < rank(B, AB, \ldots, A^{n-1}B). \qquad (6.9)$$

Then for each $T > 0$, there exists $\widehat{y}_0 \in \mathbb{R}^n \setminus \{0\}$ so that the problem $(NP)_{\mathbb{U}}^{T, \widehat{y}_0}$ has an optimal control $u^*(\cdot) = (u_1^*(\cdot), \ldots, u_m^*(\cdot))^\top$, with $u_1^*(\cdot) = 0$ over $(0, T)$, so that

$$N_{\mathbb{U}}(T, \widehat{y}_0) = 1.$$

Proof Arbitrarily fix $T > 0$. Define two subsets in $\mathbb{R}^n$ as follows:

$$\mathscr{R}^{\mathbb{U}, T} \triangleq \{y(T; 0, 0, u) \mid \|u\|_{L^\infty(0, T; \mathbb{R}_{\mathbb{U}}^m)} \leq 1\} \qquad (6.10)$$

and

$$\mathcal{R}_0^{U,T} \triangleq \left\{ y(T; 0, 0, u) \mid \|u\|_{L^\infty(0,T;\mathbb{R}_U^m)} \leq 1, u = (u_1, \ldots, u_m)^\top, u_1 = 0 \right\}. \quad (6.11)$$

From (6.10) and (6.11), one can easily check the following facts: Fact One: $\mathcal{R}_0^{U,T} \subseteq \mathcal{R}^{U,T}$; Fact Two: $\mathcal{R}_0^{U,T}$ and $\mathcal{R}^{U,T}$ are nonempty, bounded, and closed subsets of $\mathbb{R}^n$; Fact Three: For each $z \in \mathbb{R}^n$,

$$\max_{y \in \mathcal{R}^{U,T}} \langle y, z \rangle_{\mathbb{R}^n} = \sum_{j=1}^m \int_0^T a_j |b_j^\top e^{A^*(T-t)} z| dt \quad (6.12)$$

and

$$\max_{y \in \mathcal{R}_0^{U,T}} \langle y, z \rangle_{\mathbb{R}^n} = \sum_{j=2}^m \int_0^T a_j |b_j^\top e^{A^*(T-t)} z| dt. \quad (6.13)$$

We now claim that there exist $\widehat{z} \in \mathbb{R}^n$ and $\widehat{y}_1 \in \mathcal{R}_0^{U,T}$ so that

$$\max_{y \in \mathcal{R}^{U,T}} \langle y, \widehat{z} \rangle_{\mathbb{R}^n} = \max_{y \in \mathcal{R}_0^{U,T}} \langle y, \widehat{z} \rangle_{\mathbb{R}^n} = \langle \widehat{y}_1, \widehat{z} \rangle_{\mathbb{R}^n} > 0. \quad (6.14)$$

In fact, by (6.9), we can choose $\widehat{z} \in \text{Range}\,(B, AB, \ldots, A^{n-1}B) \setminus \{0\}$ so that

$$\langle \widehat{z}, A^k b_1 \rangle_{\mathbb{R}^n} = 0 \quad \text{for each} \quad k \in \{0, \ldots, n-1\}.$$

From this, we can apply the Hamilton-Caley Theorem to verify that

$$b_1^\top e^{A^* t} \widehat{z} = 0 \quad \text{for each} \quad t \in \mathbb{R}.$$

This, together with (6.12) and (6.13), leads to the first equality in (6.14). The second equality in (6.14) follows from the fact that $\mathcal{R}_0^{U,T}$ is closed and bounded.

From (6.12), we see that to show the last inequality in (6.14), it suffices to prove that

$$\sum_{j=1}^m \int_0^T a_j |b_j^\top e^{A^*(T-t)} \widehat{z}| dt > 0.$$

By contradiction, we suppose that the above was not true. Then we would have that

$$b_j^\top e^{A^*(T-t)} \widehat{z} = 0 \quad \text{over} \quad [0, T] \quad \text{for each} \quad j \in \{1, \ldots, m\},$$

which indicates that

$$\langle \widehat{z}, A^k b_j \rangle_{\mathbb{R}^n} = 0 \quad \text{for all} \quad k \in \mathbb{N}^+ \cup \{0\} \quad \text{and} \quad j \in \{1, \ldots, m\}.$$

This leads to a contradiction since $\widehat{z} \in \text{Range}(B, AB, \cdots, A^{n-1}B) \setminus \{0\}$. Hence, the last inequality in (6.14) is true. Thus, we have proved (6.14).

Let

$$\widehat{y}_0 \triangleq -e^{-AT}\widehat{y}_1 \quad (\text{with } \widehat{y}_1 \text{ given by } (6.14)). \tag{6.15}$$

By the inequality in (6.14), we see that $\widehat{y}_0 \neq 0$. The rest of the proof is carried out by the following two steps:

Step 1. We show that

$$N_{\mathbb{U}}(T, \widehat{y}_0) \leq 1. \tag{6.16}$$

Indeed, since $\widehat{y}_1 \in \mathscr{R}_0^{\mathbb{U},T}$, according to (6.11), there exists a control

$$\widehat{u}(\cdot) = (\widehat{u}_1(\cdot), \ldots, \widehat{u}_m(\cdot))^\top \in L^\infty(0, T; \mathbb{R}_{\mathbb{U}}^m) \quad \text{with } \widehat{u}_1(\cdot) = 0 \text{ over } (0, T) \tag{6.17}$$

so that

$$\widehat{y}_1 = y(T; 0, 0, \widehat{u}) \quad \text{and} \quad \|\widehat{u}\|_{L^\infty(0,T;\mathbb{R}_{\mathbb{U}}^m)} \leq 1.$$

These, together with (6.15), imply that

$$y(T; 0, \widehat{y}_0, \widehat{u}) = 0 \quad \text{and} \quad \|\widehat{u}\|_{L^\infty(0,T;\mathbb{R}_{\mathbb{U}}^m)} \leq 1. \tag{6.18}$$

From (6.18), we see that $\widehat{u}$ is an admissible control of $(NP)_{\mathbb{U}}^{T,\widehat{y}_0}$, and consequently (6.16) holds.

Step 2. We show that $N_{\mathbb{U}}(T, \widehat{y}_0) = 1$.

By contradiction, suppose that the above was not true. Then by (6.16), we would have that $N_{\mathbb{U}}(T, \widehat{y}_0) < 1$. Thus there is a control $u^* \in L^\infty(0, T; \mathbb{R}_{\mathbb{U}}^m)$ so that

$$y(T; 0, \widehat{y}_0, u^*) = 0 \quad \text{and} \quad \|u^*\|_{L^\infty(0,T;\mathbb{R}_{\mathbb{U}}^m)} < 1. \tag{6.19}$$

By the inequality in (6.19), we can find a constant $\lambda > 1$ so that

$$\|\lambda u^*\|_{L^\infty(0,T;\mathbb{R}_{\mathbb{U}}^m)} \leq 1. \tag{6.20}$$

Meanwhile, it follows from (6.15) and the equality in (6.19) that

$$\lambda \widehat{y}_1 = \lambda(-e^{AT}\widehat{y}_0) = y(T; 0, 0, \lambda u^*).$$

This, along with (6.10) and (6.20), implies that

$$\lambda \widehat{y}_1 \in \mathscr{R}^{\mathbb{U},T}.$$

From the latter and (6.14), we see that

$$\langle \widehat{y}_1, \widehat{z} \rangle_{\mathbb{R}^n} > 0 \quad \text{and} \quad \langle \widehat{y}_1, \widehat{z} \rangle_{\mathbb{R}^n} \geq \langle \lambda \widehat{y}_1, \widehat{z} \rangle_{\mathbb{R}^n}.$$

Since $\lambda > 1$, the above leads to a contradiction. Therefore, we have proved the conclusion in Step 2.

Finally, by (6.18), (6.17) and the conclusion in Step 2, we see that $\widehat{u}$ is the desired optimal control and $\widehat{y}_0$ (given by (6.15)) is the desired initial state. This completes the proof of Lemma 6.1.   □

The second lemma is concerned with some kind of controllability.

Lemma 6.2 *Let $T > t > 0$. Then for each $u \in L^\infty(0, T; \mathbb{R}^m)$, there exists $v_u \in L^\infty(0, T; \mathbb{R}^m)$ so that*

$$y(T; 0, 0, \chi_{(0,t)}u) = y(T; 0, 0, \chi_{(t,T)}v_u). \tag{6.21}$$

Conversely, for each $v \in L^\infty(0, T; \mathbb{R}^m)$, there exists $u_v \in L^\infty(0, T; \mathbb{R}^m)$ so that

$$y(T; 0, 0, \chi_{(t,T)}v) = y(T; 0, 0, \chi_{(0,t)}u_v). \tag{6.22}$$

Proof Arbitrarily fix $T > t > 0$. First of all, we show that there exists a constant $C(T, t) > 0$ so that

$$\int_0^t \|B^* e^{A^*(T-s)} z\|_{\mathbb{R}^m} ds \leq C(T, t) \int_t^T \|B^* e^{A^*(T-s)} z\|_{\mathbb{R}^m} ds \quad \text{for each } z \in \mathbb{R}^n. \tag{6.23}$$

For this purpose, we define two subspaces of $L^1(0, T; \mathbb{R}^m)$ as follows:

$$\mathcal{O}_{0,t} \triangleq \{\chi_{(0,t)}(\cdot) B^* e^{A^*(T-\cdot)} z \mid z \in \mathbb{R}^n\}$$

and

$$\mathcal{O}_{t,T} \triangleq \{\chi_{(t,T)}(\cdot) B^* e^{A^*(T-\cdot)} z \mid z \in \mathbb{R}^n\}.$$

It is clear that both $\mathcal{O}_{0,t}$ and $\mathcal{O}_{t,T}$ are of finite dimension. Meanwhile, for each fixed $z \in \mathbb{R}^n$, since the function $s \rightarrow B^* e^{A^* s} z$, $s \in \mathbb{R}$, is analytic, we can check that

$$B^* e^{A^*(T-\cdot)} z = 0 \text{ over } (0, t) \text{ if and only if } B^* e^{A^*(T-\cdot)} z = 0 \text{ over } (t, T).$$

Therefore, (6.23) is true.

Next, we arbitrarily fix $u \in L^\infty(0, T; \mathbb{R}^m)$ and define

$$\mathscr{F}(\chi_{(t,T)}(\cdot) B^* e^{A^*(T-\cdot)} z) \triangleq \int_0^t \langle B^* e^{A^*(T-s)} z, u(s) \rangle_{\mathbb{R}^m} ds \quad \text{for each } z \in \mathbb{R}^n. \tag{6.24}$$

From (6.23), we see that $\mathscr{F}$ is well defined over $\mathcal{O}_{t,T}$ and is linear bounded from $\mathcal{O}_{t,T}$ to $\mathbb{R}$. Then, according to the Hahn-Banach Theorem and the Riesz Representation Theorem (see Theorems 1.5 and 1.4), there exists $v_u \in L^\infty(0, T; \mathbb{R}^m)$ so that

$$\mathscr{F}(\chi_{(t,T)}(\cdot)B^* e^{A^*(T-\cdot)}z) = \int_t^T \langle B^* e^{A^*(T-s)}z, v_u(s)\rangle_{\mathbb{R}^m}\,ds \quad \text{for each } z \in \mathbb{R}^n.$$

This, together with (6.24), leads to (6.21).

Finally, the conclusion (6.22) can be proved by the very similar way to the above. Hence, we finish the proof of Lemma 6.2.   □

We now are in the position to prove Theorem 6.2.

Proof (Proof of Theorem 6.2) We prove the conclusions one by one.

(i) It suffices to prove this theorem in the case that both Y_S and Y_E are singleton subsets of $\mathbb{R}^n$. (The reason is similar to that explained in the proof of Theorem 6.1.) Arbitrarily fix two different vectors y_0 and y_1 in $\mathbb{R}^n$. We set

$$Y_S \triangleq \{y_0\} \quad \text{and} \quad Y_E \triangleq \{y_1\}.$$

Let $(0, y_0, t_E^*, u^*)$ be an optimal tetrad of $(TP)_{min}^{Q_S, Q_E}$. It is clear that $t_E^* > 0$. By the similar arguments to those used in the proof of Theorem 6.1, we can obtain that $(TP)_{min}^{Q_S, Q_E}$ satisfies the classical Pontryagin Maximum Principle. Then according to Definition 4.1, there exists $z^* \in \mathbb{R}^n \setminus \{0\}$ so that

$$\langle u^*(t), B^* e^{A^*(t_E^*-t)}z^*\rangle_{\mathbb{R}^m} = \max_{v \in U}\langle v, B^* e^{A^*(t_E^*-t)}z^*\rangle_{\mathbb{R}^m} \quad \text{for a.e. } t \in (0, t_E^*).$$
$$(6.25)$$

Meanwhile, since the function $t \to b_j^\top e^{A^*t}z^*$, $t \in \mathbb{R}$ (with $j \in \{1, \dots, m\}$) is analytic, it follows by (6.5) that for each $j \in \{1, \dots, m\}$,

$$b_j^\top e^{A^*t}z^* \neq 0 \quad \text{for a.e. } t \in (0, t_E^*).$$

Write $u^*(\cdot) = (u_1^*(\cdot), \dots, u_m^*(\cdot))^\top$. From the above and (6.25), we see that

$$|u_j^*(t)| = a_j \quad \text{for a.e. } t \in (0, t_E^*) \text{ and each } j \in \{1, \dots, m\}.$$

Then by Definition 6.1, $(0, y_0, t_E^*, u^*)$ satisfies the bang-bang property, consequently, $(TP)_{min}^{Q_S, Q_E}$ holds the bang-bang property.

(ii) Assume that (6.6) holds. Arbitrarily fix $T > 0$. According to Lemma 6.1, there exists $\widehat{y}_0 \in \mathbb{R}^n \setminus \{0\}$ so that $(NP)_U^{T, \widehat{y}_0}$ has an optimal control $u^*(\cdot) = (u_1^*(\cdot), \dots, u_m^*(\cdot))^\top$ satisfying that

$$y(T; 0, \widehat{y}_0, u^*) = 0, \quad \|u^*\|_{L^\infty(0,T;\mathbb{R}^m_{\mathbb{U}})} = N_{\mathbb{U}}(T, \widehat{y}_0) = 1, \quad u_1^*(\cdot) = 0 \text{ over } (0, T).$$
$$(6.26)$$

Denote $Q_S = \{0\} \times \{\widehat{y}_0\}$ and $Q_E = (0, +\infty) \times \{0\}$. We next show that the problem $(TP)_{min}^{Q_S, Q_E}$ has at least one optimal control but does not hold the bang-bang property. The proof will be carried out by the following two steps.

Step 1. We show that $(0, \widehat{y}_0, T, u^*)$ is an optimal tetrad to the problem $(TP)_{min}^{Q_S, Q_E}$.

From the first two conclusions in (6.26), we see that $(0, \widehat{y}_0, T, u^*)$ is an admissible tetrad to the problem $(TP)_{min}^{Q_S, Q_E}$. The reminder is to show that T is the optimal time of $(TP)_{min}^{Q_S, Q_E}$. By contradiction, we suppose that it was not true. Then there would exist $\widehat{t} \in (0, T)$ and $\widehat{v} \in L(0, \widehat{t}; \mathbb{U})$ so that

$$y(\widehat{t}; 0, \widehat{y}_0, \widehat{v}) = 0. \tag{6.27}$$

Meanwhile, according to Lemma 6.2, there exists $\widetilde{v} \in L^\infty(\widehat{t}, T; \mathbb{R}^m)$ so that

$$\int_0^{\widehat{t}} e^{A(T-t)} B\widehat{v}(t)dt = \int_{\widehat{t}}^T e^{A(T-t)} B\widetilde{v}(t)dt. \tag{6.28}$$

We choose a small $\lambda \in (0, 1)$ so that

$$2\lambda\widetilde{v}(t) \in \mathbb{U} \text{ for a.e. } t \in (\widehat{t}, T). \tag{6.29}$$

Define a new control as follows:

$$v_\lambda(t) \triangleq \begin{cases} (1-\lambda)\widehat{v}(t), & t \in (0, \widehat{t}), \\ \lambda\widetilde{v}(t), & t \in (\widehat{t}, T). \end{cases} \tag{6.30}$$

Then, by (6.30) and (6.28), we find that

$$y(T; 0, \widehat{y}_0, v_\lambda)$$
$$= e^{AT}\widehat{y}_0 + (1-\lambda)\int_0^{\widehat{t}} e^{A(T-t)} B\widehat{v}(t)dt + \lambda\int_{\widehat{t}}^T e^{A(T-t)} B\widetilde{v}(t)dt$$
$$= e^{AT}\widehat{y}_0 + \int_0^{\widehat{t}} e^{A(T-t)} B\widehat{v}(t)dt.$$

This, together with (6.27), implies that

$$y(T; 0, \widehat{y}_0, v_\lambda) = e^{A(T-\widehat{t})}y(\widehat{t}; 0, \widehat{y}_0, \widehat{v}) = 0.$$

Thus, v_λ is an admissible control to $(NP)_{\mathbb{U}}^{T, \widehat{y}_0}$ (see (6.8)). Then it follows from the second conclusion in (6.26) that

$$\|v_\lambda\|_{L^\infty(0,T;\mathbb{R}^m_{\mathbb{U}})} \geq 1. \tag{6.31}$$

Meanwhile, since $\widehat{v}(t) \in \mathbb{U}$ for a.e. $t \in (0, \widehat{t})$, we see from (6.7), (6.30), and (6.29) that

$$\|v_\lambda(t)\|_{\mathbb{R}_\mathbb{U}^m} \le \max\{1 - \lambda, 1/2\} < 1 \quad \text{for a.e. } t \in (0, T),$$

which contradicts (6.31). Therefore, T is the optimal time of $(TP)_{min}^{Q_S, Q_E}$.
Hence, $(0, \widehat{y}_0, T, u^*)$ is an optimal tetrad of $(TP)_{min}^{Q_S, Q_E}$.

Step 2. We prove that $(TP)_{min}^{Q_S, Q_E}$ does not hold the bang-bang property.
By Step 1, the last conclusion in (6.26) and Definition 6.1, we see that u^* does not have the bang-bang property, consequently, $(TP)_{min}^{Q_S, Q_E}$ does not hold the bang-bang property.

Hence, we finish the proof of Theorem 6.2.   □

The next two examples give auxiliary instructions on Theorem 6.2.

Example 6.4 Let

$$A = \begin{pmatrix} 0 & 1 \\ -1 & 0 \end{pmatrix}, \quad B = \begin{pmatrix} 1 & 0 \\ 0 & 1 \end{pmatrix}, \quad y_0 = \begin{pmatrix} 1 \\ 1 \end{pmatrix} \quad \text{and} \quad y_1 = \begin{pmatrix} 0 \\ 0 \end{pmatrix}.$$

Consider the problem $(TP)_{min}^{Q_S, Q_E}$, where the controlled system is as:

$$\dot{y}(t) = Ay(t) + Bu(t), \quad t \in (0, +\infty) \quad (\text{with } y(t) \in \mathbb{R}^2, u(t) \in \mathbb{R}^2)$$

and where

$$Q_S = \{0\} \times \{y_0\}, \quad Q_E = (0, +\infty) \times \{y_1\}, \quad \mathbb{U} = [-1, 1] \times [-1, 1].$$

One can easily check the following facts:

(i) The problem $(TP)_{min}^{Q_S, Q_E}$ can be put into the framework $(\mathscr{B}_1)$ (given at the beginning of Section 6.1);
(ii) The problem $(TP)_{min}^{Q_S, Q_E}$ has at least one admissible control; (This follows from Theorem 3.3 and Corollary 3.2.)
(iii) The problem $(TP)_{min}^{Q_S, Q_E}$ has at least one optimal control;
(iv) The assumptions in (i) of Theorem 6.2 hold.

Then by (i) of Theorem 6.2, $(TP)_{min}^{Q_S, Q_E}$ holds the bang-bang property. We end Example 6.4.

Example 6.5 Let

$$A = \begin{pmatrix} 0 & 0 \\ 0 & 0 \end{pmatrix}, \quad B = \begin{pmatrix} 1 & 0 \\ 0 & 1 \end{pmatrix}, \quad y_0 = \begin{pmatrix} 1 \\ 1 \end{pmatrix} \quad \text{and} \quad y_1 = \begin{pmatrix} 0 \\ 0 \end{pmatrix}.$$

Consider the problem $(TP)_{min}^{Q_S,Q_E}$, where the controlled equation is as:

$$\dot{y}(t) = Ay(t) + Bu(t), \quad t \in (0, +\infty) \quad \text{(with } y(t) \in \mathbb{R}^2 \text{ and } u(t) \in \mathbb{R}^2\text{)}$$

and where

$$Q_S = \{0\} \times \{y_0\}, \quad Q_E = (0, +\infty) \times \{y_1\}, \quad \mathbb{U} = [-1, 1] \times [-2, 2].$$

One can easily check the following facts:

(i) The problem $(TP)_{min}^{Q_S,Q_E}$ can be put into the framework $(\mathscr{B}_1)$ (given at the beginning of Section 6.1);

(ii) The assumption (6.6) in (ii) of Theorem 6.2 holds;

(iii) The tetrad $(0, y_0, 1, u^*)$ is optimal for $(TP)_{min}^{Q_S,Q_E}$, where

$$u^*(t) \triangleq \begin{cases} (-1, -2)^\top, & t \in (0, 1/2), \\ (-1, 0)^\top, & t \in [1/2, 1). \end{cases}$$

From the above, we see that u^* is not a bang-bang control. Consequently, $(TP)_{min}^{Q_S,Q_E}$ does not hold the bang-bang property. We now end Example 6.5.

6.1.2 Bang-Bang Property for Maximal Time Controls

This subsection aims to give some conditions to ensure the bang-bang property of the maximal time control problem $(TP)_{max}^{Q_S,Q_E}$. We start with the next lemma.

Lemma 6.3 *Let* $(A, B) \in \mathbb{R}^{n \times n} \times \mathbb{R}^{n \times m}$ *satisfy (6.3). Then for any* $T > t \geq 0$ *and each subset* $E \subseteq (t, T)$ *of positive measure, there exists a constant* $C \triangleq C(T, t, E) > 0$ *so that for each* $y_0 \in \mathbb{R}^n$, *there exists* $u \in L^\infty(t, T; \mathbb{R}^m)$ *satisfying that*

$$y(T; t, y_0, \chi_E u) = 0 \quad \text{and} \quad \|\chi_E u\|_{L^\infty(t,T;\mathbb{R}^m)} \leq C\|y_0\|_{\mathbb{R}^n}.$$

Proof Arbitrarily fix $T > t > 0$. First of all, we show that there exists a constant $C \triangleq C(T, t, E) > 0$ so that

$$\|z\|_{\mathbb{R}^n} \leq C \int_E \|B^* e^{A^*(T-s)} z\|_{\mathbb{R}^m} ds \quad \text{for each } z \in \mathbb{R}^n. \tag{6.32}$$

(Here and throughout the proof of this lemma, $C \triangleq C(T, t, E)$ denotes a generic positive constant depending on T, t, and E.) For this purpose, we define a subspace of $L^1(t, T; \mathbb{R}^m)$ as follows:

$$\mathscr{O} \triangleq \{\chi_E(\cdot) B^* e^{A^*(T-\cdot)} z \mid z \in \mathbb{R}^n\}.$$

It is clear that $\mathcal{O}$ is of finite dimension. Meanwhile, for each $z \in \mathbb{R}^n$, the function $s \to B^* e^{A^* s} z$, $s \in \mathbb{R}$, is analytic. Thus, by (6.3), we see that

$$B^* e^{A^*(T-\cdot)} z = 0 \text{ over } E \text{ if and only if } z = 0.$$

Therefore, (6.32) is true.

Next, we arbitrarily fix $y_0 \in \mathbb{R}^n$ and define

$$\mathcal{F}(\chi_E(\cdot) B^* e^{A^*(T-\cdot)} z) \triangleq \langle z, -e^{(T-t)A} y_0 \rangle_{\mathbb{R}^n} \text{ for each } z \in \mathbb{R}^n. \tag{6.33}$$

From (6.32), we see that $\mathcal{F}$ is well defined over $\mathcal{O}$. Besides, one can easily check that $\mathcal{F}$ is linear. Then by (6.33) and (6.32), we find that

$$|\mathcal{F}(\chi_E(\cdot) B^* e^{A^*(T-\cdot)} z)|$$
$$\leq C \|y_0\|_{\mathbb{R}^n} \int_t^T \|\chi_E(s) B^* e^{A^*(T-s)} z\|_{\mathbb{R}^m} ds \text{ for each } z \in \mathbb{R}^n,$$

which indicates that

$$\|\mathcal{F}\|_{L(\mathcal{O}, \mathbb{R})} \leq C \|y_0\|_{\mathbb{R}^n}.$$

Thus, $\mathcal{F}$ is a linear bounded functional on $\mathcal{O}$. Then, according to the Hahn-Banach Theorem and the Riesz Representation Theorem (see Theorems 1.5 and 1.4), there exists $u \in L^\infty(t, T; \mathbb{R}^m)$ so that

$$\mathcal{F}(\chi_E(\cdot) B^* e^{A^*(T-\cdot)} z) = \int_E \langle B^* e^{A^*(T-s)} z, u(s) \rangle_{\mathbb{R}^m} ds \text{ for each } z \in \mathbb{R}^n,$$

and so that

$$\|u\|_{L^\infty(t, T; \mathbb{R}^m)} = \|\mathcal{F}\|_{\mathscr{L}(\mathcal{O}, \mathbb{R})} \leq C \|y_0\|_{\mathbb{R}^n}.$$

These, together with (6.33), lead to the desired result.

Hence, we finish the proof of Lemma 6.3.   □

The first main theorem of this subsection concerns the bang-bang property for maximal time control problems where controls have the ball-type constraint.

Theorem 6.3 *Let $T > 0$ and $\mathbb{U} = B_\rho(0)$ with $\rho > 0$. Assume that (6.3) holds. Let*

$$Q_S = \{(t, e^{At} y_0) \mid 0 \leq t < T, y_0 \in Y_S\} \text{ and } Q_E = \{T\} \times Y_E,$$

where Y_S and Y_E are two nonempty subsets of $\mathbb{R}^n$ so that $(e^{AT} Y_S) \cap Y_E = \emptyset$. Then the problem $(TP)_{max}^{Q_S, Q_E}$ has the bang-bang property.

Proof By contradiction, we suppose that $(TP)_{max}^{Q_S,Q_E}$ did not hold the bang-bang property. Then by Definition 6.1, there would exist an optimal tetrad $(t_S^*, e^{At_S^*}y_0^*, T, u^*)$ of $(TP)_{max}^{Q_S,Q_E}$, a constant $\varepsilon \in (0, \rho)$ and a subset $E \subseteq (t_S^*, T)$ of positive measure so that

$$\|u^*(t)\|_{\mathbb{R}^m} < \rho - \varepsilon \quad \text{for a.e. } t \in E. \tag{6.34}$$

Meanwhile, by the optimality of the tetrad $(t_S^*, e^{At_S^*}y_0^*, T, u^*)$, we have that

$$0 \le t_S^* < T, \quad y_0^* \in Y_S, \quad y(T; t_S^*, e^{At_S^*}y_0^*, u^*) \in Y_E \tag{6.35}$$

and that

$$\|u^*(t)\|_{\mathbb{R}^m} \le \rho \quad \text{for a.e. } t \in (t_S^*, T). \tag{6.36}$$

Define a subset of E in the following manner:

$$E_1 \triangleq \{t \in E \mid t_S^* + |E|/2 < t < T\}. \tag{6.37}$$

It is obvious that

$$E_1 \subseteq (t_S^* + |E|/2, T) \quad \text{and} \quad |E_1| \ge |E|/2. \tag{6.38}$$

Let $C \triangleq C(T, t, E) > 0$ be given by Lemma 6.3, where (T, t, E) is replaced by $(T, t_S^* + |E|/2, E_1)$. We choose a small $\delta \in (0, |E|/2)$ so that

$$\|\widehat{y_0}\|_{\mathbb{R}^n} \le \varepsilon / \left(C \max_{0 \le s \le T} \|e^{As}\|_{\mathscr{L}(\mathbb{R}^n)} \right) \quad \text{with } \widehat{y_0} \triangleq -y(t_S^* + \delta; t_S^*, 0, u^*). \tag{6.39}$$

Then by Lemma 6.3, we can find a control $\widehat{u} \in L^\infty(0, +\infty; \mathbb{R}^m)$ so that

$$y(T; t_S^* + |E|/2, y(t_S^* + |E|/2; t_S^* + \delta, \widehat{y_0}, 0), \chi_{E_1}\widehat{u}) = 0 \tag{6.40}$$

and so that

$$\|\chi_{E_1}\widehat{u}\|_{L^\infty(0,+\infty;\mathbb{R}^m)} \le C\|y(t_S^* + |E|/2; t_S^* + \delta, \widehat{y_0}, 0)\|_{\mathbb{R}^n}. \tag{6.41}$$

It follows from (6.38) and (6.40) that

$$y(T; t_S^* + \delta, \widehat{y_0}, \chi_{E_1}\widehat{u}) = 0. \tag{6.42}$$

Define a new control in the following manner:

$$\widetilde{u}(t) \triangleq u^*(t) + \chi_{E_1}(t)\widehat{u}(t), \quad t \in (t_S^* + \delta, T). \tag{6.43}$$

On one hand, by (6.36), (6.34), (6.37), (6.41), and (6.39), we can directly check that

$$\|\tilde{u}(t)\|_{\mathbb{R}^m} \le \rho \quad \text{for a.e. } t \in (t_S^* + \delta, T). \tag{6.44}$$

On the other hand, it follows from (6.43) that

$$y(T; t_S^* + \delta, e^{A(t_S^*+\delta)}y_0^*, \tilde{u}) = \int_{t_S^*+\delta}^{T} e^{A(T-t)}B(\chi_{E_1}\widehat{u})(t)dt$$

$$+ e^{A(T-t_S^*-\delta)}y(t_S^* + \delta; t_S^*, e^{At_S^*}y_0^*, u^*) + \int_{t_S^*+\delta}^{T} e^{A(T-t)}Bu^*(t)dt$$

$$+ e^{A(T-t_S^*-\delta)}(e^{A(t_S^*+\delta)}y_0^* - y(t_S^* + \delta; t_S^*, e^{At_S^*}y_0^*, u^*)),$$

which, along with (6.39), yields that

$$y(T; t_S^* + \delta, e^{A(t_S^*+\delta)}y_0^*, \tilde{u})$$

$$= y(T; t_S^*, e^{At_S^*}y_0^*, u^*) + e^{A(T-t_S^*-\delta)}\widehat{y}_0 + \int_{t_S^*+\delta}^{T} e^{A(T-t)}B(\chi_{E_1}\widehat{u})(t)dt.$$

This, combined with (6.42) and the third conclusion of (6.35), indicates that

$$y(T; t_S^* + \delta, e^{A(t_S^*+\delta)}y_0^*, \tilde{u}) = y(T; t_S^*, e^{At_S^*}y_0^*, u^*) + y(T; t_S^* + \delta, \widehat{y}_0, \chi_{E_1}\widehat{u})$$

$$= y(T; t_S^*, e^{At_S^*}y_0^*, u^*) \in Y_E. \tag{6.45}$$

Since $0 \le t_S^* + \delta < t_S^* + |E|/2 < T$, it follows by the second conclusion of (6.35), (6.44), and (6.45) that $(t_S^* + \delta, e^{A(t_S^*+\delta)}y_0^*, T, \tilde{u})$ is an admissible tetrad of $(TP)_{max}^{Q_S,Q_E}$, which contradicts the optimality of t_S^*. Thus, $(TP)_{max}^{Q_S,Q_E}$ holds the bang-bang property.

This completes the proof of Theorem 6.3.   □

The next two examples give auxiliary instructions on Theorem 6.3.

Example 6.6 Let

$$A = \begin{pmatrix} 0 & 1 \\ 0 & 0 \end{pmatrix}, \quad B = \begin{pmatrix} 0 \\ 1 \end{pmatrix}, \quad y_0 = \begin{pmatrix} 0 \\ 0 \end{pmatrix} \quad \text{and} \quad y_1 = \begin{pmatrix} 1 \\ 0 \end{pmatrix}.$$

Consider the problem $(TP)_{max}^{Q_S,Q_E}$, where the controlled equation is as:

$$\dot{y}(t) = Ay(t) + Bu(t), \quad t \in (0, +\infty) \quad \text{(with } y(t) \in \mathbb{R}^2 \text{ and } u(t) \in \mathbb{R})$$

and where

$$Q_S = \{(t, e^{At}y_0) \mid 0 \le t < 2\}, \quad Q_E = \{2\} \times \{y_1\}, \quad \mathbb{U} = [-1, 1], \quad T = 2.$$

One can easily check the following facts:

(i) The problem $(TP)_{max}^{Q_S, Q_E}$ can be put into the framework $(\mathscr{B}_1)$ (given at the beginning of Section 6.1);
(ii) It holds that rank $(B, AB) = 2$;
(iii) The tetrad $(0, y_0, 2, u)$ is admissible for $(TP)_{max}^{Q_S, Q_E}$, where

$$u(t) \triangleq \begin{cases} 1, & t \in (0, 1), \\ -1, & t \in [1, 2); \end{cases}$$

(iv) The problem $(TP)_{max}^{Q_S, Q_E}$ has optimal controls. (This follows from (iii).)
Then by Theorem 6.3, $(TP)_{max}^{Q_S, Q_E}$ holds the bang-bang property.

We now end Example 6.6.

Example 6.7 Let

$$A = \begin{pmatrix} 0 & 0 \\ 0 & 1 \end{pmatrix}, \quad B = \begin{pmatrix} 1 \\ 0 \end{pmatrix}, \quad y_0 = \begin{pmatrix} 0 \\ 1 \end{pmatrix} \quad \text{and} \quad y_1 = \begin{pmatrix} 1 \\ e \end{pmatrix}.$$

Consider the problem $(TP)_{max}^{Q_S, Q_E}$, where the controlled system is as:

$$\dot{y}(t) = Ay(t) + Bu(t), \quad t \in (0, +\infty) \quad (\text{with } y(t) \in \mathbb{R}^2, u(t) \in \mathbb{R}),$$

and where

$$Q_S = \{(t, e^{At} y_0) \mid 0 \le t < 1\}, \quad Q_E = \{1\} \times \{y_1\}, \quad \mathbb{U} = [-1, 1], \quad T = 1.$$

One can easily check the following facts:

(i) The problem $(TP)_{max}^{Q_S, Q_E}$ can be put into the framework $(\mathscr{B}_1)$ (given at the beginning of Section 6.1);
(ii) It holds that rank $(B, AB) = 1 < 2$ (i.e., (A, B) does not satisfy (6.3));
(iii) The tetrad $(0, y_0, 1, 0)$ is optimal for $(TP)_{max}^{Q_S, Q_E}$.

Consequently, $(TP)_{max}^{Q_S, Q_E}$ has no bang-bang property. Now we end Example 6.9.

The second main result of this subsection concerns the bang-bang property for maximal time control problems where controls have the rectangle-type constraint.

Theorem 6.4 *Let* $m \ge 2$ *and* $B = (b_1, \ldots, b_m)$ *with* $\{b_j\}_{j=1}^m \subseteq \mathbb{R}^n$. *Let* $\mathbb{U} = \prod_{j=1}^m [-a_j, a_j]$ *with* $\{a_j\}_{j=1}^m \subseteq (0, +\infty)$. *Then the following two conclusions are true:*

(i) Assume that

$$rank(b_j, Ab_j, \ldots, A^{n-1} b_j) = n \quad \text{for each} \quad j \in \{1, \ldots, m\}.$$

Let

$$Q_S = \{(t, e^{At} y_0) \mid 0 \le t < T, y_0 \in Y_S\} \quad and \quad Q_E = \{T\} \times Y_E,$$

where $T > 0$, Y_S and Y_E are two nonempty subsets of $\mathbb{R}^n$ with $(e^{AT} Y_S) \cap Y_E = \emptyset$. Then the problem $(TP)_{max}^{Q_S, Q_E}$ holds the bang-bang property.

(ii) Assume that

$$rank(b_1, Ab_1, \ldots, A^{n-1}b_1) < rank(B, AB, \ldots, A^{n-1}B). \qquad (6.46)$$

Then there exists $\widehat{y}_0 \in \mathbb{R}^n \setminus \{0\}$ and $\widehat{T} > 0$ so that the problem $(TP)_{max}^{Q_S, Q_E}$, with

$$Q_S = \{(t, e^{At} \widehat{y}_0) : 0 \le t < \widehat{T}\} \quad and \quad Q_E = \{\widehat{T}\} \times \{0\},$$

has at least one optimal control but does not hold the bang-bang property.

Proof (i) By contradiction, we suppose that $(TP)_{max}^{Q_S, Q_E}$ did not hold the bang-bang property. Then by Definition 6.1, we could assume, without loss of generality, that there would be an optimal tetrad $(t_S^*, e^{At_S^*} y_0^*, T, u^*)$ (with $u^* = (u_1^*, \ldots, u_m^*)^\top$) of $(TP)_{max}^{Q_S, Q_E}$, an $\varepsilon \in (0, a_1)$ and a subset E (in (t_S^*, T)) of positive measure so that

$$|u_1^*(t)| < a_1 - \varepsilon \quad \text{for each } t \in E. \qquad (6.47)$$

It is clear that

$$0 \le t_S^* < T, \quad y_0^* \in Y_S, \quad y(T; t_S^*, e^{At_S^*} y_0^*, u^*) \in Y_E \qquad (6.48)$$

and

$$|u_i^*(t)| \le a_i \quad \text{for each } i \in \{1, \ldots, m\} \text{ and a.e. } t \in (t_S^*, T). \qquad (6.49)$$

Define a subset of E in the following manner:

$$E_1 \triangleq \{t \in E \mid t_S^* + |E|/2 < t < T\}. \qquad (6.50)$$

It is obvious that

$$E_1 \subseteq (t_S^* + |E|/2, T) \quad \text{and} \quad |E_1| \ge |E|/2. \qquad (6.51)$$

Let $C \triangleq C(T, t, E) > 0$ be given by Lemma 6.3, where (A, B) and (B, T, t, E) are replaced by (A, b_1) and $(b_1, T, t_S^* + |E|/2, E_1)$ respectively. We choose a small $\delta \in (0, |E|/2)$ so that

$$\|\widehat{y}_0\|_{\mathbb{R}^n} \leq \varepsilon / \left(C \max_{0 \leq s \leq T} \|e^{As}\|_{\mathscr{L}(\mathbb{R}^n)} \right) \quad \text{with} \quad \widehat{y}_0 \triangleq -y(t_S^* + \delta; t_S^*, 0, u^*).$$

$$(6.52)$$

Then according to Lemma 6.3, there is $\widehat{u} \in L^\infty(0, +\infty; \mathbb{R}^m)$, with $\widehat{u} = (\widehat{u}_1, 0, \ldots, 0)^\top$, so that

$$y(T; t_S^* + |E|/2, y(t_S^* + |E|/2; t_S^* + \delta, \widehat{y}_0, 0), \chi_{E_1}\widehat{u}) = 0 \qquad (6.53)$$

and so that

$$\|\chi_{E_1}\widehat{u}_1\|_{L^\infty(0,+\infty;\mathbb{R})} \leq C\|y(t_S^* + |E|/2; t_S^* + \delta, \widehat{y}_0, 0)\|_{\mathbb{R}^n}. \qquad (6.54)$$

It follows from (6.51) and (6.53) that

$$y(T; t_S^* + \delta, \widehat{y}_0, \chi_{E_1}\widehat{u}) = 0. \qquad (6.55)$$

Define a new control in the following manner:

$$(\widetilde{u}_1(t), \ldots, \widetilde{u}_m(t))^\top \triangleq \widetilde{u}(t) \triangleq u^*(t) + \chi_{E_1}(t)\widehat{u}(t), \quad t \in (t_S^* + \delta, T). \qquad (6.56)$$

Then it follows by (6.49), (6.47), (6.50), (6.54), and (6.52) that

$$|\widetilde{u}_j(t)| \leq a_j \text{ for a.e. } t \in (t_S^* + \delta, T) \text{ and for each } j \in \{1, \ldots, m\}. \qquad (6.57)$$

Meanwhile, from (6.56), we see that

$$y(T; t_S^* + \delta, e^{A(t_S^*+\delta)}y_0^*, \widetilde{u}) = \int_{t_S^*+\delta}^T e^{A(T-t)} B(\chi_{E_1}\widehat{u})(t) dt$$

$$+ e^{A(T-t_S^*-\delta)} y(t_S^* + \delta; t_S^*, e^{At_S^*}y_0^*, u^*) + \int_{t_S^*+\delta}^T e^{A(T-t)} B u^*(t) dt$$

$$+ e^{A(T-t_S^*-\delta)}(e^{A(t_S^*+\delta)}y_0^* - y(t_S^* + \delta; t_S^*, e^{At_S^*}y_0^*, u^*)),$$

which, together with (6.52), yields that

$$y(T; t_S^* + \delta, e^{A(t_S^*+\delta)}y_0^*, \widetilde{u})$$

$$= y(T; t_S^*, e^{At_S^*}y_0^*, u^*) + e^{A(T-t_S^*-\delta)}\widehat{y}_0 + \int_{t_S^*+\delta}^T e^{A(T-t)} B(\chi_{E_1}\widehat{u})(t) dt.$$

This, combined with (6.55) and the third conclusion of (6.48), indicates that

$$y(T; t_S^* + \delta, e^{A(t_S^*+\delta)}y_0^*, \widetilde{u}) = y(T; t_S^*, e^{At_S^*}y_0^*, u^*) + y(T; t_S^* + \delta, \widehat{y}_0, \chi_{E_1}\widehat{u})$$
$$= y(T; t_S^*, e^{At_S^*}y_0^*, u^*) \in Y_E.$$

$$(6.58)$$

Finally, since $0 \leq t_S^* + \delta < t_S^* + |E|/2 < T$, it follows by the second conclusion of (6.48), (6.57), and (6.58) that $(t_S^* + \delta, e^{A(t_S^*+\delta)}y_0^*, T, \widetilde{u})$ is an admissible tetrad of $(TP)_{max}^{Q_S,Q_E}$, which contradicts the optimality of t_S^*. Thus, $(TP)_{max}^{Q_S,Q_E}$ holds the bang-bang property.

(ii) Assume that (6.46) holds. Arbitrarily fix $T > 0$. According to (6.46) and Lemma 6.1, there exists $\widehat{y}_0 \in \mathbb{R}^n \setminus \{0\}$ so that $(NP)_{\mathbb{U}}^{T,\widehat{y}_0}$ has an optimal control $u^*(\cdot) = (u_1^*(\cdot), \ldots, u_m^*(\cdot))^\top$ satisfying that

$$y(T; 0, \widehat{y}_0, u^*) = 0, \quad \|u^*\|_{L^\infty(0,T;\mathbb{R}_{\mathbb{U}}^m)} = N_{\mathbb{U}}(T, \widehat{y}_0) = 1$$

$$\text{and } u_1^*(\cdot) = 0 \text{ over } (0, T). \tag{6.59}$$

Let

$$Q_S = \{(t, e^{At}\widehat{y}_0) : 0 \leq t < T\} \text{ and } Q_E = \{T\} \times \{0\}.$$

We next show that the problem $(TP)_{max}^{Q_S,Q_E}$ has at least one optimal tetrad but does not hold the bang-bang property. The proof will be carried out by the following two steps.

Step 1. We show that $(0, \widehat{y}_0, T, u^*)$ is an optimal tetrad to the problem $(TP)_{max}^{Q_S,Q_E}$.

From the first two conclusions in (6.59), we see that $(0, \widehat{y}_0, T, u^*)$ is an admissible tetrad to the problem $(TP)_{max}^{Q_S,Q_E}$. The reminder is to show that T is the optimal time of $(TP)_{max}^{Q_S,Q_E}$. By contradiction, we suppose that it was not true. Then there would exist $\widehat{t} \in (0, T)$ and $\widehat{v} \in L(\widehat{t}, T; \mathbb{U})$ so that

$$y(T; \widehat{t}, e^{A\widehat{t}}\widehat{y}_0, \widehat{v}) = 0. \tag{6.60}$$

Meanwhile, according to Lemma 6.2, there exists $\widetilde{v} \in L^\infty(0, \widehat{t}; \mathbb{R}^m)$ so that

$$\int_{\widehat{t}}^T e^{A(T-t)} B\widehat{v}(t)dt = \int_0^{\widehat{t}} e^{A(T-t)} B\widetilde{v}(t)dt. \tag{6.61}$$

We choose a small $\lambda \in (0, 1)$ so that

$$2\lambda\widetilde{v}(t) \in \mathbb{U} \text{ for a.e. } t \in (0, \widehat{t}). \tag{6.62}$$

Define a new control as follows:

$$v_\lambda(t) \triangleq \begin{cases} \lambda\widetilde{v}(t), & t \in (0, \widehat{t}), \\ (1 - \lambda)\widehat{v}(t), & t \in (\widehat{t}, T). \end{cases} \tag{6.63}$$

Then it follows from (6.63) and (6.61) that

$$y(T; 0, \widehat{y}_0, v_\lambda)$$

$$= e^{AT}\widehat{y}_0 + (1 - \lambda) \int_{\widehat{t}}^{T} e^{A(T-t)} B\widehat{v}(t)dt + \lambda \int_0^{\widehat{t}} e^{A(T-t)} B\widetilde{v}(t)dt$$

$$= e^{AT}\widehat{y}_0 + \int_{\widehat{t}}^{T} e^{A(T-t)} B\widehat{v}(t)dt.$$

This, together with (6.60), implies that

$$y(T; 0, \widehat{y}_0, v_\lambda) = y(T; \widehat{t}, e^{A\widehat{t}}\widehat{y}_0, \widehat{v}) = 0.$$

Thus, v_λ is an admissible control to $(NP)_{\mathbb{U}}^{T, \widehat{y}_0}$ (see (6.8)). Then it follows from the second conclusion in (6.59) that

$$\|v_\lambda\|_{L^\infty(0, T; \mathbb{R}_{\mathbb{U}}^m)} \geq 1. \tag{6.64}$$

Meanwhile, since $\widehat{v}(t) \in \mathbb{U}$ a.e. $t \in (\widehat{t}, T)$, from (6.7), (6.63), and (6.62), we see that

$$\|v_\lambda(t)\|_{\mathbb{R}_{\mathbb{U}}^m} \leq \max\{1 - \lambda, 1/2\} < 1 \quad \text{for a.e. } t \in (0, T),$$

which contradicts (6.64). Therefore, T is the optimal time of $(TP)_{max}^{Q_S, Q_E}$. Hence, $(0, \widehat{y}_0, T, u^*)$ is an optimal tetrad of $(TP)_{max}^{Q_S, Q_E}$.

Step 2. We prove that $(TP)_{max}^{Q_S, Q_E}$ does not hold the bang-bang property.

By the conclusion in Step 1, the third conclusion in (6.59), and Definition 6.1, we see that $(0, \widehat{y}_0, T, u^*)$ does not have the bang-bang property, consequently, $(TP)_{max}^{Q_S, Q_E}$ does not hold the bang-bang property.

This completes the proof of Theorem 6.4. □

The next two examples give auxiliary instructions on Theorem 6.4.

Example 6.8 Let

$$A = \begin{pmatrix} 0 & 1 \\ -1 & 0 \end{pmatrix}, \quad B = \begin{pmatrix} 1 & 0 \\ 0 & 1 \end{pmatrix}, \quad y_0 = \begin{pmatrix} 0 \\ -1 \end{pmatrix} \quad \text{and} \quad y_1 = \begin{pmatrix} 0 \\ 0 \end{pmatrix}.$$

Consider the problem $(TP)_{max}^{Q_S, Q_E}$, where the controlled equation is as:

$$\dot{y}(t) = Ay(t) + Bu(t), \quad t \in (0, +\infty), \quad \text{with } y(t) \in \mathbb{R}^2 \text{ and } u(t) \in \mathbb{R}^2,$$

and where

$$Q_S = \{(t, e^{At}y_0) \mid 0 \leq t < 1\}, \quad Q_E = \{1\} \times \{y_1\}, \quad \mathbb{U} = [-1, 1] \times [-1, 1], \quad T = 1.$$

One can easily check the following facts:

(i) The problem $(TP)_{max}^{Q_S, Q_E}$ can be put into the framework $(\mathscr{B}_1)$ (given at the beginning of Section 6.1);

(ii) The assumptions in (i) of Theorem 6.4 hold;

(iii) The tetrad $(0, y_0, 1, u)$ is admissible for $(TP)_{max}^{Q_S, Q_E}$, where

$$u(t) \triangleq \begin{pmatrix} 1 \\ 0 \end{pmatrix}, \quad t \in (0, 1);$$

(iv) The problem $(TP)_{max}^{Q_S, Q_E}$ has at least one optimal control. (This follows from the existence of admissible controls, see (iii).)

Then by (i) of Theorem 6.4, $(TP)_{max}^{Q_S, Q_E}$ holds the bang-bang property. We now end Example 6.8.

Example 6.9 Let

$$A = \begin{pmatrix} 0 & 0 \\ 0 & 0 \end{pmatrix}, \quad B = \begin{pmatrix} 1 & 0 \\ 0 & 1 \end{pmatrix}, \quad y_0 = \begin{pmatrix} 1 \\ 1 \end{pmatrix} \quad \text{and} \quad y_1 = \begin{pmatrix} 0 \\ 0 \end{pmatrix}.$$

Consider the problem $(TP)_{max}^{Q_S, Q_E}$, where the controlled system is as:

$$\dot{y}(t) = Ay(t) + Bu(t), \quad t \in (0, +\infty) \quad (\text{with } y(t) \in \mathbb{R}^2, u(t) \in \mathbb{R}^2),$$

and where

$$Q_S = \{(t, e^{At} y_0) \mid 0 \le t < 1\}, \quad Q_E = \{1\} \times \{y_1\}, \quad U = [-1, 1] \times [-2, 2], \quad T = 1.$$

One can easily check the following facts:

(i) The problem $(TP)_{max}^{Q_S, Q_E}$ can be put into the framework $(\mathscr{B}_1)$ (given at the beginning of Section 6.1);

(ii) The assumption in (ii) of Theorem 6.4 holds;

(iii) The tetrad $(0, y_0, 1, u^*)$ is optimal for $(TP)_{max}^{Q_S, Q_E}$, where

$$u^*(t) \triangleq \begin{cases} (-1, -2)^\top, & t \in (0, 1/2), \\ (-1, 0)^\top, & t \in [1/2, 1). \end{cases}$$

By (iii), we see that u^* is not a bang-bang control. Thus, $(TP)_{max}^{Q_S, Q_E}$ does not hold the bang-bang property. We now end Example 6.9.

6.2 Bang-Bang Property and Null Controllability

In this section, we will study the bang-bang properties of minimal/maximal time control problems $(TP)_{min}^{Q_S,Q_E}/(TP)_{max}^{Q_S,Q_E}$ from the perspective of the L^∞-null controllability from measurable sets in time. These problems are under the following framework ($\mathscr{B}_2$):

(i) The state space Y and the control space U are two real separable Hilbert spaces.
(ii) The controlled system is as:

$$\dot{y}(t) = Ay(t) + D(t)y(t) + B(t)u(t), \quad t > 0, \tag{6.65}$$

where $A : \mathscr{D}(A) \subseteq Y \mapsto Y$ generates a C_0 semigroup $\{e^{At}\}_{t\geq 0}$ on Y; $D(\cdot) \in L^\infty(0, +\infty; \mathscr{L}(Y))$ and $B(\cdot) \in L^\infty(0, +\infty; \mathscr{L}(U, Y))$. Given $\tau \in [0, +\infty)$, $y_0 \in Y$ and $u \in L^\infty(\tau, +\infty; U)$, we write $y(\cdot; \tau, y_0, u)$ for the solution of (6.65) over $[\tau, +\infty)$, with the initial condition that $y(\tau) = y_0$. Moreover, write $\{\Phi(t, s) : t \geq s \geq 0\}$ for the evolution system generated by $A + D(\cdot)$ over Y.

(iii) The control constraint set $\mathbb{U}$ is nonempty and closed in U.
(iv) Sets Q_S and Q_E are two disjoint nonempty subsets of $[0, +\infty) \times Y$ and $(0, +\infty) \times Y$, respectively.

As what we did in the previous section, we always assume that $(TP)_{min}^{Q_S,Q_E}$ and $(TP)_{max}^{Q_S,Q_E}$ have optimal controls.

We now give the definition of the bang-bang property for problems $(TP)_{min}^{Q_S,Q_E}$ and $(TP)_{max}^{Q_S,Q_E}$ as follows:

Definition 6.2

(i) An optimal control u^* (associated with an optimal tetrad $(t_S^*, y_0^*, t_E^*, u^*)$) to the problem $(TP)_{min}^{Q_S,Q_E}$ (or $(TP)_{max}^{Q_S,Q_E}$) is said to have the bang-bang property if

$$u^*(t) \in \partial\mathbb{U} \quad \text{for a.e. } t \in (t_S^*, t_E^*), \tag{6.66}$$

where $\partial\mathbb{U}$ denotes the boundary of $\mathbb{U}$. (We also say that u^* is a bang-bang control when (6.66) holds.)

(ii) The problem $(TP)_{min}^{Q_S,Q_E}$ (or $(TP)_{max}^{Q_S,Q_E}$) is said to have the bang-bang property, if any optimal control satisfies the bang-bang property.

Remark 6.2 In Section 6.1, we defined the bang-bang property for the finite dimension controlled system (see Definition 6.1). Definition 6.1 differs from Definition 6.2 (see "ex $\mathbb{U}$" in (6.2) and "$\partial\mathbb{U}$" in (6.66)). In general, ex $\mathbb{U}$ and $\partial\mathbb{U}$ are different. For example, when $\mathbb{U}$ is a rectangle of $\mathbb{R}^m$ (with $m \in \mathbb{N}^+$), they are different. But if $\mathbb{U}$ is a nonempty closed ball in a Hilbert space, then ex $\mathbb{U}$ and $\partial\mathbb{U}$ are the same.

For time optimal control problems $(TP)_{min}^{Q_S,Q_E}$ (or $(TP)_{max}^{Q_S,Q_E}$), where the controlled systems are in infinitely dimensional spaces, we will only study the bang-bang property given by Definition 6.2. The reason is that in infinitely dimensional cases, the study on the bang-bang property given by Definition 6.1 is very hard. (To our best knowledge, there have not been any result on this issue so far.)

We next give the definition of the L^∞-null controllability from measurable sets in time for the system (6.65).

Definition 6.3 The system (6.65) is said to be L^∞-null controllable from measurable sets in time, if for any $T > t > 0$ and each subset $E \subseteq (t, T)$ of positive measure, there exists a constant $C(T, t, E) > 0$ so that for each $y_0 \in Y$, there is a control $u \in L^\infty(0, +\infty; U)$ satisfying that

$$y(T; t, y_0, \chi_E u) = 0 \quad \text{and} \quad \|\chi_E u\|_{L^\infty(0,+\infty;U)} \leq C(T, t, E)\|y_0\|_Y. \qquad (6.67)$$

6.2.1 Bang-Bang Property of Minimal Time Control

In this subsection, we aim to use the L^∞-null controllability (from measurable sets in time) to study the bang-bang property for the minimal time control problem $(TP)_{min}^{Q_S,Q_E}$. For this purpose, we need the following assumption $(\widehat{H}1)$:

(i) $D(\cdot) \equiv 0$ and $B(\cdot) \equiv B$ with $B \in \mathscr{L}(U, H)$. (That is, the system (6.65) is time-invariant.)
(ii) The operator A generates an analytic semigroup $\{e^{At}\}_{t\geq 0}$ over Y.
(iii) There exist two constants $d > 0$ and $\sigma > 0$ so that for each $L \in (0, 1]$,

$$\|e^{A^*L}z\|_Y^2 \leq e^{d/L^\sigma} \int_0^L \|B^*e^{A^*(L-t)}z\|_U^2 dt \quad \text{for all } z \in Y.$$

Proposition 6.1 Suppose that $(\widehat{H}1)$ holds. Then the system (6.65) is L^∞-null controllable from measurable sets in time.

The proof of Proposition 6.1 will be given later. We now use it to prove the following main result of this subsection:

Theorem 6.5 Suppose that $(\widehat{H}1)$ holds. Let

$$Q_S = \{0\} \times Y_S \quad \text{and} \quad Q_E = (0, +\infty) \times Y_E,$$

where Y_S and Y_E are two disjoint nonempty subsets of Y. Then the problem $(TP)_{min}^{Q_S,Q_E}$ has the bang-bang property.

Proof By contradiction, we suppose that $(TP)_{min}^{Q_S,Q_E}$ did not hold the bang-bang property. Then according to Definition 6.2, there would exist an optimal tetrad

$(0, y_0^*, t_E^*, u^*)$ of $(TP)_{min}^{Q_S, Q_E}$ and a subset E (in $(0, t_E^*)$) of positive measure so that

$$u^*(t) \notin \partial \mathbb{U} \quad \text{for each } t \in E. \tag{6.68}$$

By the optimality of $(0, y_0^*, t_E^*, u^*)$, we find that

$$y_0^* \in Y_S, \quad y(t_E^*; 0, y_0^*, u^*) \in Y_E \quad \text{and} \quad u^*(t) \in \mathbb{U} \quad \text{for a.e. } t \in (0, t_E^*). \tag{6.69}$$

For each $k \in \mathbb{N}^+$, we define a subset of E in the following manner:

$$E_k \triangleq \{t \in E \mid d_U(u^*(t), \partial \mathbb{U}) > k^{-1}\}. \tag{6.70}$$

Here and in what follows, $d_U(\cdot, \partial \mathbb{U})$ denotes the distance between $\cdot$ and $\partial \mathbb{U}$ in U.

We first claim that for each $k \in \mathbb{N}^+$, E_k is measurable. To this end, We arbitrarily fix $k \in \mathbb{N}^+$. Since $u^*(\cdot) : E \to U$ is strongly measurable, there exists $\{n_j\}_{j \geq 1} \subseteq \mathbb{N}^+$, $\{a_{j\ell} : j \in \mathbb{N}^+, 1 \leq \ell \leq n_j\} \subseteq U$ and a sequence of measurable subsets $\{F_{j\ell} \mid j \in \mathbb{N}^+, 1 \leq \ell \leq n_j\} \subseteq E$, with

$$F_{j\ell} \cap F_{j\ell'} = \emptyset, \quad \text{when } 1 \leq \ell, \ell' \leq n_j, \ \ell \neq \ell' \text{ and } j \in \mathbb{N}^+,$$

so that

$$u_j(t) \triangleq \sum_{\ell=1}^{n_j} a_{j\ell} \chi_{F_{j\ell}}(t) \to u^*(t) \quad \text{strongly in } U \text{ for a.e. } t \in E.$$

From this it follows that for a.e. $t \in E$,

$$d_U(u^*(t), \partial \mathbb{U}) = \lim_{j \to +\infty} d_U(u_j(t), \partial \mathbb{U}). \tag{6.71}$$

Meanwhile, we find that for each $j \in \mathbb{N}^+$,

$$d_U(u_j(t), \partial \mathbb{U}) = \sum_{\ell=1}^{n_j} \chi_{F_{j\ell}}(t) d_U(a_{j\ell}, \partial \mathbb{U}) \quad \text{for each } t \in E,$$

which indicates that for each $j \in \mathbb{N}^+$, the function $t \to d_U(u_j(t), \partial \mathbb{U})$ is measurable from E to $\mathbb{R}$. This, together with (6.71), yields that the function $t \to d_U(u^*(t), \partial \mathbb{U})$ is also measurable from E to $\mathbb{R}$. From this and (6.70), we see that each E_k is measurable.

Next, from (6.68) and (6.70), one can directly check that

$$E = \bigcup_{k=1}^{+\infty} E_k \quad \text{and} \quad E_j \subseteq E_\ell \quad \text{for all } j, \ell \in \mathbb{N}^+ \text{ with } j \leq \ell. \tag{6.72}$$

Two observations are given in order: (i) There exists $\widehat{k} \in \mathbb{N}^+$ so that $|E_{\widehat{k}}| > 0$. (This follows from (6.72) and the fact that $|E| > 0$.) (ii) It follows from (6.70) that

$$d_U(u^*(t), \partial \mathbb{U}) > 1/\widehat{k} \quad \text{for each } t \in E_{\widehat{k}}. \tag{6.73}$$

We define a subset of $E_{\widehat{k}}$ in the following manner:

$$E_{\widehat{k},1} \triangleq \{t \in E_{\widehat{k}} : |E_{\widehat{k}}|/3 < t < t_E^* - |E_{\widehat{k}}|/3\}. \tag{6.74}$$

Then one can easily check that

$$E_{\widehat{k},1} \subseteq (|E_{\widehat{k}}|/3, t_E^* - |E_{\widehat{k}}|/3) \quad \text{and} \quad |E_{\widehat{k},1}| \geq |E_{\widehat{k}}|/3. \tag{6.75}$$

Now we will use the above observations (i)–(ii) and Proposition 6.1 to derive a contradiction.

Indeed, by $(\widehat{H}1)$, we can apply Proposition 6.1 to find that the system (6.65) is L^∞-null controllable from measurable sets in time. Then by Definition 6.3 (where (T, t, E) is replaced by $(t_E^* - |E_{\widehat{k}}|/3, |E_{\widehat{k}}|/3, E_{\widehat{k},1})$), for each $\delta \in (0, |E_{\widehat{k}}|/3)$ satisfying that

$$\|\widehat{y}_0\|_Y \leq 1/\left(\widehat{k}C \max_{0 \leq s \leq t_E^*} \|e^{As}\|_{\mathscr{L}(Y)}\right) \quad \text{with } \widehat{y}_0 \triangleq y_0^* - y(\delta; 0, y_0^*, u^*), \tag{6.76}$$

there is $\widehat{u} \in L^\infty(0, +\infty; U)$ so that

$$y(t_E^* - |E_{\widehat{k}}|/3; |E_{\widehat{k}}|/3, y(|E_{\widehat{k}}|/3; \delta, \widehat{y}_0, 0), \chi_{E_{\widehat{k},1}} \widehat{u}) = 0, \tag{6.77}$$

and so that

$$\|\chi_{E_{\widehat{k},1}} \widehat{u}\|_{L^\infty(0,+\infty; U)} \leq C\|y(|E_{\widehat{k}}|/3; \delta, \widehat{y}_0, 0)\|_Y. \tag{6.78}$$

(Here, C is the corresponding constant in Definition 6.3.) It follows from (6.75) and (6.77) that

$$y(t_E^* - |E_{\widehat{k}}|/3; \delta, \widehat{y}_0, \chi_{E_{\widehat{k},1}} \widehat{u}) = 0. \tag{6.79}$$

Define a new control in the following manner:

$$\widetilde{u}(t) \triangleq u^*(t + \delta) + \chi_{E_{\widehat{k},1}}(t + \delta)\widehat{u}(t + \delta), \quad 0 < t < t_E^* - \delta. \tag{6.80}$$

Then we have that

$$\widetilde{u}(t) \in \mathbb{U} \quad \text{for a.e. } t \in (0, t_E^* - \delta). \tag{6.81}$$

In fact, by the third conclusion in (6.69) and (6.73), we obtain that

$$u^*(t) + B_{1/\widehat{k}}(0) \subseteq \mathbb{U} \text{ for a.e. } t \in E_{\widehat{k}}.$$

This, together with (6.80), (6.78), (6.76), and (6.74), yields that

$$\widetilde{u}(t) = u^*(t + \delta) + \widehat{u}(t + \delta)$$

$$\in u^*(t + \delta) + B_{1/\widehat{k}}(0) \subseteq \mathbb{U} \text{ for a.e. } t \in E_{\widehat{k},1} - \{\delta\},$$

which, together with (6.80) and the third conclusion in (6.69), leads to (6.81). Meanwhile, from (6.80), (6.76), and (6.75) it follows that

$$y(t_E^* - \delta; 0, y_0^*, \widetilde{u}) = e^{A(t_E^* - \delta)} y_0^*$$

$$+ \int_0^{t_E^* - \delta} e^{A(t_E^* - \delta - t)} B\big[u^*(t + \delta) + (\chi_{E_{\widehat{k},1}} \widehat{u})(t + \delta)\big] dt$$

$$= \Big[e^{A(t_E^* - \delta)} y(\delta; 0, y_0^*, u^*) + \int_\delta^{t_E^*} e^{A(t_E^* - t)} Bu^*(t) dt\Big]$$

$$+ \Big[e^{A(t_E^* - \delta)} \widehat{y}_0 + \int_\delta^{t_E^*} e^{A(t_E^* - t)} B(\chi_{E_{\widehat{k},1}} \widehat{u})(t) dt\Big]$$

$$= y(t_E^*; 0, y_0^*, u^*) + e^{A|E_{\widehat{k}}|/3} y(t_E^* - |E_{\widehat{k}}|/3; \delta, \widehat{y}_0, \chi_{E_{\widehat{k},1}} \widehat{u}),$$

which, combined with (6.79) and the second conclusion in (6.69), indicates that

$$y(t_E^* - \delta; 0, y_0^*, \widetilde{u}) = y(t_E^*; 0, y_0^*, u^*) \in Y_E. \tag{6.82}$$

Finally, from (6.82) and (6.81), we see that $(0, y_0^*, t_E^* - \delta, \widetilde{u})$ is an admissible tetrad of $(TP)_{min}^{Qs,QE}$, which contradicts the optimality of t_E^*. Hence, $(TP)_{min}^{Qs,QE}$ holds the bang-bang property. This completes the proof of Theorem 6.5. $\qquad\square$

We are now in the position to show Proposition 6.1. For this purpose, some preliminaries are needed. The first preliminary concerns complexifications. Write Y^C and U^C for the complexifications of Y and U, respectively. Define two operators in the following manner:

$$A^C(y_1 + iy_2) \triangleq Ay_1 + iAy_2 \text{ for each } y_1, y_2 \in \mathcal{D}(A), \tag{6.83}$$

and

$$B^C(u_1 + iu_2) \triangleq Bu_1 + iBu_2 \text{ for each } u_1, u_2 \in U. \tag{6.84}$$

From (6.83) and (6.84), we see that for each $z \in Y$ and each $u \in U$,

$$e^{(A^C)^*t} z = e^{A^*t} z \text{ for all } t \geq 0 \text{ and } (B^C)^* u = Bu.$$

This, along with (ii) and (iii) of $(\widehat{H}1)$, indicates the following two properties:

(ii)′ The operator A^C generates an analytic semigroup $\{e^{A^C t}\}_{t \geq 0}$ over Y^C.

(iii)′ There exist two constants $d > 0$ and $\sigma > 0$ so that for each $L \in (0, 1]$,

$$\|e^{(A^C)^* L} z\|_{Y^C}^2 \leq e^{d/L^\sigma} \int_0^L \|(B^C)^* e^{(A^C)^*(L-t)} z\|_{U^C}^2 \, dt \quad \text{for all } z \in Y^C.$$

(Here (d, τ) is given by (iii) of $(\widehat{H}1)$.)

Thus we conclude that

$$(ii) \text{ and } (iii) \text{ of } (H1) \Rightarrow (ii)' \text{ and } (iii)'. \tag{6.85}$$

The second preliminary is the L^1-observability inequality from measurable sets in time, presented in the next Lemma 6.4, which will be proved after the proof of Proposition 6.1.

Lemma 6.4 *Assume that $(\widehat{H}1)$ holds. Let $T > t > 0$ and let $E \subseteq (t, T)$ be a subset of positive measure. Then there exists a constant $C_0 \triangleq C_0(A, B, T, t, E) > 0$ so that*

$$\|e^{(A^C)^*(T-t)} z\|_{Y^C} \leq C_0 \int_E \|(B^C)^* e^{(A^C)^*(T-s)} z\|_{U^C} \, ds \quad \text{for all } z \in Y^C. \tag{6.86}$$

Now we give the proof of Proposition 6.1.

Proof (Proof of Proposition 6.1) Arbitrarily fix $T > t > 0$ and a subset E (in (t, T)) of positive measure. According to Lemma 6.4, (6.83), and (6.84), there is $C_0 \triangleq C_0(A, B, T, t, E) > 0$ so that

$$\|e^{A^*(T-t)} z\|_Y \leq C_0 \int_E \|B^* e^{A^*(T-s)} z\|_U \, ds \quad \text{for all } z \in Y.$$

Then by the similar arguments to those used in the proof of Theorem 1.20, we can obtain (6.67). Thus, the system (6.65) is L^∞-null controllable from measurable sets in time (see Definition 6.3). This ends the proof of Proposition 6.1. □

Next we go back to the proof of Lemma 6.4. Two other lemmas (Lemma 6.5 and Lemma 6.6) are needed. We will prove them after the proof of Lemma 6.4.

Lemma 6.5 *Suppose that $(\widehat{H}1)$ holds. Let $0 < t_1 < t_2$ with $0 < t_2 - t_1 \leq 1$. Let $\eta \in (0, 1)$ and E be a measurable subset so that*

$$|E \cap (t_1, t_2)| \geq \eta(t_2 - t_1).$$

Then there exist two constants $\widehat{C} \triangleq \widehat{C}(A, B, \eta) > 0$ and $\theta \triangleq \theta(A, B, \eta) \in (0, 1)$ so that for each $z \in Y^C$,

$$\|e^{(A^C)^*t_2}z\|_{Y^C} \leq \left(\widehat{C}e^{\frac{\widehat{c}}{(t_2-t_1)^\sigma}} \int_{E\cap(t_1,t_2)} \|(B^C)^*e^{(A^C)^*t}z\|_{U^C}\mathrm{d}t\right)^\theta \|e^{(A^C)^*t_1}z\|_{Y^C}^{1-\theta}.$$
(6.87)

Lemma 6.6 *Let $E \subseteq (0, +\infty)$ be a subset of positive measure. Let $\ell \in (0, +\infty)$ be a Lebesgue density point of E. Then for each $\lambda \in (0, 1)$, there exists an increasing sequence $\{\ell_m\}_{m\geq 1} \subseteq (0, +\infty) \cap (\ell - 1, \ell)$ so that for each $m \in \mathbb{N}^+$,*

$$\ell - \ell_m = \lambda^{m-1}(\ell - \ell_1) \quad and \quad |E \cap (\ell_m, \ell_{m+1})| \geq \frac{1}{3}|\ell_{m+1} - \ell_m|.$$

(The above Lemma 6.6 is quoted from [20, Proposition 2.1].) We now give the proof of Lemma 6.4.

Proof (Proof of Lemma 6.4.) First of all, by (6.85), we have (ii)′ and (iii)′. Let $\ell \in (t, T)$ be a Lebesgue density point of E. Let σ be given by (iii)′ and let $(\widehat{C}, \theta) \triangleq (\widehat{C}(A, B, 3^{-1}), \theta(A, B, 3^{-1}))$ be given by Lemma 6.5. We choose $\lambda \in \left((1-\theta)^{1/\sigma}, 1\right)$. According to Lemma 6.6, there exists an increasing sequence $\{\ell_m\}_{m\geq 1} \subseteq (t, \ell) \cap (\ell - 1, \ell)$ so that for each $m \in \mathbb{N}^+$,

$$\ell - \ell_m = \lambda^{m-1}(\ell - \ell_1) \quad and \quad |E \cap (\ell_m, \ell_{m+1})| \geq \frac{1}{3}|\ell_{m+1} - \ell_m|. \qquad (6.88)$$

Arbitrarily fix $z \in Y^C$. Define

$$h(t) \triangleq e^{(A^C)^*(T-t)}z, \quad t \in [0, T]. \qquad (6.89)$$

For each $m \in \mathbb{N}^+$, we apply Lemma 6.5, where

$$(E, t_1, t_2, \eta) \text{ is replaced by } (\{T\} - E \cap (\ell_m, \ell_{m+1}), T - \ell_{m+1}, T - \ell_m, 1/3),$$

to find that

$$\|h(\ell_m)\|_{Y^C} \leq \left(\widehat{C}e^{\widehat{C}/(\ell_{m+1}-\ell_m)^\sigma} \int_{E\cap(\ell_m, \ell_{m+1})} \|(B^C)^*h(s)\|_{U^C}\mathrm{d}s\right)^\theta \|h(\ell_{m+1})\|_{Y^C}^{1-\theta}.$$

By the latter inequality and the Young inequality:

$$ab \leq \varepsilon a^p + \varepsilon^{-\frac{r}{p}}b^r \text{ when } a > 0, b > 0, \varepsilon > 0 \text{ with } \frac{1}{p} + \frac{1}{r} = 1, p > 1, r > 1,$$

we see that for each $m \in \mathbb{N}^+$ and each $\varepsilon > 0$,

$$\|h(\ell_m)\|_{Y^C} \leq \varepsilon\|h(\ell_{m+1})\|_{Y^C}$$
$$+ \frac{\widehat{C}}{\varepsilon^{(1-\theta)/\theta}}e^{\widehat{C}/(\ell_{m+1}-\ell_m)^\sigma} \int_{E\cap(\ell_m, \ell_{m+1})} \|(B^C)^*h(s)\|_{U^C}\mathrm{d}s,$$

which implies that for each $m \in \mathbb{N}^+$ and each $\varepsilon > 0$,

$$\varepsilon^{(1-\theta)/\theta} e^{-\frac{\widehat{c}}{(\ell_{m+1}-\ell_m)^\sigma}} \|h(\ell_m)\|_{YC} - \varepsilon^{1/\theta} e^{-\frac{\widehat{c}}{(\ell_{m+1}-\ell_m)^\sigma}} \|h(\ell_{m+1})\|_{YC}$$

$$\leq \widehat{C} \int_{E \cap (\ell_m, \ell_{m+1})} \|(B^C)^* h(s)\|_{U^C} ds.$$

In the above, we choose, for each $m \in \mathbb{N}^+$,

$$\varepsilon \triangleq \varepsilon_m \triangleq \left[e^{-\frac{\widehat{c}}{(\ell_{m+1}-\ell_m)^\sigma}} \right]^{\frac{1-\lambda^\sigma}{\lambda^\sigma + (1-\theta)(\lambda^\sigma - 1)/\theta}}.$$

Then we obtain that for each $m \in \mathbb{N}^+$,

$$e^{-\frac{\widehat{C}\lambda^\sigma}{[\lambda^\sigma + (1-\theta)(\lambda^\sigma - 1)/\theta](\ell_{m+1}-\ell_m)^\sigma}} \|h(\ell_m)\|_{YC}$$

$$-e^{-\frac{\widehat{C}\lambda^\sigma}{[\lambda^\sigma + (1-\theta)(\lambda^\sigma - 1)/\theta](\ell_{m+2}-\ell_{m+1})^\sigma}} \|h(\ell_{m+1})\|_{YC}$$

$$\leq \widehat{C} \int_{E \cap (\ell_m, \ell_{m+1})} \|(B^C)^* h(s)\|_{U^C} ds.$$

Summing the above over all m, we see that

$$\|h(\ell_1)\|_{YC} \leq \widehat{C} e^{\frac{\widehat{C}\lambda^\sigma}{[\lambda^\sigma + (1-\theta)(\lambda^\sigma - 1)/\theta](\ell_2-\ell_1)^\sigma}} \int_{E \cap (\ell_1, \ell)} \|(B^C)^* h(s)\|_{U^C} ds.$$

This, along with (6.88) and (6.89), leads to (6.86).

Hence, we finish the proof of Lemma 6.4.   □

The remainder on Proposition 6.1 is to give proofs of Lemma 6.5 and Lemma 6.6. To show Lemma 6.5, we need a propagation of smallness estimate from measurable sets for real analytic functions, built up in [25] (see also [1, Lemma 2] or [2, Lemma 13]). This estimate will be given in Lemma 6.7, without proof.

Lemma 6.7 *Let* $f : [a, a+s] \mapsto \mathbb{R}$, *where* $a \in \mathbb{R}$ *and* $s > 0$, *be an analytic function satisfying that*

$$|f^{(\beta)}(x)| \leq M\beta!(s\rho)^{-\beta} \quad \text{for all } x \in [a, a+s] \text{ and } \beta \in \mathbb{N}^+ \cup \{0\}$$

with some constants $M > 0$ *and* $\rho \in (0, 1]$. *Assume that* $\widehat{E} \subseteq [a, a+s]$ *is a subset of positive measure. Then there are two constants* $C \triangleq C(\rho, |\widehat{E}|/s) \geq 1$ *and* $\vartheta \triangleq \vartheta(\rho, |\widehat{E}|/s) \in (0, 1)$ *so that*

$$\|f\|_{L^\infty(a, a+s)} \leq C M^{1-\vartheta} \left(\frac{1}{|\widehat{E}|} \int_{\widehat{E}} |f(x)| dx \right)^\vartheta.$$

We now give the proof of Lemma 6.5.

Proof (Proof of Lemma 6.5) First of all, we recall (6.85). Let $0 < t_1 < t_2$ with $0 < t_2 - t_1 \leq 1$. Let $\eta \in (0, 1)$ and E be a measurable subset so that

$$|E \cap (t_1, t_2)| \geq \eta(t_2 - t_1).$$

Arbitrarily fix $z \in Y^C$. The rest of the proof is carried out by the following two steps:

Step 1. We show two facts. Fact One: The following function is real analytic:

$$g(t; z) \triangleq \|(B^C)^* e^{(A^C)^* t} z\|_{U^C}^2, \quad t > 0;$$

Fact Two: There is $\rho \in (0, 1)$ (only depending on A^C) so that for all $t, s > 0$ with $0 < t - s \leq 1$,

$$|g^{(\beta)}(t; z)| \leq \|B^C\|_{\mathscr{L}(U^C, Y^C)}^2 \frac{\beta!}{[\rho(t-s)]^\beta} \|e^{(A^C)^* s} z\|_{Y^C}^2 \text{ for all } \beta \in \mathbb{N}^+.$$

$$\text{(6.90)}$$

Recall properties $(ii)'$ and $(iii)'$ before (6.85). Fact One follows from $(ii)'$. We now show Fact Two. According to $(ii)'$, there exist three positive constants M, ω and δ (only depending on A^C) so that

$$\|e^{A^C \tau}\|_{\mathscr{L}(Y^C, U^C)} \leq M e^{\omega \tau} \text{ for each } \tau \in \Delta_\delta \triangleq \{\tau \in \mathbb{C} : |\arg \tau| < \delta\}. \quad \text{(6.91)}$$

We claim that there exists $\rho_1 \in (0, 1)$ (only depending on A^C) so that

$$\|[(A^C)^*]^\alpha e^{(A^C)^* \tau} y\|_{Y^C}$$
$$\leq \frac{\alpha!}{(\rho_1 \tau)^\alpha} \|y\|_{Y^C} \text{ for each } \tau \in (0, 1], \alpha \in \mathbb{N}^+ \cup \{0\} \text{ and } y \in Y^C. \quad \text{(6.92)}$$

Indeed, by $(ii)'$ and (6.91), we see that $A^C - \omega I$ generates a bounded analytic semigroup over Y^C. (Here and in what follows, I denotes the identity operator from Y^C to Y^C.) This, together with Theorem 5.2 in Chapter 2 of [19], implies that for each $\tau > 0$,

$$\|(A^C - \omega I)e^{(A^C - \omega I)\tau}\| \leq C_1/\tau \text{ (with } C_1 > 0 \text{ depending only on } A^C).$$

From the above inequality, $(ii)'$ and (6.91), we see that for each $\tau > 0$,

$$\|(A^C)^* e^{(A^C)^* \tau}\|_{\mathscr{L}(Y^C, Y^C)} = \|A^C e^{A^C \tau}\|_{\mathscr{L}(Y^C, Y^C)}$$
$$\leq e^{\omega \tau} \left(\|(A^C - \omega I)e^{(A^C - \omega I)\tau}\|_{\mathscr{L}(Y^C, Y^C)} \right.$$
$$\left. + \omega \|e^{(A^C - \omega I)\tau}\|_{\mathscr{L}(Y^C, Y^C)} \right)$$
$$\leq e^{\omega \tau} (C_1/\tau + \omega M),$$

which indicates that

$$\|(A^C)^* e^{(A^C)^* \tau}\|_{\mathscr{L}(Y^C, Y^C)} \le e^{\omega}(C_1 + \omega M)/\tau \triangleq C_2/\tau \quad \text{for all } \tau \in (0, 1].$$

This yields that for each $\alpha \in \mathbb{N}^+ \cup \{0\}$ and $\tau \in (0, 1]$,

$$\|[(A^C)^*]^\alpha e^{(A^C)^* \tau} y\|_{Y^C} = \|[(A^C)^* e^{(A^C)^* \tau/\alpha}]^\alpha y\|_{Y^C} \le \left(\frac{C_2 \alpha}{\tau}\right)^\alpha \|y\|_{Y^C}.$$

From this and the Stirling formula: $\alpha^\alpha \lesssim e^\alpha \alpha!$ for each $\alpha \in \mathbb{N}^+ \cup \{0\}$, we obtain (6.92).

Next, we arbitrarily fix $t, s > 0$ with $0 < t - s \le 1$. For each $\beta \in \mathbb{N}^+ \cup \{0\}$, since

$$g^{(\beta)}(t; z) = \sum_{\beta_1 + \beta_2 = \beta} \frac{\beta!}{\beta_1! \beta_2!} \left\langle \frac{d^{\beta_1}}{dt^{\beta_1}} ((B^C)^* e^{(A^C)^* t} z), \frac{d^{\beta_2}}{dt^{\beta_2}} ((B^C)^* e^{(A^C)^* t} z) \right\rangle_{U^C},$$

it follows that

$$|g^{(\beta)}(t; z)|$$
$$\le \sum_{\beta_1 + \beta_2 = \beta} \frac{\beta!}{\beta_1! \beta_2!} \|(B^C)^* [(A^C)^*]^{\beta_1} e^{(A^C)^* t} z\|_{U^C} \|(B^C)^* [(A^C)^*]^{\beta_2} e^{(A^C)^* t} z\|_{U^C}$$
$$\le \|(B^C)^*\|^2_{\mathscr{L}(Y^C, U^C)} \sum_{\beta_1 + \beta_2 = \beta} \frac{\beta!}{\beta_1! \beta_2!} \|[(A^C)^*]^{\beta_1} e^{(A^C)^* t} z\|_{Y^C} \|[(A^C)^*]^{\beta_2} e^{(A^C)^* t} z\|_{Y^C}.$$
$$(6.93)$$

Finally, from (6.93) and (6.92), we see that for each $\beta \in \mathbb{N}^+ \cup \{0\}$,

$$|g^{(\beta)}(t; z)| \le \|(B^C)^*\|^2_{\mathscr{L}(Y^C, U^C)} \sum_{\beta_1 + \beta_2 = \beta} \frac{\beta!}{\beta_1! \beta_2!} \|[(A^C)^*]^{\beta_1} e^{(A^C)^*(t-s)} (e^{(A^C)^* s} z)\|_{Y^C}$$

$$\|[(A^C)^*]^{\beta_2} e^{(A^C)^*(t-s)} (e^{(A^C)^* s} z)\|_{Y^C}$$

$$\le \|(B^C)^*\|^2_{\mathscr{L}(Y^C, U^C)} \sum_{\beta_1 + \beta_2 = \beta} \frac{\beta!}{\beta_1! \beta_2!} \frac{\beta_1! \beta_2!}{(\rho_1(t-s))^{\beta_1 + \beta_2}} \|e^{(A^C)^* s} z\|^2_{Y^C}$$

$$= \left[\|(B^C)^*\|_{\mathscr{L}(Y^C, U^C)} \|(e^{(A^C)^* s}) z\|_{Y^C} \right]^2 \frac{(\beta + 1)!}{(\rho_1(t-s))^\beta}.$$

Since $\beta + 1 \le 2^\beta$ for each $\beta \in \mathbb{N}^+ \cup \{0\}$, the above leads to (6.90). So Fact Two is true.

Step 2. We prove (6.87).
Set

$$\tau \triangleq t_1 + \frac{\eta}{2}(t_2 - t_1) \quad \text{and} \quad \widehat{E} \triangleq E \cap [\tau, t_2].$$

It is clear that

$$|\widehat{E}| \geq |E \cap (t_1, t_2)| - (\tau - t_1) \geq \frac{\eta}{2}(t_2 - t_1).$$

Let $\widehat{E}_1 \subseteq \widehat{E}$ be a measurable subset so that

$$|\widehat{E}_1| = \frac{\eta}{2}(t_2 - t_1).$$

By Step 1, there exist two constants $K \geq 1$ and $\rho \in (0, 1)$ (only depending on A^C and B^C) so that for each $t \in [\tau, t_2]$ and $\beta \in \mathbb{N}^+ \cup \{0\}$,

$$|g^{(\beta)}(t; z)| \leq K \frac{\beta!}{[\rho(t - t_1)]^\beta} \|e^{(A^C)^*t_1} z\|_{YC}^2$$

$$\leq K \frac{\beta!}{[\rho\eta(t_2 - t_1)/2]^\beta} \|e^{(A^C)^*t_1} z\|_{YC}^2.$$

Then according to Lemma 6.7, where

$$(\widehat{E}, M, \rho, s, a) \text{ is replaced by } \left(\widehat{E}_1, K\|e^{(A^C)^*t_1} z\|_{YC}^2, \frac{\rho\eta}{2 - \eta}, \left(1 - \frac{\eta}{2}\right)(t_2 - t_1), \tau\right),$$

there exists $C_3 > 0$ and $\nu \in (0, 1)$ (only depending on A^C, B^C and η) so that for each $t \in [\tau, t_2]$,

$$|g(t; z)| \leq C_3 \left(K\|e^{(A^C)^*t_1} z\|_{YC}^2\right)^{1-\nu} \left(\frac{1}{|\widehat{E}_1|} \int_{\widehat{E}_1} |g(s; z)| ds\right)^\nu.$$

Thus, for each $t \in [\tau, t_2]$, we have that

$$\|(B^C)^* e^{(A^C)^*t} z\|_{UC}^2 \leq \frac{2C_3 K}{\eta(t_2 - t_1)} \|e^{(A^C)^*t_1} z\|_{YC}^{2(1-\nu)} \left(\int_{\widehat{E}_1} \|(B^C)^* e^{(A^C)^*s} z\|_{UC}^2 ds\right)^\nu.$$

From this and (iii)′, we see that

$$\|e^{(A^C)^*t_2} z\|_{YC}^2 = \|e^{(A^C)^*(t_2 - \tau)} (e^{(A^C)^*\tau} z)\|_{YC}^2$$

$$\leq e^{d/(t_2 - \tau)^\sigma} \int_0^{t_2 - \tau} \|(B^C)^* e^{(A^C)^*(t_2 - \tau - s)} (e^{(A^C)^*\tau} z)\|_{UC}^2 ds$$

$$= e^{d/(t_2-\tau)^\sigma} \int_\tau^{t_2} \|(B^C)^* e^{(A^C)^* t} z\|_{U^C}^2 dt$$

$$\leq e^{d/(t_2-\tau)^\sigma} (t_2 - \tau) \frac{2C_3 K}{\eta(t_2 - t_1)} \|e^{(A^C)^* t_1} z\|_{Y^C}^{2(1-\nu)} \left(\int_{\widehat{E}_1} \|(B^C)^* e^{(A^C)^* t} z\|_{U^C}^2 dt \right)^\nu.$$

This implies that

$$\|e^{(A^C)^* t_2} z\|_{Y^C}^2 \leq e^{d/(t_2-\tau)^\sigma} \frac{(2-\eta)C_3 K}{\eta} \|e^{(A^C)^* t_1} z\|_{Y^C}^{2(1-\nu)} \times$$

$$\left(\sup_{s \in \widehat{E}_1} \|(B^C)^* e^{(A^C)^* s} z\|_{U^C} \int_E \|(B^C)^* e^{(A^C)^* t} z\|_{U^C} dt \right)^\nu$$

$$\leq \left[e^{d/(t_2-\tau)^\sigma} \frac{(2-\eta)C_3 K}{\eta} \right]$$

$$\left[\|(B^C)^*\|_{\mathscr{L}(U^C, Y^C)} \sup_{s \in [0, t_2-t_1]} \|e^{(A^C)^* s}\|_{\mathscr{L}(Y^C)} \right]^\nu$$

$$\times \|e^{(A^C)^* t_1} z\|_{Y^C}^{2-\nu} \left(\int_E \|(B^C)^* e^{(A^C)^* t} z\|_{U^C} dt \right)^\nu,$$

which leads to (6.87). This completes the proof of Lemma 6.5.   □

Finally, we prove Lemma 6.6.

Proof (Proof of Lemma 6.6.) Arbitrarily fix $\lambda \in (0, 1)$. Write E^c for the complement subset of E. Since ℓ is a Lebesgue density point of E, we have that

$$\lim_{h \to 0+} \frac{1}{h} |E^c \cap (\ell - h, \ell + h)| = 0.$$

This yields that

$$\lim_{m \to +\infty} \frac{1}{\lambda^m \ell - \lambda^{m+1} \ell} |E \cap (\ell - \lambda^m \ell, \ell - \lambda^{m+1} \ell)|$$

$$= 1 - \lim_{m \to +\infty} \frac{1}{\lambda^m \ell - \lambda^{m+1} \ell} |E^c \cap (\ell - \lambda^m \ell, \ell - \lambda^{m+1} \ell)|$$

$$= 1 - \frac{2}{1 - \lambda} \lim_{m \to +\infty} \frac{1}{2\lambda^m \ell} |E^c \cap (\ell - \lambda^m \ell, \ell + \lambda^m \ell)| = 1.$$

Thus, there exists $m_0 \in \mathbb{N}^+$ so that for each $m \geq m_0$,

$$\frac{1}{\lambda^m \ell - \lambda^{m+1} \ell} |E \cap (\ell - \lambda^m \ell, \ell - \lambda^{m+1} \ell)| \geq 1/3 \quad \text{and} \quad \lambda^{m_0} \ell < 1. \qquad (6.94)$$

Let $\ell_m \triangleq \ell - \lambda^{m+m_0}\ell$ for each $m \in \mathbb{N}^+$. We get the desired result from (6.94) immediately. This completes the proof of Lemma 6.6. □

We end this subsection with the next example which gives auxiliary instructions on Theorem 6.5.

Example 6.10 Consider the problem $(TP)_{min}^{Q_S,Q_E}$, where the controlled system is the heat equation (3.59) (with $f = 0$); and where

$$Q_S = \{0\} \times \{y_0\}, \quad Q_E = (0, +\infty) \times B_r(0) \text{ and } \mathbb{U} = B_\rho(0),$$

with $y_0 \in L^2(\Omega) \setminus B_r(0)$, $r > 0$ and $\rho > 0$. One can easily check the following facts:

(i) The problem $(TP)_{min}^{Q_S,Q_E}$ can be put into the framework $(\mathscr{B}_2)$ (given at the beginning of Section 6.2);
(ii) The problem $(TP)_{min}^{Q_S,Q_E}$ has at least one admissible control (see Remark 3.1 after Theorem 3.5), which leads to the existence of optimal controls;
(iii) The assumption $(\widehat{H}1)$ is satisfied for this case (see [5, 20] and [21]).

Then according to Theorem 6.5, $(TP)_{min}^{Q_S,Q_E}$ holds the bang-bang property. Now we end Example 6.10.

6.2.2 Bang-Bang Property of Maximal Time Control

In this subsection, we aim to use the L^∞-null controllability (from measurable sets in time) to study the bang-bang property for the maximal control problem $(TP)_{max}^{Q_S,Q_E}$. For this purpose, we need the following assumption:

$(\widehat{H}2)$ The system (6.65) is L^∞-null controllable from measurable sets in time.

The main result of this subsection is stated as follows:

Theorem 6.6 *Suppose that the assumption $(\widehat{H}2)$ holds. Let $T > 0$ and let*

$$Q_S = \{(t, \Phi(t,0)z) \mid 0 \le t < T, z \in Y_S\} \text{ and } Q_E = \{T\} \times Y_E,$$

where Y_S and Y_E are two nonempty subsets of Y with $(\Phi(T,0)Y_S) \cap Y_E = \emptyset$. Then the maximal time control problem $(TP)_{max}^{Q_S,Q_E}$ holds the bang-bang property.

Proof By contradiction, we suppose that $(TP)_{max}^{Q_S,Q_E}$ did not hold the bang-bang property. Then by Definition 6.2, there would exist an optimal tetrad $(t_S^*, \Phi(t_S^*, 0)y_0^*, T, u^*)$ of $(TP)_{max}^{Q_S,Q_E}$ and a subset $E \subseteq (t_S^*, T)$ of positive measure so that

$$u^*(t) \notin \partial \mathbb{U} \text{ for a.e. } t \in E. \tag{6.95}$$

It is clear that

$$0 \leq t_S^* < T, \quad y_0^* \in Q_S, \quad y(T; t_S^*, \Phi(t_S^*, 0)y_0^*, u^*) \in Y_E \tag{6.96}$$

and

$$u^*(t) \in \mathbb{U} \text{ for a.e. } t \in (t_S^*, T). \tag{6.97}$$

For each $k \in \mathbb{N}^+$, we define a subset of E in the following manner:

$$E_k \triangleq \{t \in E \mid d_U(u^*(t), \partial\mathbb{U}) > k^{-1}\}, \tag{6.98}$$

where $d_U(\cdot, \partial\mathbb{U})$ denotes the distance between $\cdot$ and $\partial\mathbb{U}$ in U. By the same way as that used in the proof of Theorem 6.5, we can prove that for each $k \in \mathbb{N}^+$, E_k is measurable. Meanwhile, from (6.95) and (6.98), one can directly check that

$$E = \bigcup_{k=1}^{+\infty} E_k \text{ and } E_j \subseteq E_\ell \text{ for all } j, \ell \in \mathbb{N}^+ \text{ with } j \leq \ell. \tag{6.99}$$

Two observations are given in order:

(i) There exists $\widehat{k} \in \mathbb{N}^+$ so that $|E_{\widehat{k}}| > 0$; (This follows from (6.99) and the fact that $|E| > 0$.)

(ii) It follows from (6.98) that

$$d_U(u^*(t), \partial\mathbb{U}) > 1/\widehat{k} \text{ for each } t \in E_{\widehat{k}}. \tag{6.100}$$

Define a subset of $E_{\widehat{k}}$ in the following manner:

$$E_{\widehat{k},1} \triangleq \{t \in E_{\widehat{k}} \mid t_S^* + |E_{\widehat{k}}|/2 < t < T\}. \tag{6.101}$$

It is obvious that

$$E_{\widehat{k},1} \subseteq (t_S^* + |E_{\widehat{k}}|/2, T) \text{ and } |E_{\widehat{k},1}| \geq |E_{\widehat{k}}|/2. \tag{6.102}$$

We now use the above observations (i) and (ii) to obtain a contradiction. First, by $(\widehat{H}2)$ and Definition 6.3 (where (T, t, E) is replaced by $(T, t_S^* + |E_{\widehat{k}}|/2, E_{\widehat{k},1})$), we see that for each $\delta \in (0, |E_{\widehat{k}}|/2)$ satisfying that

$$\|\widehat{y}_0\|_Y \leq 1 / \left(\widehat{k}C \sup_{0 \leq s \leq t \leq T} \|\Phi(t, s)\|_{\mathscr{L}(Y)} \right) \text{ with } \widehat{y}_0 \triangleq -y(t_S^* + \delta; t_S^*, 0, u^*), \tag{6.103}$$

there is a control $\widehat{u} \in L^\infty(0, +\infty; U)$ so that

$$y(T; t_S^* + |E_{\widehat{k}}|/2, y(t_S^* + |E_{\widehat{k}}|/2; t_S^* + \delta, \widehat{y}_0, 0), \chi_{E_{\widehat{k},1}} \widehat{u}) = 0 \qquad (6.104)$$

and so that

$$\|\chi_{E_{\widehat{k},1}} \widehat{u}\|_{L^\infty(0,+\infty;U)} \leq C \|y(t_S^* + |E_{\widehat{k}}|/2; t_S^* + \delta, \widehat{y}_0, 0)\|_Y. \qquad (6.105)$$

(Here, C is the corresponding constant given by Definition 6.3.) Then, it follows from (6.102) and (6.104) that

$$y(T; t_S^* + \delta, \widehat{y}_0, \chi_{E_{\widehat{k},1}} \widehat{u}) = 0. \qquad (6.106)$$

Define a new control in the following manner:

$$\widetilde{u}(t) \triangleq u^*(t) + \chi_{E_{\widehat{k},1}}(t)\widehat{u}(t), \quad t \in (t_S^* + \delta, T). \qquad (6.107)$$

We claim that

$$\widetilde{u}(t) \in \mathbb{U} \text{ for a.e. } t \in (t_S^* + \delta, T). \qquad (6.108)$$

Indeed, by (6.97) and (6.100), we obtain that

$$u^*(t) + B_{1/\widehat{k}}(0) \subseteq \mathbb{U} \text{ for a.e. } t \in E_{\widehat{k}}.$$

From the above, (6.107), (6.101), (6.105), and (6.103), we find that

$$\widetilde{u}(t) = u^*(t) + \widehat{u}(t) \in u^*(t) + B_{1/\widehat{k}}(0) \subseteq \mathbb{U} \text{ for a.e. } t \in E_{\widehat{k},1}.$$

This, along with (6.107) and (6.97), leads to (6.108).

Next, by (6.107) and (6.103), one can directly check that

$$y(T; t_S^* + \delta, \Phi(t_S^* + \delta, 0)y_0^*, \widetilde{u})$$

$$= y(T; t_S^*, \Phi(t_S^*, 0)y_0^*, u^*) + \Phi(T, t_S^* + \delta)\widehat{y}_0 + \int_{t_S^*+\delta}^{T} \Phi(T, t)B(t)(\chi_{E_{\widehat{k},1}} \widehat{u})(t)dt,$$

which, combined with (6.106) and the third conclusion of (6.96), indicates that

$$\begin{aligned}
y(T; t_S^* + \delta, \Phi(t_S^* + \delta, 0)y_0^*, \widetilde{u}) \\
= y(T; t_S^*, \Phi(t_S^*, 0)y_0^*, u^*) + y(T; t_S^* + \delta, \widehat{y}_0, \chi_{E_{\widehat{k},1}} \widehat{u}) \\
= y(T; t_S^*, \Phi(t_S^*, 0)y_0^*, u^*) \in Y_E.
\end{aligned} \qquad (6.109)$$

Finally, since $0 \leq t_S^* + \delta < T$, it follows by the second conclusion of (6.96), (6.109) and (6.108) that $(t_S^* + \delta, \Phi(t_S^* + \delta, 0)y_0^*, T, \widetilde{u})$ is an admissible tetrad of $(TP)_{max}^{Q_S,Q_E}$, which contradicts the optimality of t_S^*. Thus, $(TP)_{max}^{Q_S,Q_E}$ holds the bang-bang property. This completes the proof of Theorem 6.6. $\qquad \square$

We end this subsection with an example which gives auxiliary instructions on Theorem 6.6.

Example 6.11 Consider the maximal time control problem $(TP)_{max}^{Q_S,Q_E}$, where the controlled system is the heat equation (3.107) (with $a \in L^\infty(0, +\infty; L^\infty(\Omega)))$; and where

$$Q_S = \{(t, \Phi(t,0)y_0) \mid 0 \le t < T\}, \quad Q_E = \{T\} \times B_r(0) \text{ and } \mathbb{U} = B_\rho(0),$$

with $\Phi(T,0)y_0 \in L^2(\Omega) \setminus B_r(0)$, $r > 0$ and $\rho > 0$. One can easily check the following facts:

(i) $(TP)_{max}^{Q_S,Q_E}$ can be put into the framework of $(\mathscr{B}_2)$ (given at the beginning of Section 6.2);
(ii) The assumption $(\widehat{H2})$ (given at the beginning of Section 6.2.2) holds in this case. (Here, we used Theorems 1.22 and 1.21.)

If we further assume that $(TP)_{max}^{Q_S,Q_E}$ has optimal controls, then according to Theorem 6.6, $(TP)_{max}^{Q_S,Q_E}$ holds the bang-bang property. We end Example 6.11.

6.3 Bang-Bang Property and Maximum Principle

In this section, we will study the bang-bang property of the minimal time control problem $(TP)_{min}^{Q_S,Q_E}$ from the perspective of the Pontryagin Maximum Principle. This problem is under the framework $(\mathscr{B}_3)$:

(i) The state space Y and the control space U are two real separable Hilbert spaces.
(ii) The controlled system is as:

$$\dot{y}(t) = Ay(t) + D(t)y(t) + B(t)u(t), \quad t \in (0, +\infty), \tag{6.110}$$

where $A : \mathscr{D}(A) \subseteq Y \mapsto Y$ generates a C_0 semigroup $\{e^{At}\}_{t\ge 0}$ over Y; $D(\cdot) \in L^\infty(0, +\infty; \mathscr{L}(Y))$ and $B(\cdot) \in L^\infty(0, +\infty; \mathscr{L}(U, Y))$. Given $\tau \in [0, +\infty)$, $y_0 \in Y$ and $u \in L^\infty(\tau, +\infty; U)$, we write $y(\cdot; \tau, y_0, u)$ for the solution of (6.110) over $[\tau, +\infty)$ with the initial condition that $y(\tau) = y_0$. Moreover, write $\{\Phi(t,s) : t \ge s \ge 0\}$ for the evolution system generated by $A + D(\cdot)$ over Y.
(iii) The control constraint set $\mathbb{U}$ is bounded, convex, and closed in U, and has a nonempty interior in U.
(iv) Sets Q_S and Q_E are disjoint nonempty subsets of $[0, +\infty) \times Y$ and $(0, +\infty) \times Y$, respectively.

As what we did in the previous two sections, we always assume that $(TP)_{min}^{Q_S,Q_E}$ has optimal controls.

We now give the definition of the bang-bang property for the problem $(TP)_{min}^{Q_S, Q_E}$.

Definition 6.4

(i) An optimal control u^* (associated with an optimal tetrad $(0, y_0^*, t_E^*, u^*)$) to the problem $(TP)_{min}^{Q_S, Q_E}$ is said to have the bang-bang property if

$$u^*(t) \in \partial \mathbb{U} \quad \text{for a.e. } t \in (0, t_E^*),$$

where $\partial \mathbb{U}$ denotes the boundary of $\mathbb{U}$.

(ii) The problem $(TP)_{min}^{Q_S, Q_E}$ is said to have the bang-bang property, if any optimal control satisfies the bang-bang property.

In this section, the following unique continuation property $(\widehat{H}3)$ plays an important role:

$(\widehat{H}3)$ If there is a measurable subset $E \subseteq (0, t_0)$ (with $t_0 > 0$ and $|E| > 0$) and $z \in Y$ so that $B(t)^* \Phi(T, t)^* z = 0$ for each $t \in E$, then $z = 0$ in Y.

The first main result of this section is stated as follows:

Theorem 6.7 *Suppose that $(\widehat{H}3)$ holds. Let $Y_S \subseteq Y$ be a nonempty subset and $Y_E \subseteq Y$ be a bounded, convex, and closed subset with a nonempty interior. Assume that $Y_S \cap Y_E = \emptyset$. Let*

$$Q_S = \{0\} \times Y_S \quad \text{and} \quad Q_E = (0, +\infty) \times Y_E.$$

Then the problem $(TP)_{min}^{Q_S, Q_E}$ holds the bang-bang property.

Proof It suffices to prove this theorem in the case that Y_S is a singleton subset of Y. The reason is the same as that stated in the proof of Theorem 6.1. Thus, we can let $Y_S = \{y_0\}$ with $y_0 \in Y$.

Let $(0, y_0, t_E^*, u^*)$ be an optimal tetrad to $(TP)_{min}^{Q_S, Q_E}$. Since $Y_S \cap Y_E = \emptyset$, we have that $t_E^* > 0$. According to Theorem 4.1 and Theorem 4.2, the problem $(TP)_{min}^{Q_S, Q_E}$ satisfies the classical Pontryagin Maximum Principle. Then by Definition 4.1, there exists $z^* \in Y \setminus \{0\}$ so that

$$\langle u^*(t), B(t)^* \Phi(t_E^*, t)^* z^* \rangle_U = \max_{v \in U} \langle v, B(t)^* \Phi(t_E^*, t)^* z^* \rangle_U \quad \text{for a.e. } t \in (0, t_E^*).$$

$$(6.111)$$

We now are going to use (6.111) and $(\widehat{H}3)$ to derive that

$$u^*(t) \in \partial \mathbb{U} \quad \text{for a.e. } t \in (0, t_E^*). \qquad (6.112)$$

By contradiction, we suppose that (6.112) was not true. Then there would exist a subset $E \subseteq (0, t_E^*)$ of positive measure so that

$$u^*(t) \in \text{Int} \, \mathbb{U} \quad \text{for a.e. } t \in E.$$

This, together with (6.111), yields that

$$B(t)^* \Phi(t_E^*, t)^* z^* = 0 \quad \text{for a.e. } t \in E,$$

which, combined with $(\widehat{H}3)$, indicates that $z^* = 0$ in Y. This leads to a contradiction since $z^* \neq 0$. Therefore, (6.112) is true, i.e., u^* holds the bang-bang property, consequently, so does $(TP)_{min}^{Q_S, Q_E}$. Thus, we end the proof of Theorem 6.7.   □

The next example gives auxiliary instructions on Theorem 6.7.

Example 6.12 Consider the controlled system:

$$\dot{y}(t) = D(t)y(t) + B(t)u(t), \quad t \in (0, +\infty), \quad (\text{with } y(t) \in \mathbb{R}^n \text{ and } u(t) \in \mathbb{R}^m) \tag{6.113}$$

where $(D(\cdot), B(\cdot)) \in L^\infty(0, +\infty; \mathbb{R}^{n \times n}) \times L^\infty(0, +\infty; \mathbb{R}^{n \times m})$ (with $n, m \in \mathbb{N}^+$). Consider the problem $(TP)_{min}^{Q_S, Q_E}$, where the controlled system is (6.113) and where

$$Q_S = \{0\} \times \{y_0\}, \quad Q_E = (0, +\infty) \times B_r(0) \quad \text{and} \quad \mathbb{U} = B_\rho(0),$$

with $y_0 \in \mathbb{R}^n \setminus B_r(0)$ and $r, \rho > 0$. Then one can check that $(TP)_{min}^{Q_S, Q_E}$ can be put into the framework $(\mathscr{B}_3)$ (given at the beginning of Section 6.3).

If we further assume that $(TP)_{min}^{Q_S, Q_E}$ has optimal controls and that $(\widehat{H}3)$ holds for this case, then according to Theorem 6.7, $(TP)_{min}^{Q_S, Q_E}$ has the bang-bang property. We now end Example 6.12.

The second main result of this section is as follows:

Theorem 6.8 *Let*

$$D(\cdot) \equiv 0, \quad B(\cdot) \equiv B \in \mathscr{L}(U, Y),$$

and let

$$Q_S = \{0\} \times Y_S \quad \text{and} \quad Q_E = (0, +\infty) \times \{0\},$$

where Y_S is a nonempty subset of Y (with $0 \notin Y_S$). Suppose that $(\widehat{H}3)$ holds and that the system (6.110) is L^∞-null controllable over any interval. Assume that $0 \in Int \mathbb{U}$. Then the corresponding minimal time control problem $(TP)_{min}^{Q_S, Q_E}$ holds the bang-bang property.

Proof It suffices to prove this theorem in the case that Y_S is a singleton subset of Y. The reason is the same as that stated in the proof of Theorem 6.1. Thus, we can let $Y_S = \{y_0\}$ with $y_0 \neq 0$.

Let $(0, y_0, t_E^*, u^*)$ be an optimal tetrad to the problem $(TP)_{min}^{Q_S, Q_E}$. According to Theorem 4.4 and Theorem 4.3, $(TP)_{min}^{Q_S, Q_E}$ satisfies the local Pontryagin Maximum

Principle. Then by Definition 4.2, for each $T \in (0, t_E^*)$, there exists $z_T^* \in Y \setminus \{0\}$ so that

$$\langle u^*(t), B^* e^{A^*(T-t)} z_T^* \rangle_U = \max_{v \in \mathbb{U}} \langle v, B^* e^{A^*(T-t)} z_T^* \rangle_U \quad \text{for a.e. } t \in (0, T).$$

From this and using the similar arguments as those used in the proof of Theorem 6.7, we obtain that for each $T \in (0, t_E^*)$,

$$u^*(t) \in \partial \mathbb{U} \quad \text{for a.e. } t \in (0, T),$$

which leads to that

$$u^*(t) \in \partial \mathbb{U} \quad \text{for a.e. } t \in (0, t_E^*),$$

i.e., $(0, y_0, t_E^*, u^*)$ holds the bang-bang property, consequently, $(TP)_{min}^{Q_S, Q_E}$ has the bang-bang property. This ends the proof of Theorem 6.8. $\qquad \square$

Before presenting the last main result of this section, we introduce the following hypothesis:

$(\widehat{H4})$ $Y_T = Z_T$ for each $T \in (0, +\infty)$, where

$$Y_T \triangleq \overline{X_T}^{\|\cdot\|_{L^1(0,T;U)}},$$

$$X_T \triangleq \{B(\cdot)^* \Phi(T, \cdot)^* z \mid z \in Y\},$$

$$Z_T \triangleq \{B(\cdot)^* \varphi \in L^1(0, T; U) \mid \text{for each } s \in (0, T), \text{ there exists}$$
$$z_s \in Y \text{ so that } \varphi(\cdot) = \Phi(s, \cdot)^* z_s \text{ over } [0, s]\}.$$

Theorem 6.9 *Let $B(\cdot) \equiv B$ and let*

$$Q_S = \{0\} \times Y_S, \quad Q_E = (0, +\infty) \times \{0\} \quad \text{and} \quad \mathbb{U} = B_\rho(0),$$

where Y_S is a nonempty subset of Y (with $0 \notin Y_S$) and $\rho > 0$. Suppose that $(\widehat{H3})$ and $(\widehat{H4})$ hold and that the system (6.110) is L^∞-null controllable over each interval. Then the corresponding minimal time control problem $(TP)_{min}^{Q_S, Q_E}$ has the bang-bang property.

Proof It suffices to prove this theorem in the case that Y_S is a singleton subset of Y. The reason is the same as that stated in the proof of Theorem 6.1. Thus, we can let $Y_S = \{\widetilde{y}_0\}$ with $\widetilde{y}_0 \neq 0$.

In order to show this theorem, we only need to prove (4.138) and (4.139) of Theorem 4.5. Indeed, when they are proved, it follows from Theorem 4.5 and Theorem 4.3 that $(TP)_{min}^{Q_S, Q_E}$ satisfies the local Pontryagin Maximum Principle. Then by the same arguments as those used in the proof of Theorem 6.8, we can verify the bang-bang property for $(TP)_{min}^{Q_S, Q_E}$.

We now turn to the proof of (4.138) of Theorem 4.5. Since $Y_E = \{0\}$ and the system (6.110) is L^∞-null controllable over each interval, it follows from Theorem 1.20 and the definition of $Y_C(t_1, t_2)$ (see (4.82)) that $0 \in Int Y_C(t_1, t_2)$. Hence, (4.138) is true for the current case.

We next show (4.139) of Theorem 4.5. By Proposition 4.7, we see that in order to show the above-mentioned (4.139), we only need to prove that for each $t \in [0, +\infty)$ and each $y_0 \in Y$, the function $T \to N(t, T; y_0)$ (see (4.155)) is left continuous from $(t, +\infty)$ to $[0, +\infty]$. To this end, we arbitrarily fix $t \geq 0$. Using the similar arguments as those used in the proof of Proposition 5.3, we can see that in order to show the above-mentioned left continuity, it suffices to verify the following condition:

$$Y_{t,T} = Z_{t,T} \quad \text{for each } T > t, \tag{6.114}$$

where $Y_{t,T}$ and $Z_{t,T}$ are given by

$$
\begin{aligned}
Y_{t,T} &\triangleq \overline{X_{t,T}}^{\|\cdot\|_{L^1(t,T;U)}}, \\
X_{t,T} &\triangleq \{B^*\Phi(T, \cdot)^*z|_{(t,T)} \mid z \in Y\}, \\
Z_{t,T} &\triangleq \big\{B^*\varphi \in L^1(t, T; U) \mid \text{for each } s \in (t, T), \text{ there exists} \\
&\quad z_s \in Y \text{ so that } \varphi(\cdot) = \Phi(s, \cdot)^*z_s \text{ over } [t, s]\big\}.
\end{aligned}
\tag{6.115}
$$

Since $Y_T = Z_T$ (see $(\widehat{H4})$), $Y_{0,T} = Y_T$ and $Z_{0,T} = Z_T$, we find that it suffices to show (6.114) for the case that $t > 0$. The proof will be carried out by the following two steps:

Step 1. We show that $Y_{t,T} \subseteq Z_{t,T}$.

To this end, we arbitrarily take $f \in Y_{t,T}$. According to the definition of $Y_{t,T}$, there is a sequence of $\{z_n\}_{n\geq 1} \subseteq Y$ so that

$$B^*\Phi(T, \cdot)^*z_n \to f(\cdot) \quad \text{strongly in } L^1(t, T; U). \tag{6.116}$$

Since the system (6.110) is L^∞-null controllable over each interval, by Theorem 1.20, there exists a positive constant $C \triangleq C(T, t)$ so that

$$\|\Phi(T, t)^*z\|_Y \leq C \int_t^T \|B^*\Phi(T, s)^*z\|_U ds \quad \text{for each } z \in Y.$$

This, together with (6.116) and the definition of Y_T (see $(\widehat{H4})$), yields that there exists $g \in Y_T$ so that

$$B^*\Phi(T, \cdot)^*z_n \to g(\cdot) \quad \text{strongly in } L^1(0, T; U) \text{ and } g|_{(t,T)} = f.$$

From this, $(\widehat{H4})$ and the definition of $Z_{t,T}$ (see (6.115)), we find that $f \in Z_{t,T}$, which leads to $Y_{t,T} \subseteq Z_{t,T}$.

Step 2. We prove that $Z_{t,T} \subseteq Y_{t,T}$.

Arbitrarily fix $f \in Z_{t,T}$. We first claim that there is $\tilde{f} \in Z_T$ so that

$$\tilde{f}|_{(t,T)} = f. \tag{6.117}$$

For this purpose, we use the definition of $Z_{t,T}$ (given by (6.115)) to obtain the next two facts. Fact One:

$$f \in L^1(t, T; U); \tag{6.118}$$

Fact Two: For each $s \in (t, T)$, there exists $z_s \in Y$ so that

$$f(\tau) = B^* \Phi(s, \tau)^* z_s \quad \text{for a.e. } \tau \in (t, s). \tag{6.119}$$

Meanwhile, from the observability estimate (i) of Theorem 1.20, we see from (6.119) that for each s_1, s_2 with $t < s_1 \leq s_2 < T$,

$$B^* \Phi(s_1, \tau)^* z_{s_1} = B^* \Phi(s_2, \tau)^* z_{s_2} \quad \text{for each } \tau \in (0, s_1). \tag{6.120}$$

Now, we arbitrarily take $\hat{s} \in (t, T)$ and define the following function:

$$\tilde{f}(\tau) \triangleq \begin{cases} B^* \Phi(\hat{s}, \tau)^* z_{\hat{s}}, & \tau \in (0, t], \\ f(\tau), & \tau \in (t, T). \end{cases} \tag{6.121}$$

Then from (6.118)–(6.120) and the definition of Z_T (given by $(\widehat{H}4)$), we can easily verify (6.117).

Next, since $Y_T = Z_T$ (see $(\widehat{H}4)$) and $\tilde{f} \in Z_T$ (see (6.117)), we obtain that $\tilde{f} \in Y_T$. This, along with the definition of Y_T (see $(\widehat{H}4)$), indicates that there is $\{\eta_n\}_{n \geq 1} \subseteq Y$ so that

$$B^* \Phi(T, \cdot)^* \eta_n \to \tilde{f}(\cdot) \quad \text{strongly in } L^1(0, T; U).$$

This, together with the definition of $Y_{t,T}$ (see (6.115)) and (6.117), yields that

$$f = \tilde{f}|_{(t,T)} \in Y_{t,T},$$

which leads to that $Z_{t,T} \subseteq Y_{t,T}$.

Finally, (6.114) follows from the conclusions in Step 1 and Step 2. Hence, we finish the proof of Theorem 6.9. □

We next give three examples, which may help us to understand Theorems 6.9, 6.7, and 6.8 better.

Example 6.13 Let $\Omega \subseteq \mathbb{R}^d$ ($d \geq 1$) be a bounded domain with a C^2 boundary $\partial \Omega$. Let $\omega \subseteq \Omega$ be a nonempty and open subset with its characteristic function χ_ω. Let $a_1 \in L^\infty(\Omega)$ and $a_2 \in L^\infty(0, +\infty)$. Consider the controlled heat equation:

$$\begin{cases} \partial_t y - \Delta y + (a_1(x) + a_2(t))y = \chi_\omega u & \text{in } \Omega \times (0, +\infty), \\ y = 0 & \text{on } \partial\Omega \times (0, +\infty), \\ y(0) = y_0 \in L^2(\Omega), \end{cases} \qquad (6.122)$$

where $u \in L^\infty(0, +\infty; L^2(\Omega))$. Let $Y = U \triangleq L^2(\Omega)$; Let $A = \Delta$ with its domain $H^2(\Omega) \cap H_0^1(\Omega)$; Let $D(\cdot) = -a_1 - a_2(\cdot)$; Let $B = \chi_\omega$, where χ_ω is treated as a linear and bounded operator on U. Thus, the equation (6.122) can be rewritten as:

$$\begin{cases} \dot{y}(t) = Ay(t) + D(t)y(t) + Bu(t), & t \in (0, +\infty), \\ y(0) = y_0. \end{cases}$$

Consider the problem $(TP)_{min}^{Q_S, Q_E}$, where the controlled system is as:

$$\dot{y}(t) = Ay(t) + D(t)y(t) + Bu(t), \quad t \in (0, +\infty) \qquad (6.123)$$

and where

$$Q_S = \{0\} \times \{y_0\}, \quad Q_E = (0, +\infty) \times \{0\} \text{ and } \mathbb{U} = B_\rho(0),$$

with $y_0 \in L^2(\Omega) \setminus \{0\}$ and $\rho > 0$. *Notice that this problem is exactly the problem given in Example 5.4.*
   One can easily check the following facts:

(i) $(TP)_{min}^{Q_S, Q_E}$ can be put into the framework $(\mathcal{B}_3)$, given at the beginning of Section 6.3 (see Example 5.4);
(ii) Both $(\widehat{H}3)$ and $(\widehat{H}4)$ hold in the current case (see Theorem 1.22, Remark 1.5 (after Theorem 1.22), and Example 5.4);
(iii) The system (6.123) is L^∞-null controllable over each interval (see Theorems 1.22 and 1.21).

We further assume that $(TP)_{min}^{Q_S, Q_E}$ has optimal controls. Notice that it has optimal controls when ρ is large enough (see Theorem 3.7).
   Now, according to Theorem 6.9, $(TP)_{min}^{Q_S, Q_E}$ holds the bang-bang property. We end Example 6.13.

Example 6.14 Consider the problem $(TP)_{min}^{Q_S, Q_E}$ in Example 6.13, where we replace respectively Q_S and Q_E by $\{0\} \times \{y_0\}$ and $(0, +\infty) \times B_r(0)$ (with $r > 0$ and $y_0 \in L^2(\Omega) \setminus B_r(0)$). One can easily check the following facts:

(i) $(TP)_{min}^{Q_S, Q_E}$ can be put into the framework $(\mathcal{B}_3)$, given at the beginning of Section 6.3 (see Example 5.4);
(ii) $(\widehat{H}3)$ holds (see Theorem 1.22, Remark 1.5 after Theorem 1.22).

Furthermore, we assume that $(TP)_{min}^{Q_S, Q_E}$ has optimal controls. Then, according to Theorem 6.7, $(TP)_{min}^{Q_S, Q_E}$ holds the bang-bang property. We now end Example 6.14.

Example 6.15 For the problem $(TP)_{min}^{Q_S,Q_E}$ in Example 6.13, we choose $a_1(\cdot) = a_2(\cdot) = 0$ in (6.122). Then $D(\cdot) = 0$ in (6.123).

One can easily check what follows:

(i) $(TP)_{min}^{Q_S,Q_E}$ can be put into the framework of $(\mathscr{B}_3)$ (given at the beginning of Section 6.3);
(ii) $(\widehat{H}3)$ holds (see Theorem 1.22 and Remark 1.5 after Theorem 1.22);
(iii) The system (6.123) is L^∞-null controllable over each interval (see Theorems 1.22 and 1.21);
(iv) $(TP)_{min}^{Q_S,Q_E}$ has at least one admissible control (see Remark 3.1 after Theorem 3.5), which leads to the existence of optimal controls.

Then, according to Theorem 6.8, $(TP)_{min}^{Q_S,Q_E}$ holds the bang-bang property. We now end Example 6.15.

We have introduced two different methods to derive the bang-bang property in Section 6.2 and Section 6.3 respectively. In the next Remark 6.3, we will compare these two methods.

Remark 6.3 Denote by (M1) and (M2) the methods introduced in Section 6.2 and Section 6.3, respectively. Several notes on (M1)and (M2) are given as follows:

(i) For $(TP)_{min}^{Q_S,Q_E}$ in finitely dimensional cases, (M1) and (M2) are essentially the same. On one hand, the key in (M1) is the L^∞-null controllability from measurable sets in time, while the keys in (M2) are the property $(\widehat{H}3)$ and maximum principles. Maximum principles always hold for $(TP)_{min}^{Q_S,Q_E}$ in finitely dimensional cases, provided that $\mathbb{U}$, Y_S and Y_E are nonempty convex sets (see Theorem 4.1 and Theorem 4.2). On the other hand, the aforementioned controllability is equivalent to $(\widehat{H}3)$ in finitely dimensional cases. We explain this by the following controlled system:

$$\dot{y}(t) = D(t)y(t) + Bu(t), \quad t \in (0, +\infty) \quad (\text{with } y(t) \in \mathbb{R}^n, \ u(t) \in \mathbb{R}^m),$$
(6.124)

where $D(\cdot) \in L^\infty(0, +\infty; \mathbb{R}^{n\times n})$ and $B \in \mathbb{R}^{n\times m}$. Indeed, we arbitrarily fix $T > t_0 \geq 0$ and a measurable subset $E \subseteq (t_0, T)$ with positive measure. Define the following map:

$$\mathscr{F}(B^*\Phi(T, \cdot)^*z|_E) \triangleq \Phi(T, t_0)^*z \quad \text{for each } z \in \mathbb{R}^n.$$
(6.125)

If $(\widehat{H}3)$ holds, then by (6.125), we see that the map $\mathscr{F}$ is well defined. Since the domain of $\mathscr{F}$ is of finite dimension, there exists a constant $C \triangleq C(T, t_0, E) > 0$ so that

$$\|\Phi(T, t_0)^*z\|_{\mathbb{R}^n} \leq C \int_E \|B^*\Phi(T, t)^*z\|_{\mathbb{R}^m} dt \quad \text{for each } z \in \mathbb{R}^n.$$

Then using the similar arguments as those used in the proof of Theorem 1.20, we can obtain that the system (6.124) is L^∞-null controllable from measurable subsets in time. Conversely, if the system (6.124) is L^∞-null controllable from measurable subsets in time, then using the similar arguments to those used in the proof of Theorem 1.20, we can see that $(\widehat{H}3)$ holds.

(ii) For $(TP)_{min}^{Q_S, Q_E}$ in infinitely dimension cases, (M1) and (M2) are not the same in general. (M1) works for the case when both $\mathbb{U}$ and Y_E are not convex and the controlled system is time-invariant, while (M2) works for some time-varying controlled systems, provided that both $\mathbb{U}$ and Y_E are convex.

Miscellaneous Notes

The studies on the bang-bang property of time optimal control problems started with finite dimensional systems in 1950s (see [4] and [22]), and then extended to infinite dimensional systems in 1960s. To our best knowledge, [9] seems to be the first work dealing with the bang-bang property for abstract evolution equations in infinite dimensional spaces.

One of the usual methods to derive the bang-bang property is the use of the controllability from measurable sets in time. (Simply, we call it as the E-controllability.) The original idea of this method invisibly arose from [9]. This method was first presented formally in [18], which was partially inspired by [24]. More about this method, we would like to mention what follows:

- This method works well for the minimal time control problems governed by time invariant systems, i.e., the E-controllability and the time-invariance of the system imply the bang-bang property for the corresponding minimal time control problem.
- This method doesn't seem to work for the minimal time control problems governed by time-varying systems.
- This method works for the maximal time control problems governed by time-varying systems (see [18] and [20]).
- The E-controllability, combined with a fixed point argument, implies the bang-bang property of minimal time control problems governed by some semi-linear heat equations (see [21] and [29]).

About studies on the E-controllability, we would like to give the following notes:

- The E-controllability was first proved for the boundary controlled 1-dimension heat equation in [18].
- The E-controllability was showed for the internally controlled n-dimension heat equation in [26], with the aid of Lebeau and Robbiano's spectrum inequality (see [13] and [14]).

- The E-controllability was built up for internally controlled n-dimension heat equations with lower terms (where the physical domain is bounded and convex) in [20], via the frequency function method used in [6] and [23]. Then the convexity condition was dropped in [21].
- The E-controllability was obtained for some evolution equations in [30].
- A much stronger version of the E-controllability for the internally (or boundary) controlled heat equation was built up in [2] (see also [7] and [8]). The corresponding minimal time control problem has the control constraint: $|u(x, t)| \leq M$. This strong version implies that any optimal control u^* satisfies that $|u^*(x, t)| = M$ for a.e. $(x, t) \in \Omega \times (0, T^*)$ (or $|u^*(x, t)| = M$ for a.e. $(x, t) \in \partial\Omega \times (0, T^*))$, where Ω is the physical domain and T^* is the minimal time.

Another usual method to derive the bang-bang property is the use of the Pontryagin Maximum Principle, together with some unique continuation of the adjoint equation. This method was used in [4] for finite dimensional cases. Then it was extended to infinite dimensional cases (see, for instance, [10, 11, 16, 31], and [27]). As what we have seen in Chapter 4, for infinite dimensional cases, it is difficult to derive the Pontryagin Maximum Principle for the cases where controlled systems are infinitely dimensional and the target sets have empty interiors, such as points in the state spaces. Besides, it is not easy to obtain the required unique continuation for infinite dimensional cases, in general. Therefore, the studies on the bang-bang property for infinite dimensional cases via this way are much harder, compared with finite dimensional cases.

In [15], the author studied the controlled system: $y' = Ay + u$, where A is the generator of a group, and the state and the control spaces are the same. For such a system, the bang-bang property of a kind of minimal time control problem is established. More about the bang-bang property, we would like to mention the works [3, 17, 28, 32], and [12].

Materials in Section 6.1 are partially summarized and developed from the related materials in [10, 22] and [33], while most examples in Section 6.1 are given by us. Materials in Section 6.2 are taken from [30] with a bit of modification. Materials in Section 6.3 are developed from [31] and [32].

Finally, we give some open problems which might interest readers:

- Can we have the bang-bang property for infinite dimensional controlled systems with general control constraints, especially with the rectangular-type control constraint?
- Does the bang-bang property hold for controlled systems with some state constraints? (For instance, heat equations with nonnegative states.)
- Does the bang-bang property hold for general time-varying controlled systems in infinite dimensional spaces?
- How to estimate switching times of a bang-bang control?
- Do controlled wave equations (or controlled wave-like equations) have the bang-bang property? (We believe that the key to answer this question is to understand the relation between the bang-bang property and the propagation.)
- What happens for bang-bang controls from the numerical perspective?

References

1. J. Apraiz, L. Escauriaza, Null-control and measureable sets. ESAIM Control Optim. Calc. Var. **19**, 239–254 (2013)
2. J. Apraiz, L. Escauriaza, G. Wang, C. Zhang, Observability inequalities and measurable sets. J. Eur. Math. Soc. **16**, 2433–2475 (2014)
3. V. Barbu, The time optimal control of Navier-Stokes equations. Syst. Control Lett. **30**, 93–100 (1997)
4. R. Bellman, I. Glicksberg, O. Gross, On the "bang-bang" control problem. Q. Appl. Math. **14**, 11–18 (1956)
5. T. Duyckaerts, X. Zhang, E. Zuazua, On the optimality of the observability inequalities for parabolic and hyperbolic systems with potentials. Ann. Inst. H. Poincaré, Anal. Non Linéaire **25**, 1–41 (2008)
6. L. Escauriaza, F.J. Fernández, S. Vessella, Doubling properties of caloric functions. Appl. Anal. **85**, 205–223 (2006)
7. L. Escauriaza, S. Montaner, C. Zhang, Observation from measurable sets for parabolic analytic evolutions and applications. J. Math. Pures et Appl. **104**, 837–867 (2015)
8. L. Escauriaza, S. Montaner, C. Zhang, Analyticity of solutions to parabolic evolutions and applications. SIAM J. Math. Anal. **49**, 4064–4092 (2017)
9. H.O. Fattorini, Time-optimal control of solution of operational differential equations. J. SIAM Control **2**, 54–59 (1964)
10. H.O. Fattorini, *Infinite Dimensional Linear Control Systems, the Time Optimal and Norm Optimal Problems*, North-Holland Mathematics Studies, vol. 201 (Elsevier Science B.V., Amsterdam, 2005)
11. K. Kunisch, L. Wang, Time optimal control of the heat equation with pointwise control constraints. ESAIM Control Optim. Calc. Var. **19**, 460–485 (2013)
12. K. Kunisch, L. Wang, Bang-bang property of time optimal controls of semilinear parabolic equation. Discrete Contin. Dyn. Syst. **36**, 279–302 (2016)
13. G. Lebeau, L. Robbiano, Contrôle exacte l'équation de la chaleur (French). Comm. Partial Differ. Equ. **20**, 335–356 (1995)
14. G. Lebeau, E. Zuazua, Null-controllability of a system of linear thermoelasticity. Arch. Rational Mech. Anal. **141**, 297–329 (1998)
15. J.-L. Lions, *Optimal Control of Systems Governed by Partial Differential Equations* (Springer, New York, 1971)
16. J. Lohéac, M. Tucsnak, Maximum principle and bang-bang property of time optimal controls for Schrodinger-type systems. SIAM J. Control Optim. **51**, 4016–4038 (2013)
17. Q. Lü, Bang-bang principle of time optimal controls and null controllability of fractional order parabolic equations. Acta Math. Sin. (Engl. Ser.) **26**, 2377–2386 (2010)
18. V.J. Mizel, T.I. Seidman, An abstract bang-bang principle and time optimal boundary control of the heat equation. SIAM J. Control Optim. **35**, 1204–1216 (1997)
19. A. Pazy, *Semigroups of Linear Operators and Applications to Partial Differential Equations*. Applied Mathematical Sciences, vol. 44 (Springer, New York, 1983)
20. K.D. Phung, G. Wang, An observability estimate for parabolic equations from a measurable set in time and its application. J. Eur. Math. Soc. **15**, 681–703 (2013)
21. K.D. Phung, L. Wang, C. Zhang, Bang-bang property for time optimal control of semilinear heat equation. Ann. Inst. H. Poincaré, Anal. Non Linéaire **31**, 477–499 (2014)
22. L.S. Pontryagin, V.G. Boltyanskii, R.V. Gamkrelidze, E.F. Mischenko, *The Mathematical Theory of Optimal Processes*, Translated from the Russian by K. N. Trirogoff, ed. by L. W. Neustadt, (Interscience Publishers Wiley, New York, London, 1962)
23. C.C. Poon, Unique continuation for parabolic equations. Commun. Partial Differ. Equ. **21**, 521–539 (1996)
24. E.J.P.G. Schmidt, The "bang-bang" principle for the time optimal problem in boundary control of the heat equation. SIAM J. Control Optim. **18**, 101–107 (1980)

25. S. Vessella, A continuous dependence result in the analytic continuation problem. Forum Math. **11**, 695–703 (1999)
26. G. Wang, L^∞−null controllability for the heat equation and its consequence for the time optimal control problem. SIAM J. Control Optim. **47**, 1701–1720 (2008)
27. G. Wang, L. Wang, The bang-bang principle of time optimal controls for the heat equation with internal controls. Syst. Control Lett. **56**, 709–713 (2007)
28. L. Wang, Q. Yan, Time optimal controls of semilinear heat equation with switching control. J. Optim. Theory Appl. **165**, 263–278 (2015)
29. L. Wang, Q. Yan, Bang-bang property of time optimal null controls for some semilinear heat equation. SIAM J. Control Optim. **54**, 2949–2964 (2016)
30. G. Wang, C. Zhang, Observability inequalities from measurable sets for some abstract evolution equations. SIAM J. Control Optim. **55**, 1862–1886 (2017)
31. G. Wang, Y. Zhang, Decompositions and bang-bang problems. Math. Control Relat. Fields **7**, 73–170 (2017)
32. G. Wang, Y. Xu, Y. Zhang, Attainable subspaces and the bang-bang property of time optimal controls for heat equations. SIAM J. Control Optim. **53**, 592–621 (2015)
33. J. Yong, H. Lou, *A Concise Course on Optimal Control Theory (Chinese)* (Higher Education Press, Beijing, 2006)

Index

© Springer International Publishing AG, part of Springer Nature 2018
G. Wang et al., *Time Optimal Control of Evolution Equations*, Progress
in Nonlinear Differential Equations and Their Applications 92,
https://doi.org/10.1007/978-3-319-95363-2

Printed in the United States
By Bookmasters